THE ATLAS OF THE REAL WORLD

Daniel Dorling, Mark Newman and Anna Barford

THE ATLAS OF THE REAL WORLD
Mapping the Way We Live
REVISED AND EXPANDED EDITION

 Thames & Hudson

First published in the United Kingdom in 2008 by
Thames & Hudson Ltd, 181A High Holborn,
London WC1V 7QX

thamesandhudson.com

This revised and expanded edition 2010

British Library Cataloguing-in-Publication Data
A catalogue record for this book is available from the
British Library

ISBN 978-0-500-28853-5

Printed and bound in China by Toppan Leefung
Printing Limited

Contents

The Social World

Introduction

We know a lot more about the world than we used to. Even ten years ago our knowledge was a good deal less complete than it is today. If at any time in the last fifty years you had wanted to know the population of Brazil or the number of televisions in France, you could probably, with a little determination, have found an answer. But until recently there were very few such questions you could have answered for every country in the world. If you had wanted to know the number of televisions in Burkina Faso, for example, you would probably have been out of luck.

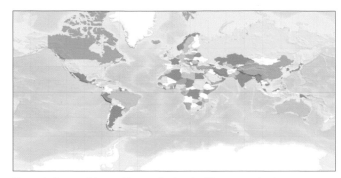

Figure 1: A map of the world created using the Mercator projection

Figure 2: A map of the world created using the Gall–Peters projection

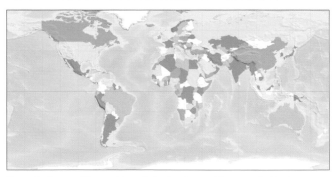

Figure 3: A map of the world created using the equidistant cylindrical projection or *plate carrée*

But as we move into the twenty-first century, things are changing. The changes are partly technological: the arrival of the internet and the widespread availability of computers have made it much easier to gather reliable data. And they are partly political: a number of major new projects have been set in motion by the United Nations and others with the aim of amassing a broader range of global statistics than was previously available. Today we have good worldwide statistics on such diverse topics as income, literacy rates, numbers of doctors, nurses and dentists, numbers of teenage mothers, how many people smoke, military spending, endangered species, greenhouse gas emissions and mobile phones.

All of these data are freely available on the web, but for the most part in the form of plain rows of numbers in tables, which are difficult to read and more difficult still to understand. Japan has a domestic water consumption of 17 billion cubic metres a year. Is that a lot or a little? How does it compare to the total amount of fresh water in Japan? How does it compare to what other countries consume? It is hard to answer such questions without spending a long time poring over the tables. A much better way to show what is going on in the world is to make a map – which is what this book does.

The Atlas of the Real World contains 382 maps showing all sorts of geographical and social statistics, ranging from basic data on population, health, wealth and occupation to how many toys we import and who's eating their vegetables. Open this book at almost any page and you will learn something you never knew about the world.

Maps and Mapmaking

Cartography, the art of mapmaking, is as old as recorded history. People were drawing views of the land as if seen from above thousands of years before they ever had such views in real life, even from balloons, let alone from aeroplanes and satellites. The Babylonians had produced maps by at least 2500 BC. Papyrus maps from the time of the pharaohs have been recovered in Egypt. The ancient Greeks were particularly adept at mapmaking and it is even believed that they knew the Earth to be round – a discovery often attributed to Aristotle, although the real credit probably goes to Eratosthenes, who was the first to measure the curvature of the Earth's surface and hence make an estimate of the size of the globe. (His estimate was remarkably accurate given the equipment he was using. He got the circumference of the world somewhat too large, but only by about 20%.)

In the second century AD Ptolemy, another Greek, took mapmaking to new heights, compiling a multi-volume atlas based on the work of several previous surveyors and mapmakers. Of the many innovations introduced by Ptolemy, perhaps the most important was the use of latitude and longitude coordinates to locate positions on his maps, a practice still in use today. On the other hand, Ptolemy's maps were still quite inaccurate, largely because of limitations in surveying.

Scholars in China and the Middle East made significant contributions to cartography during the middle ages, and had been mapping much earlier too, but an accurate picture of the Earth's surface did not emerge until the beginning of the sixteenth century, when extensive exploration of the globe and improved techniques for navigation and surveying resulted in the first virtually complete maps of the world, including the known portions of the Americas (but still omitting Antarctica and Australasia).

One of the things mapmakers quickly discovered when they started making maps of the world is that it is quite difficult to draw a picture of a spherical world on a flat piece of paper. Imagine a mapmaker's globe of the Earth and think of the map on its surface as a rubber sheet, unwrapped from the sphere on which it lives and squashed down flat on a page. Inevitably we have to bend or stretch parts of the map to make it flat; and the end result is that, no matter how we do this, any flat depiction of the world must distort some regions, and often all of them.

There are a variety of different ways of doing this 'flattening' – different 'projections', in cartographic terminology – and they distort different regions by different amounts. The most commonly used projections distort the areas near the North and South Poles quite severely but represent the areas nearer the equator relatively accurately. In many cases this is a reasonable compromise, since relatively few people live near the poles and so for most human purposes it is less important to represent the polar areas exactly.

Perhaps the most famous projection of the world is the Mercator projection, created by the Flemish cartographer Gheert Cremer (in Latin, Gerardus Mercator) in 1569 (Figure 1). This projection has the appealing property of faithfully representing the *shapes* of regions on the map, but it greatly exaggerates the *sizes* of features near the poles. An alternative is the Gall–Peters projection, first described by clergyman James Gall in 1855 (Figure 2). This projection depicts sizes accurately – the area covered by a feature on the map is proportional to the true land area of the feature in real life – but now the shapes are distorted. And again, this distortion is greatest near the poles.

In between the Mercator and Gall–Peters projections lies another common choice, the equidistant cylindrical projection (sometimes also called by its French name *plate carrée*, which means 'flat square'). This projection gets neither the areas nor the shapes of map features correct, but adopts a compromise solution that distorts neither areas as badly as the Mercator projection nor shapes as badly as the Gall–Peters (Figure 3).

There are many other projections of the world, including ones that are not rectangular, ones that come in several disconnected parts and even ones on which the same features appear more than once in different parts of the map. All of them have their uses, but all of them, inevitably, distort our view of the world.

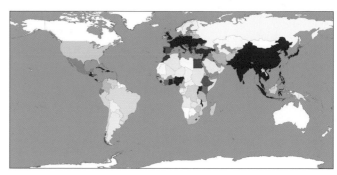

Figure 4: Map of world population density 2002

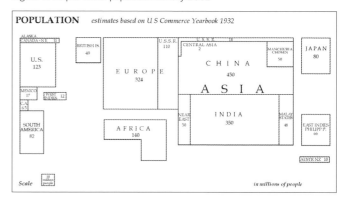

Figure 5a: 1936 population cartogram of the world by Erwin Raisz

Using Maps to Represent Data

Maps are often used to represent geographical data in visual form. For example, a standard way to depict the population of the world is to colour countries in different shades according to how many people live there. Figure 4 shows countries coloured according to population density – the number of people per square kilometre; the darker the colour the denser the population.

There are some problems with maps of this kind, however. In particular, it is very hard to tell when two countries have similar populations if they are of different sizes. Russia and Nigeria, for example, have roughly equal populations numerically, but appear on the map in completely different colours because their land areas and hence population densities are so different.

In this book we take a different approach to representing data, changing not the colour of a country but its size. Since the sizes and shapes of countries are inevitably distorted by map projections, why not make the most of it? On a map of population, for example, we draw a country with twice as many people as another twice as large; on a map of wealth, a country twice as rich as another is twice the size.

Maps of this kind are called 'cartograms'. Cartograms are a far more recent invention than maps of other kinds; a few were drawn in the nineteenth century but the earliest well-known examples date from the first half of the twentieth century (see Figures 5a and 5b), and their use has become common only since the 1960s, when

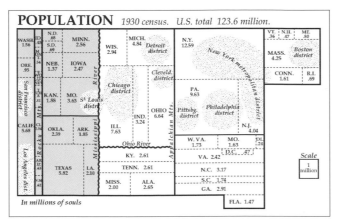

Figure 5b: 1934 population cartogram of the United States by Erwin Raisz

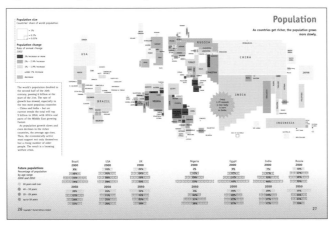

Figure 6: 2002 population cartogram of the world by Daniel Smith

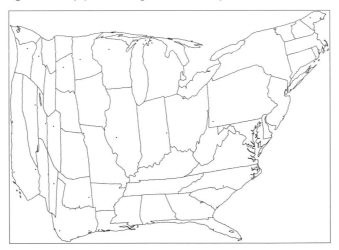

Figure 7: Population cartogram of the United States by Waldo Tobler, 1974

computers started to become widely available (Figure 6). For while the making of cartograms by hand is a laborious and difficult job, it is perfectly suited to computers.

Mapmakers started making computer-generated cartograms almost as soon as they had access to general-purpose computers. In the 1960s Waldo Tobler, a geographer at the University of Michigan in the United States, wrote what was almost certainly the first computer program to this end; one of his cartograms is shown in Figure 7. Such early efforts were limited by the slowness of the available computers, and the maps produced, though groundbreaking, were in some cases quite difficult to read. But computers have since become faster and cartograms better. The cartograms in this book were generated using a new method that uses ideas borrowed from theoretical physics. For those interested, a technical description of how they were made is given on page 12.

How to Read the Maps in this Atlas

One of the wonderful features of contemporary cartograms is that they are very easy to read, requiring no special training or expertise. Larger countries have more of whatever it is the map is showing; smaller countries have less. On a population cartogram, for instance, the size of a country represents how many people live there – see Figure 8. China and India appear large on this map: between them they are home to more than 2 billion people, over a third of the population of the planet. Canada, on the other hand, which is the third largest country in the world in terms of land area, is almost invisible because its population is relatively small, only about 30 million.

One of the tricks to making a good cartogram is to ensure that countries retain roughly their correct shapes and positions so that they are still recognizable after their sizes have changed. In some cases, however, the changes in size are so extreme that no matter how the map is drawn it can be difficult to recognize certain countries. The cartogram of yellow fever cases shown in Figure 9 is an example. To make the maps more legible we have presented the countries in the same colours on every cartogram in this book. If the identity or position of a particular country is unclear, check the colour of that country on the population or land area maps (which are maps 001–008 of this book). Using the colour and the general shape of the country, you should be able to track it down on another cartogram with relatively little effort. Copies of maps 001 (Land Area) and 002 (Total Population) also appear in the fold-out flaps of this book for convenient comparison.

What Does It All Mean?

Many of the maps in this atlas paint a worrying picture of the world. Billions of people are living in slums, in poverty and without clean water, adequate medical attention or shelter. Worldwide inequality is on the increase, with the rich getting richer and the poor poorer. The maps of life expectancy show significant improvements in the wealthier countries of the world between 1972 and the present day, but essentially none in many parts of Africa. The map of airline travel shows where people fly most often, but also shows that most people in the world never fly at all. More people do not have access to a car

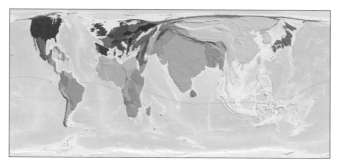

Figure 8: Total Population

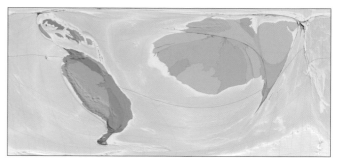

Figure 9: Yellow Fever cases 1995–2004

with every year that passes. In this atlas we show where the cars are (Map 033), so you have to work out for yourself where they are least common. The maps of trade flows show that production is concentrated on one side of the planet while consumption is concentrated on the other. And population is increasing in the poorest parts of the world while at the same time it is roughly static or even decreasing in the richer ones, thereby concentrating wealth in the hands of an ever smaller proportion of the world's population.

The picture is not entirely bleak, however. For instance, as a result of rapid growth in production and exports, China has gained near-universal access to electricity and many other benefits of the modern world. Nevertheless, it is unlikely that the average member of the Chinese population will achieve the affluent living standards typical of the richest nations any time soon. For the moment at least, high living standards tend to be associated not with industrial production but with business, finance, and intellectual property such as books, music and software.

So turn the page, and take a look at the first of the many maps in this collection. Strange though they seem at first, these maps are a thought-provoking way to learn about the world around us and understand our place within it. They can help us to see our life in relation to the more than 6 billion other humans living on the planet, in social, economic, educational, environmental and geographical terms. In a sense these maps are doing just what maps have always done: showing us where we are now, allowing us to navigate our way through the world. But the shapes and sizes of the countries shown on these cartograms are not fixed, as they are on other maps. If the distribution of population or wealth or televisions or bicycles changes in the future, then a future cartogram will look different. And though there are some things we cannot control, the future shape of the world is largely in human hands. We hope that the contents of this book will help readers to think about what shape we want that to be.

How to Use this Book

In this atlas each country is depicted in the same colour on every map to make identification easier. The countries are grouped into twelve regions, also colour-coded, allowing the reader easily to discern large-scale geographical patterns. (A list of the countries / territories in each region can be found at the end of this introduction.)

Each map is accompanied by a commentary and by a table that typically shows the ten highest-scoring and the ten lowest-scoring territories in the subject matter of the map; to avoid confusion, results in the table are almost always ranked according to the 200 territories covered by the maps, though data may not be available for all of them (see the Worldmapper website for rankings according to the exact number of territories for which data or estimates are available). On some pages graphs are included as well; these present the data depicted in the map in a different form, often aggregated by region.

This book can be read cover to cover; the order of the pages and chapters tells a story about the state of the world today. But you can also refer to individual maps as questions arise, or simply open the book at random and see what you find. A contents list is provided to help you find your way around, and to identify the locations of each territory in a way which is familiar, simply refer to map 001. Concise information about data sources appears after the 382 maps.

Each map in this atlas tells its own story, but one can learn even more by comparing maps to each other. Related maps are arranged in pairs on facing pages of the atlas so that, for instance, maps showing exports and imports are presented together, as are those showing the poorest and richest countries in the world. These juxtapositions offer many insights into the state of world trade, economics and politics, as well as social and environmental issues. And there are many other interesting comparisons to be made in addition to those suggested by the map pairs.

Main Sources and Calculation of Data

The data on which our maps are based come from a variety of sources, including the World Bank's *World Development Indicators*, the United Nations Environment Programme's *Global Environment Outlook*, the United Nations Development Programme's *Human Development Report*, the Central Intelligence Agency's *World Factbook*, the World Health Organization's *World Health Report*, Angus Maddison's historical statistics on the world economy, the United Nations Population Division's *World Population Prospects*, the United Nations Conference on Trade and Development's *Handbook of Statistics*, the United Nations Children's Fund (UNICEF) *Innocenti Report Cards* and the United Nations Statistics Division's *Millennium Development Goals Indicators*.

These organizations and projects provide some of the most complete sets of global statistics available, but unfortunately they divide the world into countries, regions or territories in different ways. For the sake of consistency, we have standardized on a fixed set of 200 territories for all maps, converting the original data to fit this template where necessary.

For cosmetic reasons, the maps are drawn using 204 resizeable areas. The four extra areas are Alaska (part of the United States), the Falkland Islands / Islas Malvinas (under United Kingdom sovereignty) and New Caledonia and French Guiana (both parts of France). Resizing these four areas according to the data of their sovereign territory was found to result in misleading maps in some cases, so instead they are resized on the basis of statistics for the surrounding region, except for Alaska, which is resized according to United States data applied to the small proportion of the country's population who live there.

Some of the territories shown are not technically countries. We show all countries recognized by the United Nations, plus a few other major territories such as Hong Kong, Greenland, Puerto Rico and the Western Sahara, which are included either because they appear as separate entities in our data sources or because there are current disputes over their sovereignty. The territories shown are home to 99.95% of the world's population so, although not perfect, the maps give effectively complete coverage.

Many of the quantities we examine, such as population, vary within territories as well as between them – most territories have areas of higher (urban) and lower (rural) population – and this internal variation is not shown on our maps. It would certainly be possible to make maps that show individual territories in more detail, but we leave that for another day.

Although the data at our disposal today are far more complete than those available even just a few years ago, there are still gaps. On the maps of the early explorers such unknown regions might have been labelled 'here be dragons', but we don't have the luxury of such whimsicality today. If our cartograms are to be reasonably accurate, we need to make some estimate of the missing numbers. Our strategy has been to assume that territories have numbers based on rates that are similar to those for neighbouring territories. If a number is missing for Peru, say, then it makes sense to estimate that number by looking at other South American countries, rather than at North America or Europe or Africa. So we have divided the world into a number of regions on a socio-economic and geographical basis – South America, North America, Eastern Europe, Western Europe, Northern Africa and so forth – and filled in the gaps for territories lacking data by giving them the average value for their region, normally scaled by the population or area of the territory. Thus, for example, a missing data point for the annual number of influenza cases in the Former Yugoslav Republic of Macedonia, which is in Eastern Europe, would be filled in by calculating the average number of such cases per member of the population for other territories in Eastern Europe and then scaling that figure to match the known population of FYR Macedonia. This technique usually provides estimates that are roughly correct, so that the final map gives a reasonable picture of the global situation. Ultimately, however, we are always dependent on the availability of good data to make these maps – if you think that an important map is missing from this series, it is probably because good data were not available. But new information is always being collected, giving ever-increasing opportunities for further mapmaking.

How the Cartograms Are Made

The cartograms in this book were created using a method proposed in 2004 by two physicists, Michael Gastner and Mark Newman (the latter being one of the authors of this book). In this method the population or other density function of interest is treated as a diffusing fluid, which spreads out from the areas where it is initially most dense into areas of lower density. As an analogy, imagine a bottle of ink emptied into a swimming pool: the ink is initially densest at the point where it is tipped into the water but over time will spread out until it is distributed uniformly throughout the pool. This process can be modelled mathematically using differential equations that allow us to predict the motion of the fluid at any time. In the cartogram calculation we then allow the features of the map to be carried along with the diffusing ink until it comes to rest at its final position, which defines the cartogram transformation. It can be shown mathematically that this process does indeed produce a map in which the final areas of territories are proportional to their initial population or other property of interest. In practice, the maps produced are also attractive and readable, with relatively minor distortion of territory shapes in most cases.

At a technical level, the solution of the differential equations is carried out in Fourier space, where it takes a particularly simple form. The initial density function is evaluated on a 4096 × 2048 lattice, transformed using a two-dimensional fast Fourier transform, multiplied by a Gaussian kernel and then back-transformed to give the diffusion field at an arbitrary later time. Since a map of the world wraps around on itself from left to right but not from top to bottom, we must apply periodic boundary conditions in the horizontal direction but closed boundary conditions in the vertical direction. This means that the Fourier transform we use is of mixed type, consisting of a complex transform in the horizontal direction and a cosine transform in the vertical direction.

The diffusion field is then used to calculate the diffusion velocity as a function of position and the velocity integrated over time to give the displacement of the map features. The integration is performed using a fourth-order Runge–Kutta integrator with an adaptive step size. With careful control of integration errors we are able to calculate vertex positions for our cartograms to an accuracy of about 1 part in 10,000 in reasonable computation times, typically on the order of a few minutes on a standard desktop computer. Although in principle one could do better than this by spending longer on the calculation, there is little point since the differences would be indiscernible to the eye.

12

The Regions and Territories

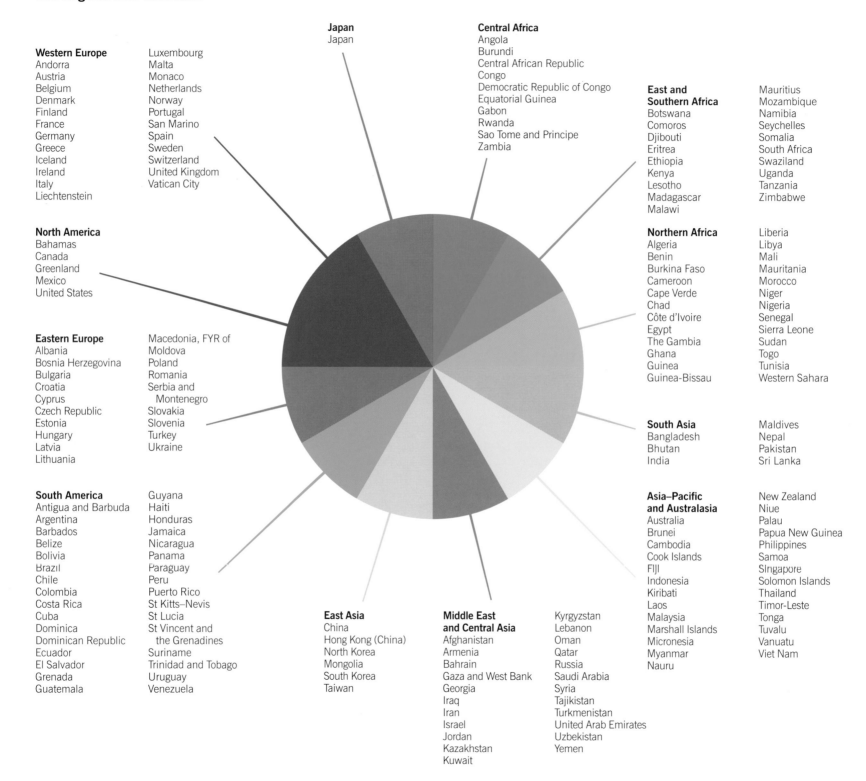

Japan
Japan

Central Africa
Angola
Burundi
Central African Republic
Congo
Democratic Republic of Congo
Equatorial Guinea
Gabon
Rwanda
Sao Tome and Principe
Zambia

Western Europe
Andorra
Austria
Belgium
Denmark
Finland
France
Germany
Greece
Iceland
Ireland
Italy
Liechtenstein
Luxembourg
Malta
Monaco
Netherlands
Norway
Portugal
San Marino
Spain
Sweden
Switzerland
United Kingdom
Vatican City

**East and
Southern Africa**
Botswana
Comoros
Djibouti
Eritrea
Ethiopia
Kenya
Lesotho
Madagascar
Malawi
Mauritius
Mozambique
Namibia
Seychelles
Somalia
South Africa
Swaziland
Uganda
Tanzania
Zimbabwe

North America
Bahamas
Canada
Greenland
Mexico
United States

Northern Africa
Algeria
Benin
Burkina Faso
Cameroon
Cape Verde
Chad
Côte d'Ivoire
Egypt
The Gambia
Ghana
Guinea
Guinea-Bissau
Liberia
Libya
Mali
Mauritania
Morocco
Niger
Nigeria
Senegal
Sierra Leone
Sudan
Togo
Tunisia
Western Sahara

Eastern Europe
Albania
Bosnia Herzegovina
Bulgaria
Croatia
Cyprus
Czech Republic
Estonia
Hungary
Latvia
Lithuania
Macedonia, FYR of
Moldova
Poland
Romania
Serbia and
    Montenegro
Slovakia
Slovenia
Turkey
Ukraine

South Asia
Bangladesh
Bhutan
India
Maldives
Nepal
Pakistan
Sri Lanka

South America
Antigua and Barbuda
Argentina
Barbados
Belize
Bolivia
Brazil
Chile
Colombia
Costa Rica
Cuba
Dominica
Dominican Republic
Ecuador
El Salvador
Grenada
Guatemala
Guyana
Haiti
Honduras
Jamaica
Nicaragua
Panama
Paraguay
Peru
Puerto Rico
St Kitts–Nevis
St Lucia
St Vincent and
    the Grenadines
Suriname
Trinidad and Tobago
Uruguay
Venezuela

**Asia–Pacific
and Australasia**
Australia
Brunei
Cambodia
Cook Islands
Fiji
Indonesia
Kiribati
Laos
Malaysia
Marshall Islands
Micronesia
Myanmar
Nauru
New Zealand
Niue
Palau
Papua New Guinea
Philippines
Samoa
Singapore
Solomon Islands
Thailand
Timor-Leste
Tonga
Tuvalu
Vanuatu
Viet Nam

East Asia
China
Hong Kong (China)
North Korea
Mongolia
South Korea
Taiwan

**Middle East
and Central Asia**
Afghanistan
Armenia
Bahrain
Gaza and West Bank
Georgia
Iraq
Iran
Israel
Jordan
Kazakhstan
Kuwait
Kyrgyzstan
Lebanon
Oman
Qatar
Russia
Saudi Arabia
Syria
Tajikistan
Turkmenistan
United Arab Emirates
Uzbekistan
Yemen

13

The Resourceful World

Land Area and Population

Travel and Transport

Natural Resources and Energy

LARGEST AND SMALLEST LAND AREAS

Rank	Territory	Hectares
1	Russia	1,689,000,000
2	China	933,000,000
3	Canada	922,000,000
4	United States	916,000,000
5	Brazil	846,000,000
6	Australia	768,000,000
7	India	297,000,000
8	Argentina	274,000,000
9	Kazakhstan	270,000,000
10	Algeria	238,000,000
191	St Kitts–Nevis	36,000
192	Niue	26,000
193	Cook Islands	23,800
194	Marshall Islands	18,000
195	Liechtenstein	16,000
196	San Marino	6,000
197	Tuvalu	3,000
198	Nauru	2,000
199	Monaco	200
200	Vatican City	40

001 Land Area

The size of each territory represents exactly its land area in proportion to that of the others, giving a strikingly different perspective from the Mercator projection most commonly used.

Maps based on the Mercator projection enormously distort the size of land masses at the poles, making Greenland and Antarctica disproportionately large by comparison with Africa and South America. This map uses land area data for each of the 200 territories shown throughout the atlas, scaled to represent accurately in two dimensions the actual relative sizes of the territories.

The total land area of the 200 territories is 13,056 million hectares. A hectare is the area of a square 100 metres on each side. If all this land were divided up equally there would be 2.1 hectares for each person on the Earth. Land, however, is not divided equally: Australia, for example, has 21 times the land area of Japan, although the population of Japan is more than 6 times that of Australia.

LAND DISTRIBUTION

Japan 0.2%
Western Europe 2.4%
Central Africa 3.7%
North America 13.8%
E. and S. Africa 18.5%
Eastern Europe 1.9%
South America 12.1%
Northern Africa 10.7%
South Asia 2.7%
East Asia 7.4%
Asia–Pacific 8.5%
Middle East 18.0%

'Secure access to land remains essential for diverse land-based livelihoods and is a precondition for sustainable agriculture, economic growth and poverty reduction.' Oxfam, 2006

LARGEST AND SMALLEST POPULATIONS

Rank	Territory	Population
1	China	1,295,000,000
2	India	1,050,000,000
3	United States	291,000,000
4	Indonesia	217,000,000
5	Brazil	176,000,000
6	Pakistan	150,000,000
7	Bangladesh	144,000,000
8	Russia	144,000,000
9	Japan	128,000,000
10	Nigeria	121,000,000
191	St Kitts–Nevis	42,000
192	Monaco	34,000
193	Liechtenstein	33,000
194	San Marino	27,000
195	Palau	20,000
196	Cook Islands	18,000
197	Nauru	13,000
198	Tuvalu	10,000
199	Niue	2,000
200	Vatican City	1,000

MOST AND LEAST LAND AREA PER PERSON

Rank	Territory	Hectares per person
1	Greenland	1,821
2	Western Sahara	97
3	Mongolia	60
4	Namibia	41
5	Australia	39
6	Suriname	39
7	Mauritania	37
8	Iceland	33
9	St Vincent and the Grenadines	33
10	Botswana	31
191	Nauru	0.154
192	Barbados	0.143
193	Bahrain	0.101
194	Maldives	0.100
195	Bangladesh	0.091
196	Malta	0.080
197	Vatican City	0.044
198	Singapore	0.016
199	Hong Kong (China)	0.015
200	Monaco	0.006

002 Total Population

The size of each territory indicates the proportion of the world's population living there, showing how the Earth's population is distributed over the planet's surface.

In the spring of 2000, the population of the world passed 6 billion people for the first time. On this map, India, China and Japan appear large because they have large populations; Panama, Namibia and Guinea-Bissau have small populations and so are barely visible.

Population is only weakly related to land area. Sudan, for example, is the largest country in Africa in terms of land area but has a smaller population than many other African countries, including Nigeria, Egypt, Ethiopia, the Democratic Republic of Congo, South Africa and Tanzania.

'Out of every 100 persons added to the population in the coming decade, 97 will live in developing countries.'
Hania Zlotnik, UNFPA, 2005

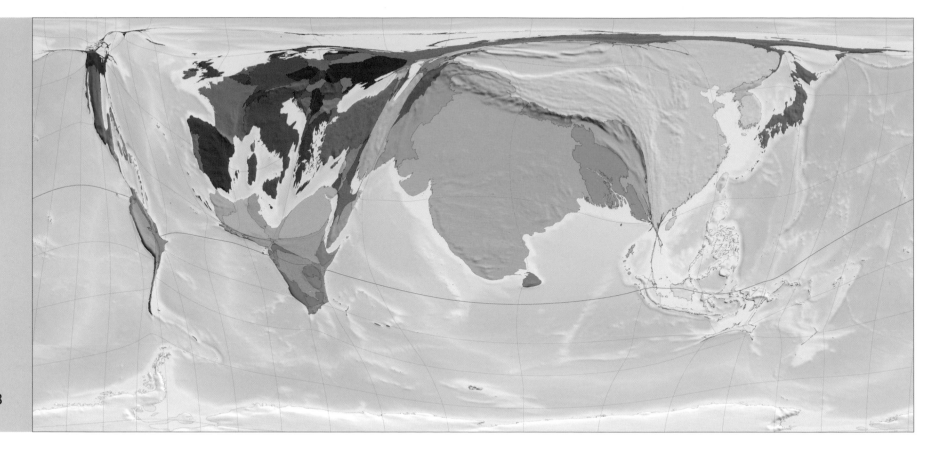

003 Population in the Year 1

This map shows the distribution of the world's population in AD 1. At that time it was estimated at 231 million – under 4 per cent of what it is today.

This map and the five that follow it form a series showing the historical distribution of the world's population, changing over time as a result of differing rates of population growth or decline in different territories. You might like to compare these maps with one another and with Map 002, which shows the population of the world as it is today.

The population of the Earth 2,000 years ago is estimated to have been 231 million. These people were spread across the planet in a way quite similar to today's population – the total numbers of people have changed more than their distribution – but there are some significant differences. North and South America, for instance, were sparsely populated at that time, as was the Asia–Pacific and Australasia region; New Zealand is thought to have been entirely uninhabited. Colder latitudes were generally less populated. South Asia, Northern Africa, China and southern Europe, on the other hand (which are all parts of the same land mass), had relatively high populations. The largest numbers of people lived in or near the Ganges, Tigris, Yangtze, Nile and Po river valleys.

POPULATION DISTRIBUTION YEAR 1

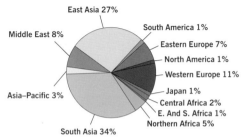

East Asia 27%
South America 1%
Middle East 8%
Eastern Europe 7%
North America 1%
Western Europe 11%
Japan 1%
Asia–Pacific 3%
Central Africa 2%
E. And S. Africa 1%
Northern Africa 5%
South Asia 34%

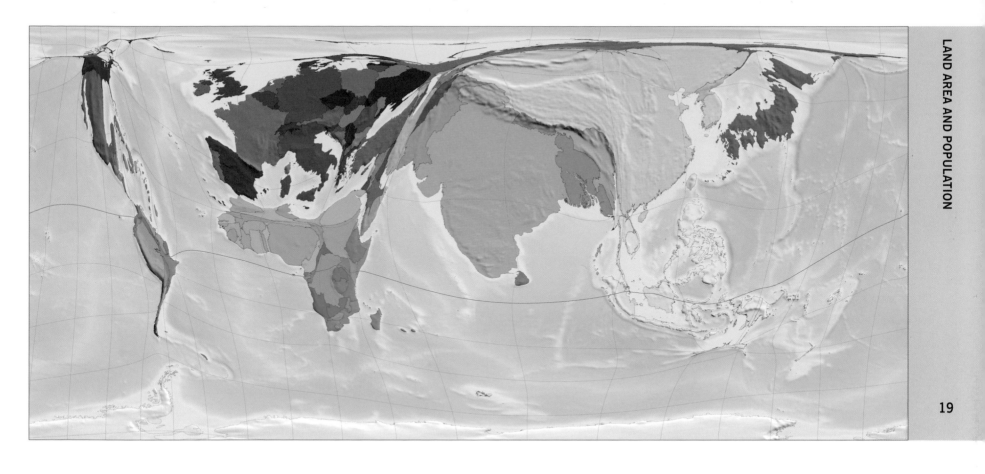

20 HIGHEST POPULATIONS IN 1500

Rank	Territory	Million people
1	China	103
2	India	90
3	Japan	15
4	France	15
5	Germany	12
6	Bangladesh	11
7	Russia	11
8	Indonesia	11
9	Italy	11
10	Pakistan	10
11	Nigeria	8
12	Philippines	8
13	Mexico	7
14	Spain	7
15	Ukraine	7
16	Turkey	6
17	DR Congo	5
18	Sudan	4
18	Egypt	4
18	Iran	4

004 Population in the Year 1500

This map shows the distribution of the world's population in the year 1500.
The pattern remains broadly similar, though the numbers have increased.

The distribution of population over the planet was roughly similar in 1500 to that in the year 1, although the total number of people had almost doubled. The largest populations lived in South Asia and East Asia. Together these regions contained more than half of the world's population at the time. The Americas were still relatively sparsely inhabited and the people who did live there were concentrated primarily in Central America and the northern parts of South America; the combined population of Mexico and Peru in the time of Christopher Columbus and the Spanish conquest was greater than the total of all the other American territories put together.

Even though the world population doubled between the years 1 and 1500, population growth was slow compared to recent times: the number of humans on the Earth has doubled in just the last 50 years.

'The population was decimated…Spanish colonizers treated the native population brutally and the European diseases which they brought to the country were fatal to indigenous people.'

'Mexico', BBC Scotland Education website, 2006

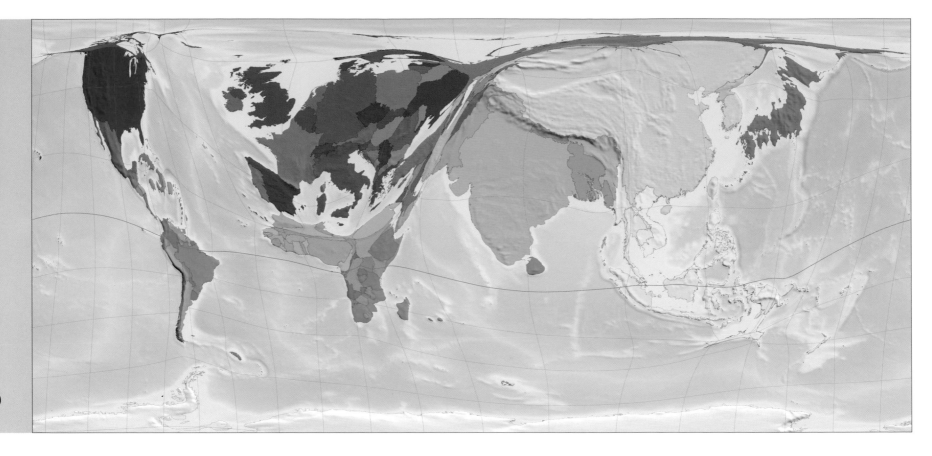

005 Population in the Year 1900

This map shows the distribution of the world's population in the year 1900, after 400 years of rapid increase and geographical spread.

The world population tripled between 1500 and 1900 to reach an estimated 1,564 million just over a century ago. During the same period the populations of the United Kingdom and the United States increased more than tenfold and the population of the Netherlands increased fivefold, although in 1500 none of these states actually existed as a unified country. By 1900 much of the world was under imperial rule and new territorial boundaries were being defined and contested. Most of the modern-day borders of the African territories on this map, for instance, were drawn as a result of the Treaty of Versailles in 1919.

20 HIGHEST POPULATIONS IN 1900

Rank	Territory	Million people
1	China	400
2	India	234
3	United States	76
4	Germany	54
5	Russia	50
6	Japan	44
7	Indonesia	43
8	France	41
9	United Kingdom	37
10	Italy	34
11	Ukraine	28
12	Bangladesh	28
13	Poland	25
14	Pakistan	24
15	Spain	19
16	Brazil	18
17	Viet Nam	16
18	Nigeria	16
19	Turkey	14
20	Mexico	14

'It has been stated that, as men progress, they shall be able to travel in airships and reach any part of the world in a few hours.' Mahatma Gandhi, *Hind Swaraj*, 1909

20 HIGHEST POPULATIONS IN 1960

Rank	Territory	Million people
1	China	667
2	India	434
3	United States	181
4	Russia	120
5	Indonesia	95
6	Japan	94
7	Germany	72
8	Brazil	72
9	Bangladesh	55
10	United Kingdom	52
11	Pakistan	50
12	Italy	50
13	France	46
14	Ukraine	43
15	Nigeria	40
16	Mexico	38
17	Viet Nam	32
18	Spain	31
19	Poland	30
20	Philippines	29

006 Population in the Year 1960

This map shows the distribution of the world's population in the year 1960.
By now some of the trends visible today are becoming clearly apparent.

By 1960 the world population was
3,039 million and China and India each
had more than twice the number of
inhabitants of the next most populous
territory, as they still do today. This map
also shows that the proportion of the world
population living in South America
increased substantially between 1900
and 1960, while the proportion living in
Western Europe declined from 15% in
1900 to 11% in 1960. By the year 2000
it had fallen still further to just 6%.

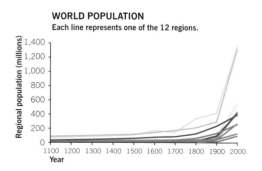

WORLD POPULATION
Each line represents one of the 12 regions.

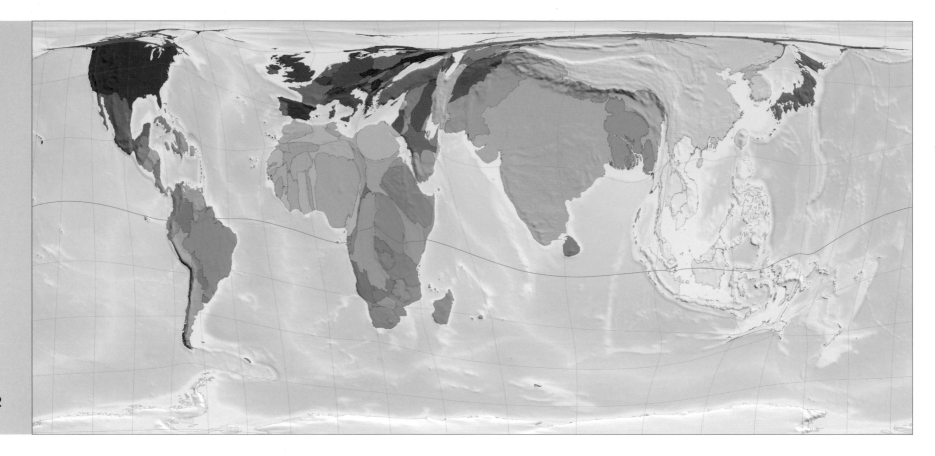

007 Projected Population in the Year 2050

The size of each territory indicates the projected proportion of the world's population that will be living there in the year 2050. The numbers used to create this map are estimates based on current trends.

It is estimated that by the year 2050 the human population of the Earth will be 9.07 billion (9,070 million), with three people alive for every two today. Of those, 62% will live in Africa, South Asia and East Asia, and the combined populations of these three regions will by then be equal to the entire population of the world today.

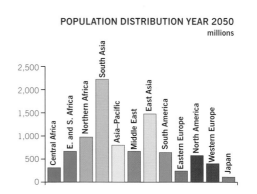

POPULATION DISTRIBUTION YEAR 2050
millions

'The choices that today's generation of young people aged 15–24 years make about the size and spacing of their families will determine whether Planet Earth will have 8, 9 or 11 billion people in the year 2050.' UNFPA, 2005

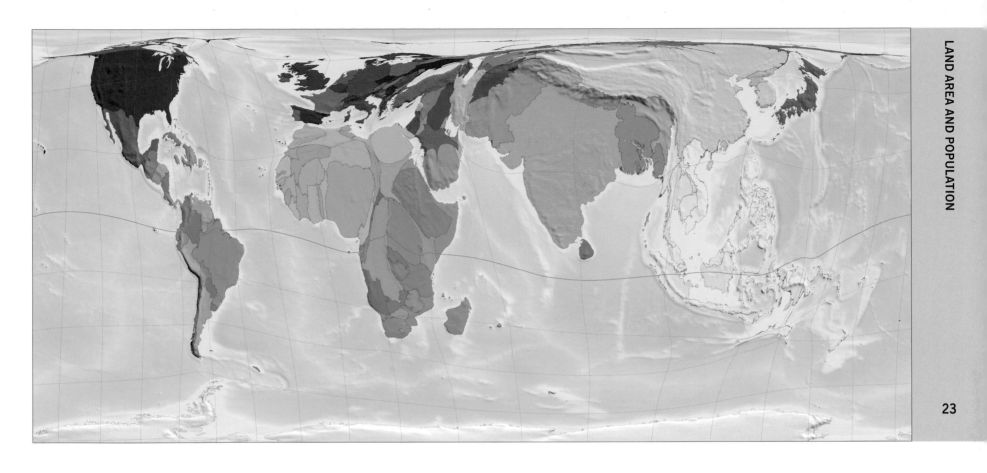

HIGHEST PREDICTED POPULATIONS IN 2300

Rank	Territory	Million people
1	India	1,372
2	China	1,285
3	United States	493
4	Pakistan	359
5	Nigeria	283
6	Indonesia	276
7	Bangladesh	243
8	Brazil	223
9	Ethiopia	207
10	DR Congo	183
11	Uganda	155
12	Yemen	130
13	Mexico	127
14	Philippines	126
15	Egypt	125
16	Viet Nam	114
17	Iran	101
17	Japan	101
19	Niger	94
20	Russia	92

008 Projected Population in the Year 2300

The size of each territory indicates the projected proportion of the world's population that will be living there in the year 2300. Inevitably some speculation is involved, but current trends suggest the outcomes shown here.

The UN forecasts that the world population will be just below 9 billion by the year 2300. World population is expected to grow, peak and then decline slightly between 2050 and 2300. The greatest long-term population growth is predicted to occur in Africa, which is currently underpopulated compared to other continents and has the lowest average life expectancy. The populations of other regions are predicted to remain roughly the same between 2050 and 2300 or else to decline, with India, China, the United States and Pakistan remaining the most populous countries (in that order).

The numbers used to create this map are estimates based on the way populations are expected to evolve in the future, but predictions this far ahead are inevitably based in part on guesswork. It is certainly possible that the future could be quite different from what is shown here.

POPULATION DISTRIBUTION YEAR 2300
millions

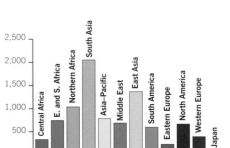

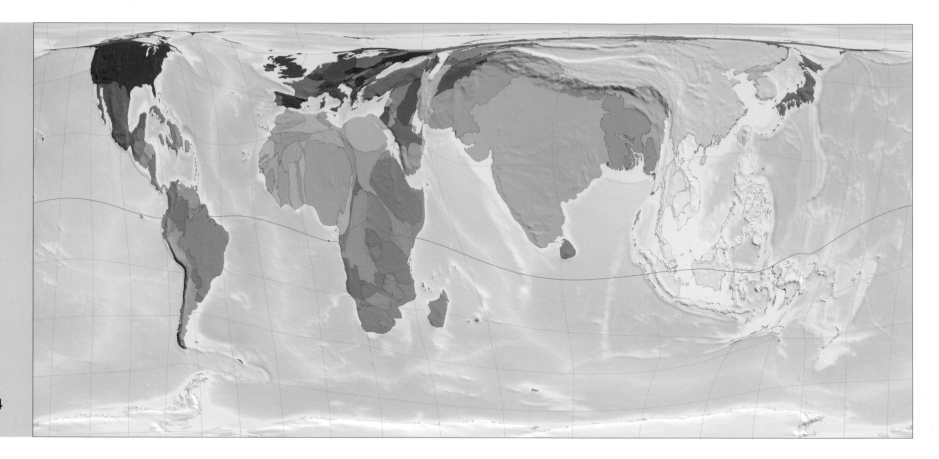

009 Births

Although the world's population is increasing overall, in some territories it is declining. In this map, the size of each territory indicates the number of births occurring there in a single year.

133,121,000 babies were born in the year 2000. This map shows how many of those births occurred in each territory. As with all population statistics, the figures are only estimates, although probably quite accurate estimates in many cases.

More babies are born each year in Africa than are born in the Americas, all of Europe and Japan put together. Worldwide, more than a third of a million people will be (or have been) born on your birthday this year. In some territories with low birth rates, on the other hand, more people are dying than are being born.

'The birth of a baby is an occasion for weaving hopeful dreams about the future.' Aung San Suu Kyi, *Letters from Burma*, 1997

BIRTH RATES
births per 1,000 people per year

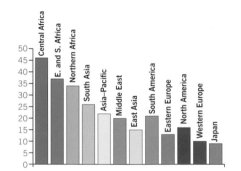

LOWEST RATES OF BIRTH ATTENDANCE

Rank	Territory	Attended births[a]
1	Ethiopia	6.0
2	Nepal	11.0
3	Bangladesh	13.0
4	Afghanistan	14.0
5	Chad	16.0
5	Niger	16.0
7	Laos	19.0
8	Pakistan	20.0
9	Yemen	22.0
10	Bhutan	24.0
10	Timor-Leste	24.0
10	Haiti	24.0
13	Eritrea	24.5
14	Burundi	25.0
15	Rwanda	31.0
16	Cambodia	32.0
17	Somalia	34.0
18	Burkina Faso	34.5
19	Guinea-Bissau	35.0
19	Guinea	35.0

[a] births attended as % of all births

010 Attended Births

The size of each territory indicates the number of births that are attended by healthcare workers. In many parts of the world, qualified help is the exception rather than the rule.

Worldwide, 62% of births in the year 2000 were attended by skilled healthcare personnel; 38% were not. The lowest rate of attended birth was 6%, in Ethiopia; the highest was in Japan, where essentially 100% of births are attended. The chance of a birth being attended in Cambodia is twice what it is in Chad.

'...how many of us realize that, in much of the world, the act of giving life to a child is still the biggest killer of women of child-bearing age?' Liya Kibede, goodwill ambassador for the WHO, 2005

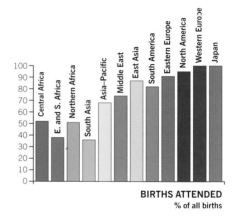

BIRTHS ATTENDED
% of all births

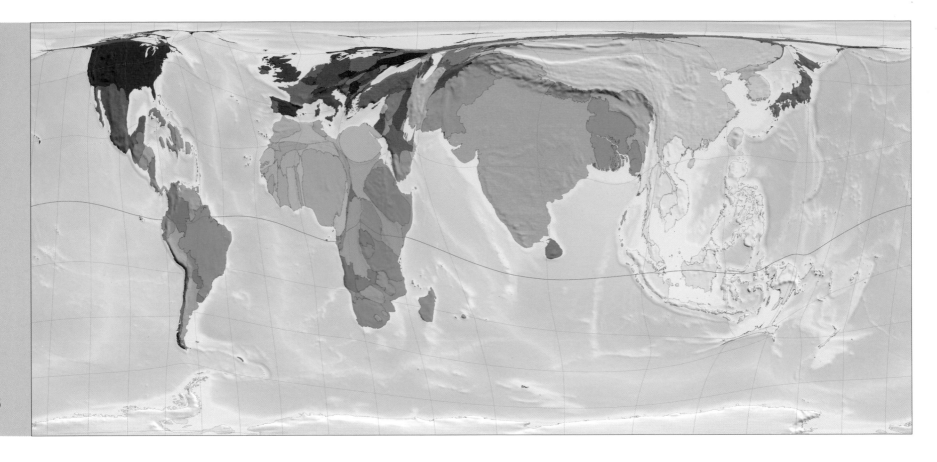

HIGHEST AND LOWEST PROPORTIONS OF POPULATION AGED 15 AND UNDER

Rank	Territory	Children[a]
1	Uganda	50.1
2	Niger	50.0
3	Mali	49.2
4	Burkina Faso	48.9
5	Yemen	48.7
6	Angola	47.5
7	Guinea-Bissau	47.1
8	Burundi	46.9
9	DR Congo	46.8
9	Congo	46.8
191	Switzerland	16.2
192	Czech Republic	15.7
192	Hong Kong (China)	15.7
194	Germany	15.2
195	Slovenia	15.0
196	Bulgaria	14.8
197	Greece	14.7
198	Japan	14.3
198	Spain	14.3
200	Italy	14.1

[a] children as a % of total population

011 Number of Children

The size of each territory indicates the number of children aged 15 and under, who overall make up a third of the world's population.

This map and the accompanying graph show statistics for children aged 15 and under only. Those aged 16 and 17 are still considered children under international law but are often excluded from international statistics on childhood. In 2002 there were 1,826 million children aged 15 and under in the world.

Among the continents, Africa has the highest percentage of children; in Uganda and Niger half the population is 15 or younger. In Italy, Spain and Japan, by contrast, only 14% of the population is aged 15 or under. The territory with the largest total number of children is India, with 349 million aged 15 or under in 2002.

POPULATION AGED 15 AND UNDER
% of total population

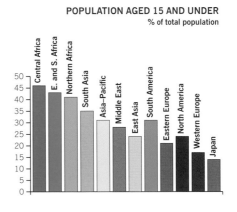

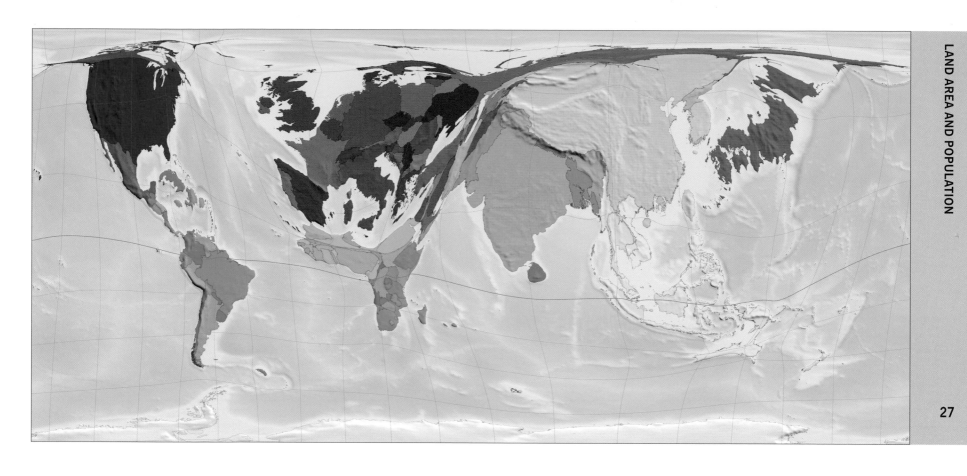

Rank	Territory	Elderly people[a]
1	Italy	18.7
2	Japan	18.2
2	Greece	18.2
4	Sweden	17.4
5	Belgium	17.3
6	Germany	17.1
7	Spain	17.0
8	Switzerland	16.4
9	Bulgaria	16.3
9	Croatia	16.3
191	Mali	2.4
191	Senegal	2.4
193	Tanzania	2.3
193	Yemen	2.3
195	Eritrea	2.1
195	Oman	2.1
197	Niger	2.0
198	Qatar	1.5
199	Kuwait	1.4
200	United Arab Emirates	1.3

[a] elderly as % of total population

012 Number of Elderly

The size of each territory indicates the number of people aged 65 and over. The proportion of elderly people in a country's population depends on political and social as well as physical factors.

In 2002 7% of the world population was 65 years old or over. The proportion of elderly worldwide and in individual countries depends on longevity and also on birth rates (since low birth rates reduce the size of the younger population).

The territory with the largest number of elderly in 2002 was China, with 92 million. Moreover, the proportion of elderly in China is expected to grow substantially in the near future because of the long-term effects of the 'one child' policy there, which limits most couples to having only a single child. Africa, by contrast, with its low life expectancy and higher birth rates, is home to only 6% of the world's population aged 65 or over.

'We live in an era of unprecedented, rapid and inexorable global ageing.' HelpAge International, 2002

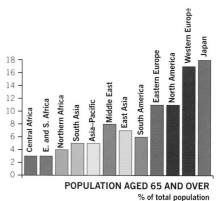

POPULATION AGED 65 AND OVER
% of total population

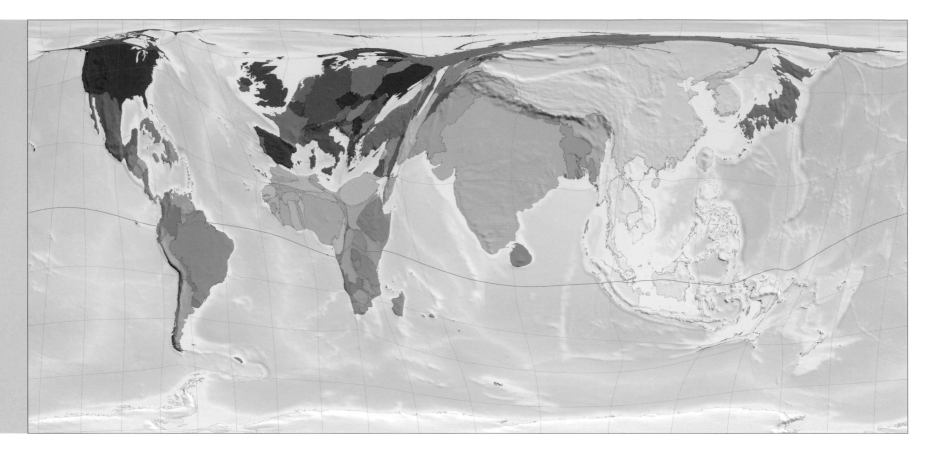

013 Right to Vote

The size of each territory indicates the collective amount of time in person-years that all adults in a territory have had the right to vote to elect those who govern them.

This map shows the total numbers of person-years lived under universal suffrage – the right of every adult to vote – calculated as the number of years of universal suffrage prior to 2004 multiplied by the total population. A person-year in this context represents one person living for one year in a territory where everyone may vote. Countries can appear large on this map because they have large populations or because they have had universal suffrage for a long time.

Universal adult suffrage was first introduced in New Zealand, in 1893;

the next territories to follow suit were the Scandinavian countries of Finland (1906), Norway (1913) and Denmark (1915). By 1994, almost all adults in 190 of the 200 territories could vote.

Some of the territories on the map have not existed for long as political entities. In some cases the inhabitants of the previous political entity to occupy the same region also had the right to vote. Thus the data on which the map are based do not always reflect the true length of time for which inhabitants of an area have been able to vote.

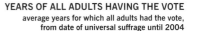

YEARS OF ALL ADULTS HAVING THE VOTE
average years for which all adults had the vote, from date of universal suffrage until 2004

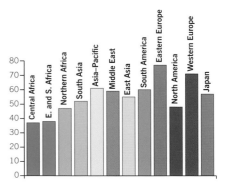

LARGEST VOTER TURNOUTS

Rank	Territory	% [a]
1	Italy	91
2	Seychelles	91
3	Angola	88
4	Cambodia	88
5	Austria	88
6	Estonia	87
7	Germany	86
8	Hungary	86
9	Kyrgyzstan	86
10	Latvia	85
11	Timor-Leste	85
12	Albania	85
12	Cook Islands	85
14	Australia	85
15	Netherlands	85
16	Malawi	85
17	Denmark	84
18	South Africa	84
19	Czech Republic	83
20	Sweden	83

[a] average % of voters who voted in all national elections for which there were data, 1945–98

014 Voter Turnout

The size of each territory indicates the estimated numbers of people who voted in national elections between 1945 and 1998. High turnout can indicate greater faith in the political process as much as greater interest in the outcome.

Voter turnout is the percentage of those eligible to vote who actually do so. Between 1945 and 1998 the average voter turnout in elections worldwide was 64%. In most territories, having the right to vote also means having the right not to vote, so that turnout is typically less than 100%, sometimes much less. At some times and in some places abstaining from voting is as strong a statement as voting.

One must, however, exercise caution in comparing the voter turnout of different territories, since political systems and types of elections vary widely and often are not directly comparable. The highest voter turnouts recorded have been in Italy and the Seychelles, both at 91%. In general, high turnout is observed in territories characterized by widespread belief in the value of political processes, credible voter anonymity and accessible ballot boxes.

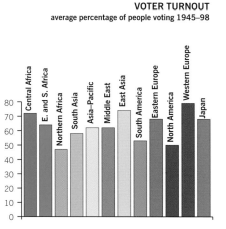

VOTER TURNOUT
average percentage of people voting 1945–98

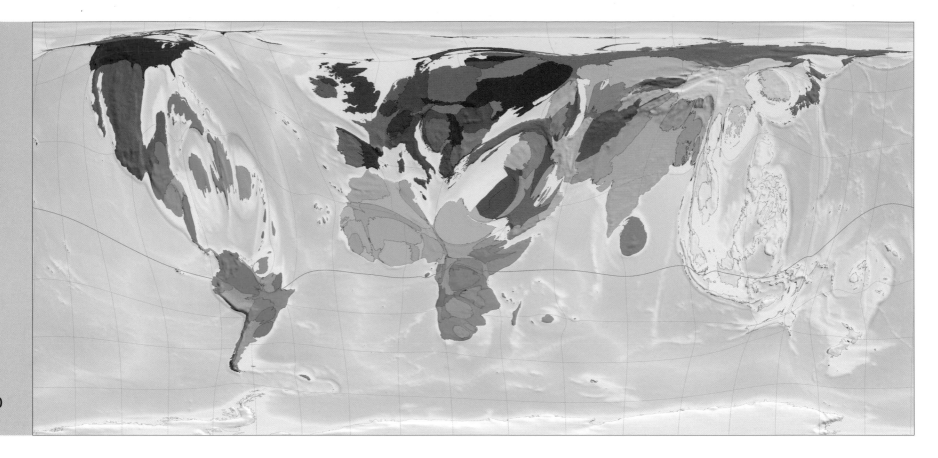

**HIGHEST AND LOWEST LEVELS
OF EMIGRATION**

Rank	Territory	Emigrants[a]
1	Andorra	76
2	Tonga	69
3	St Lucia	67
4	St Vincent and the Grenadines	66
5	Monaco	65
6	Gaza and West Bank	60
7	Samoa	55
8	Suriname	51
9	Guyana	50
10	Jamaica	38
191	Venezuela	0.64
192	Tanzania	0.64
193	Taiwan	0.60
194	Kenya	0.54
195	Brazil	0.53
196	Libya	0.49
197	China	0.43
198	Bahamas	0.36
199	Central African Republic	0.17
200	North Korea	0.16

[a] total emigrants as % of resident population

015 International Emigrants

The size of each territory indicates the number of international emigrants giving it as their place of origin. The perspective is influenced by the absolute size of populations in the source countries.

As we can see from this map, patterns of emigration are not dominated by a single region, but some variations exist nonetheless. Regional averages for the percentage of the population who emigrate range from roughly 1% in South Asia, East Asia and Japan to 8% in Eastern Europe and 9% in the Middle East and Central Asia. Despite their relatively low percentage emigration rates, South and East Asia are still easily visible on this map because they have very large total populations, and hence the total numbers of emigrants are still quite large, even if they represent only a small fraction of the population.

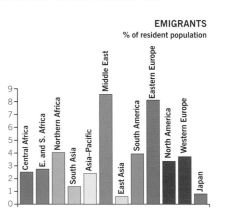

EMIGRANTS
% of resident population

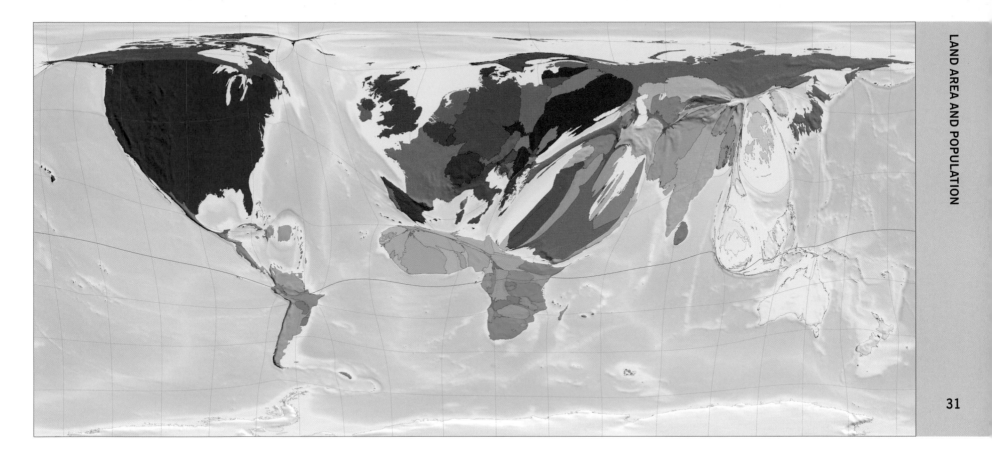

016 International Immigrants

The size of each territory indicates the number of international immigrants living there. Some of the smallest territories show both the highest and the lowest proportions of immigration.

By the year 2000, 3% of the world population – 174 million people – were living in a territory other than the one in which they were born. The United States receives the highest number of international immigrants, while Andorra has the highest proportion of immigrants: 4 out of every 5 people in Andorra are international immigrants. In the Philippines and Guyana, which have among the lowest rates of immigration, only about 1 person in every 500 is an immigrant.

IMMIGRANTS
% of resident population

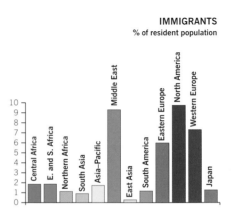

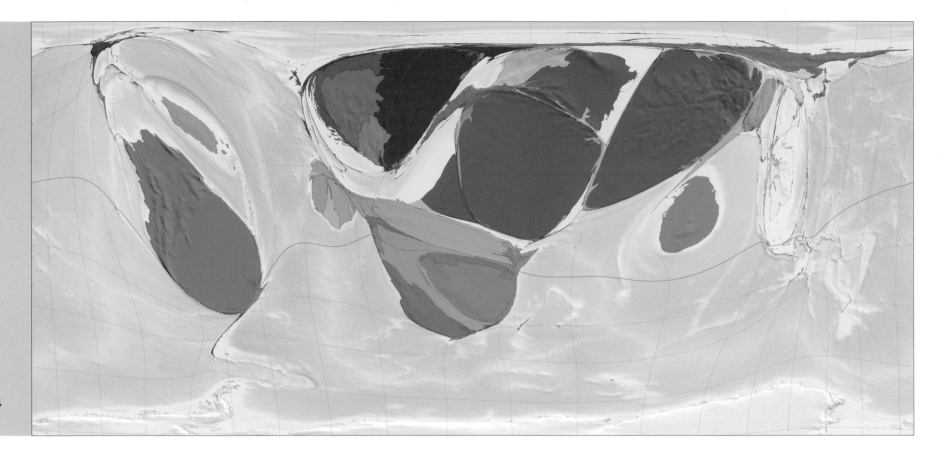

**ORIGIN OF REFUGEES AND INTERNALLY
DISPLACED PERSONS**

Rank	Territory	%ᵃ
1	Bosnia Herzegovina	12.0
2	Serbia and Montenegro	12.0
3	Afghanistan	9.9
4	Azerbaijan	9.9
5	Iraq	9.9
6	Gaza and West Bank	9.6
7	Burundi	9.5
8	Bhutan	5.2
9	Croatia	5.2
10	Georgia	5.2
11	Colombia	4.7
12	Puerto Rico	4.7
13	Sri Lanka	2.5
14	Angola	2.4
15	Liberia	1.7
16	Sudan	1.7
17	Western Sahara	1.7
18	Sierra Leone	1.6
19	Tajikistan	1.0
20	Mauritania	0.9

ᵃ number of refugees and internally
displaced persons as % of resident
population in territory of origin

019 Refugee Origins

The size of each territory indicates the proportion of all refugees
and internally displaced persons who give it as their place of origin.

Traditionally refugees (those fleeing to
another territory) are counted separately
from internally displaced persons (those
fleeing to a different part of the same
territory). This map shows the territories
of origin of both combined.

In 2003 there were an estimated 15
million refugees and internally displaced
persons in the world. Because people move
within as well as between territories, some
territories are simultaneously a major origin
and a major destination of displaced
people. Examples include Iraq and
Afghanistan.

ORIGINS
thousands of internally displaced persons (bottom)
and refugees (top)

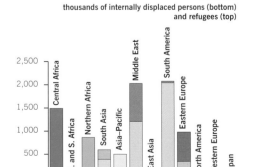

The data in this graph represent the numbers that
were officially reported. The map above includes
estimates, which particularly affects the Middle East.

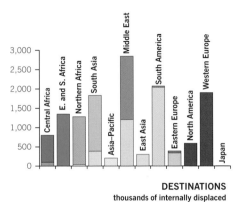

DESTINATIONS FOR REFUGEES AND INTERNALLY DISPLACED PERSONS

Rank	Territory	%[a]
1	Bosnia Herzegovina	8.5
2	Armenia	7.7
3	Azerbaijan	6.9
4	Georgia	5.1
5	Colombia	4.7
6	Djibouti	3.9
7	Serbia and Montenegro	2.6
8	Congo	2.5
9	Afghanistan	2.4
10	Iraq	2.4
11	Guinea-Bissau	2.2
12	Burundi	2.1
13	Zambia	2.1
14	Sri Lanka	2.0
15	Chad	1.8
16	Tanzania	1.8
17	Sweden	1.6
18	Puerto Rico	1.5
19	Denmark	1.4
19	Iran	1.4

[a] number of refugees and internally displaced persons as % of resident population in destination territory

020 Refugee Destinations

The size of each territory indicates the proportion of all refugees and internally displaced persons living there.

This map shows the destination territories of refugees and internally displaced persons combined. Following the definition agreed by the UN in 1951, a refugee is someone fleeing a country for 'fear of being persecuted for reasons of race, religion, nationality, membership of a particular social group or political opinion'. Someone who is in similar circumstances, or has been forced to leave their home to avoid disaster or armed conflict, but who does not cross an international border, is an internally displaced person.

In 2003 the highest numbers of refugees and IDPs combined went to the Middle East and Central Asia and to South America, while Pakistan, Iran and Germany provided asylum to the largest numbers of people from outside their own borders.

'I miss my country, the sunshine of my country, its soil, my friends, my [family], the way of life and its incredible simplicity.'
Habib Souaïdia, author of *La Sale Guerre* (*The Dirty War*), 2001

DESTINATIONS
thousands of internally displaced persons (bottom), and refugees (top)

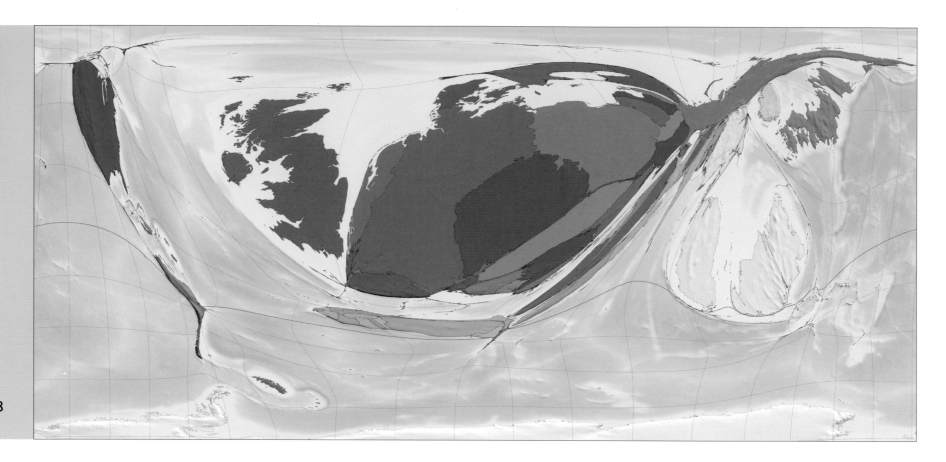

023 **Net Outgoing Tourism**

The size of each territory indicates the number of tourist trips from that territory minus the number to the territory from elsewhere.

The map is dominated by Germany, the United Kingdom, the Czech Republic, Malaysia and Poland: countries whose residents frequently travel abroad, but which are less common destinations for travellers from elsewhere. German residents, for example, make 70.7 million tourist trips a year, but only 18.4 million tourist trips a year are made to Germany.

The net outgoing tourism for Germany is thus 52.3 million visits. Together, residents of Western and Eastern European territories constitute over two-thirds of net outgoing tourism in the world, although this is partly because there are so many international borders within Europe and much of the tourism taking place is between different European countries.

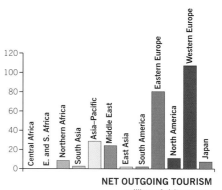

NET OUTGOING TOURISM
millions of visits per year

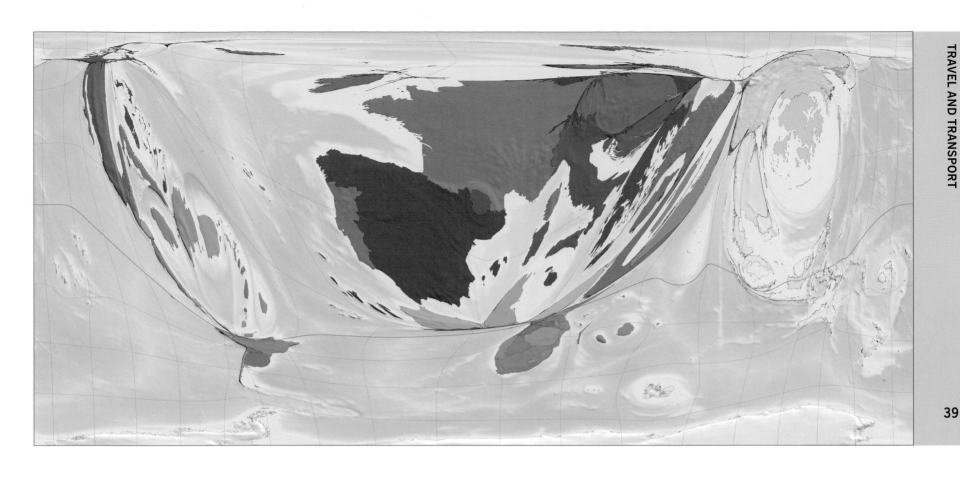

024 Net Incoming Tourism

The size of each territory indicates the number of tourist trips made to that territory minus the number made from that territory to elsewhere.

France and Spain between them account for over a third of net incoming world tourism. Each is the destination of more than three times the number of tourist trips to the next three territories combined (Austria, Italy and China).

'For my part, I travel not to go anywhere, but to go. I travel for travel's sake. The great affair is to move.'

Robert Louis Stevenson, *Travels with a Donkey*, 1879

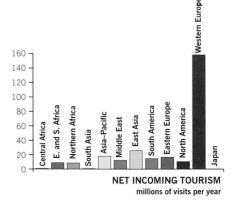

NET INCOMING TOURISM
millions of visits per year

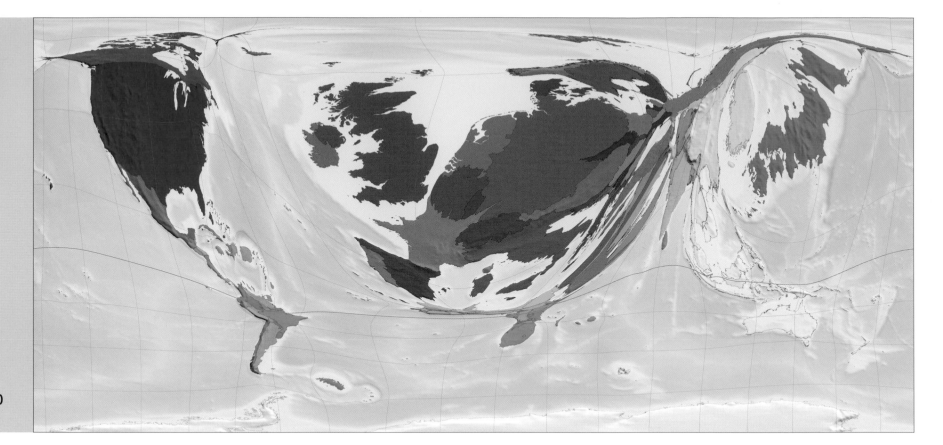

HIGHEST AND LOWEST SPENDING ON INTERNATIONAL TOURISM

Rank	Territory	US$ª
1	Luxembourg	6,005
2	Kuwait	1,563
3	Austria	1,562
4	Norway	1,499
5	Iceland	1,467
6	United Arab Emirates	1,365
7	Bahamas	1,347
8	Switzerland	1,312
9	Belgium	1,281
10	Ireland	1,239
191	Malawi	4.03
192	Rwanda	3.98
193	Angola	3.94
194	Laos	3.09
195	Bangladesh	2.81
196	Burkina Faso	2.78
197	Niger	1.39
198	Ethiopia	0.80
199	Myanmar	0.74
200	Afghanistan	0.04

ª **tourist spending abroad per person per year, US$**

025 Spending on Tourism

In 2003 each member of the world's population spent an average of US$92 on tourism. In this map, the size of each territory indicates the total amount, in dollars, spent on tourism by its residents.

Like income from tourism (Map 026), spending on tourism is very unevenly distributed, ranging from US$6,005 per person in the highest-spending territory to US$0.04 in the lowest. The highest spenders are the residents of Luxembourg, Kuwait and Austria. The lowest are the residents of Afghanistan, Myanmar and Ethiopia. As a nation, the United States spent the highest total amount of money on tourism, followed by Germany, the United Kingdom and Japan.

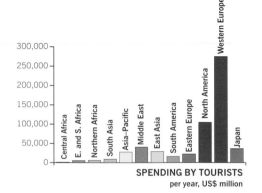

SPENDING BY TOURISTS
per year, US$ million

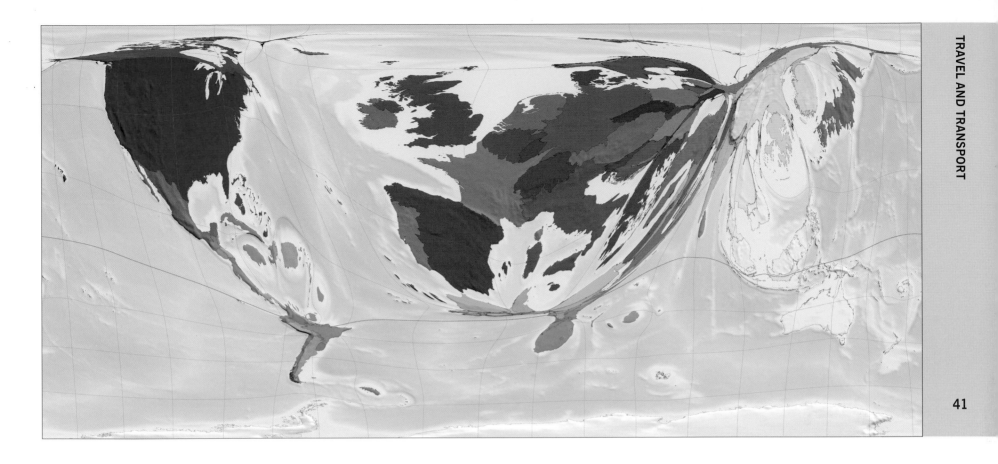

HIGHEST AND LOWEST TOURIST RECEIPTS

Rank	Territory	US$ᵃ
1	Luxembourg	7,036
2	Bahamas	5,686
3	Palau	2,803
4	Cyprus	2,665
5	Barbados	2,429
6	Seychelles	2,300
7	Malta	2,033
8	Austria	1,906
9	Iceland	1,539
10	Switzerland	1,497
191	Nigeria	2.07
192	Burkina Faso	1.89
193	Uzbekistan	1.77
194	Myanmar	1.32
195	Gaza and West Bank	1.12
196	Tajikistan	1.07
197	Guinea	0.92
198	Central African Republic	0.75
199	Bangladesh	0.39
200	Burundi	0.17

ᵃ tourism earnings per person per year, US$

026 Income from Tourism

The size of each territory indicates the total amount of income, in dollars, earned from tourism. Poorer areas earn less in part because a dollar buys more there than in richer areas.

In 2003 total income across the world from tourism came to US$573 billion. Receipts are quite unevenly distributed, however, with 72.7% going to just the top 10% of territories. Tourist income is also directed disproportionately towards developed countries such as the United States, Spain, Italy and France. Most less well-off territories have low total tourist receipts.

TOURIST RECEIPTS
per year, US$ million

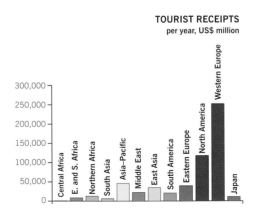

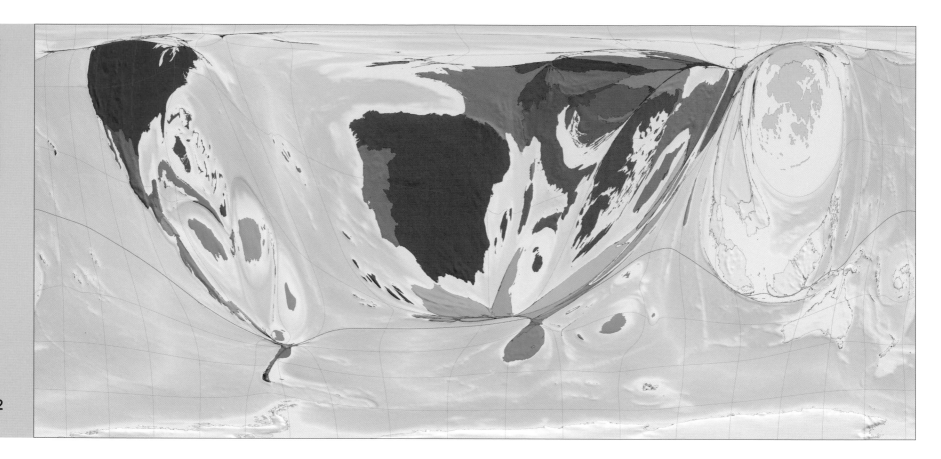

027 Net Income from Tourism

The size of each territory indicates the total income from tourism minus
the amount spent on foreign tourism by its residents travelling elsewhere.

This map shows the net income gained
when foreign tourists spend more in a
territory than its residents spend abroad
as tourists themselves. The highest figure
is for Spain, with a net income of US$33
billion in 2003, more than twice that of the
second-placed earner, the United States.
Interestingly, the territories with the highest
net income per person are all islands: the
Bahamas, Palau, Barbados, the Seychelles,
Cyprus, Malta and Hong Kong (which is
part of China).

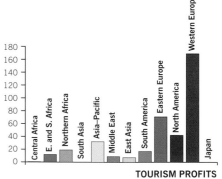

TOURISM PROFITS
US$ per person per year

43

HIGHEST NET LOSES FROM INTERNATIONAL TOURISM LOSS

Rank	Territory	US$ᵃ
1	Kuwait	1,433
2	United Arab Emirates	894
3	Norway	848
4	Germany	529
5	United Kingdom	499
6	Belgium	473
7	Sweden	354
8	Japan	201
9	Maldives	195
10	Netherlands	189
11	Israel	172
12	Canada	148
13	Oman	142
14	Suriname	127
15	Libya	99
16	Tajikistan	95
17	Gaza and West Bank	95
18	Uzbekistan	94
19	South Korea	92
20	Serbia and Montenegro	78

ᵃ **annual net loss from tourism per person, US$**

028 Net Loss from Tourism

The size of each territory indicates the total amount spent on tourism by its residents travelling abroad, minus tourist income from incoming visitors.

This map shows which territories' residents spend more on trips abroad than is spent by foreign tourists visiting. For example, Japanese tourists spent a total of US$36,506 million on holidays abroad during 2003, while foreign tourists in Japan spent only US$10,904 million. Thus Japan had a net financial loss from tourism of US$25,602 million in 2003. The largest net losses were in Germany (US$44 billion), the United Kingdom (US$29 billion) and

Japan (US$26 billion). The largest net losses per person were in Kuwait and the United Arab Emirates, which respectively lost US$1,433 and US$894 per person.

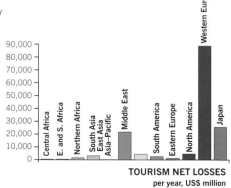

TOURISM NET LOSSES
per year, US$ million

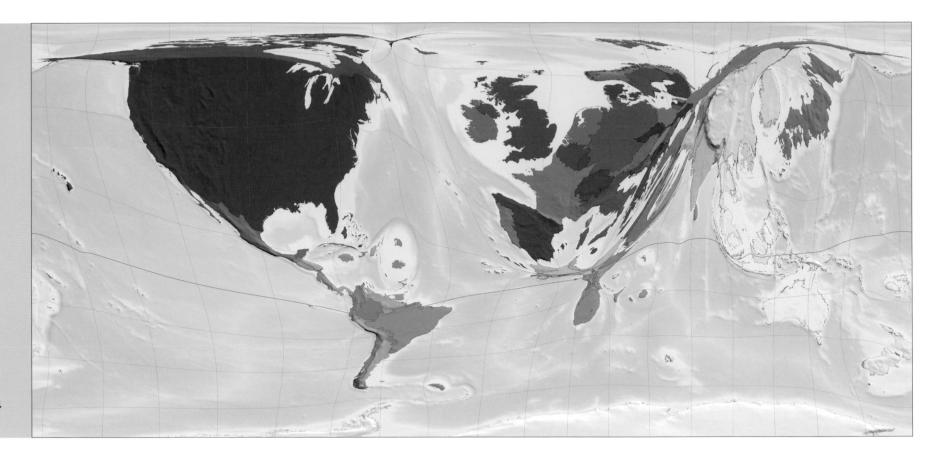

029 Aircraft Departures

The size of each territory indicates the number of civilian aircraft journeys beginning there. As it shows where aeroplanes are registered, rather than where they take off, there are one or two anomalies.

There are over 21 million civilian aircraft departures worldwide every year, or an average of 40 every minute. On this map departures are attributed to the territories in which aircraft are registered rather than the territories where they physically take off. This doesn't affect the map greatly, although it does produce a few odd results. Monaco, for example, has the second largest number of registered departures per person in the world but has no international airport.

North America and Western Europe are responsible for two-thirds of aircraft departures on this map. Africa, by contrast, accounts for only 2.5% of all departures.

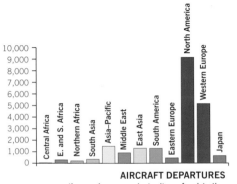

AIRCRAFT DEPARTURES
thousands per year, by territory of registration

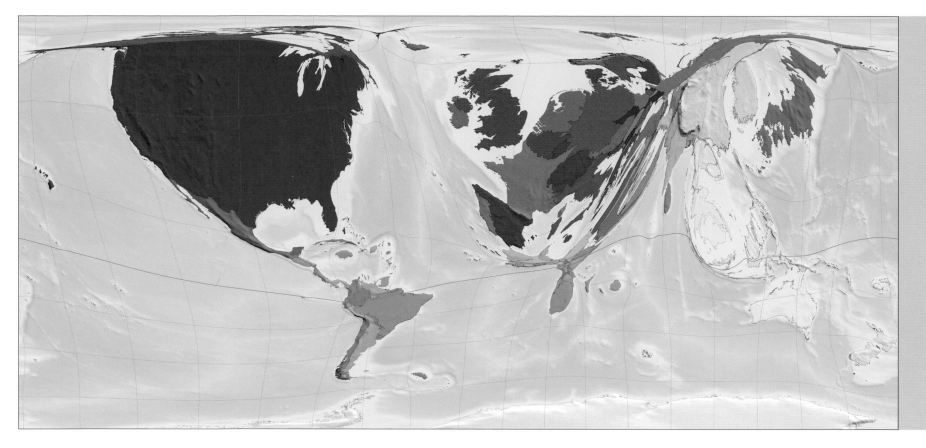

[a] km per head of population flown per year by aircraft registered in that territory

030 Aircraft Travel

Some people fly hundreds of thousands of kilometres a year while others have never been in an aeroplane. In this map, the size of each territory indicates the total distance flown by aircraft registered there.

Civilian aircraft currently fly 25 billion kilometres (16 billion miles) a year – the equivalent of going around the world 630,000 times – and the figure is on the increase. If the total distance flown by all aircraft passengers were divided equally among everyone in the world, each of us would fly 317 kilometres a year.

The people who fly most tend to be from island territories. There are exceptions, however: people from Haiti and the Dominican Republic, for instance, both parts of the Caribbean island of Hispaniola, are among those who fly the least.

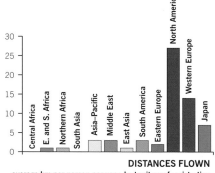

DISTANCES FLOWN
average km per person per year, by territory of registration

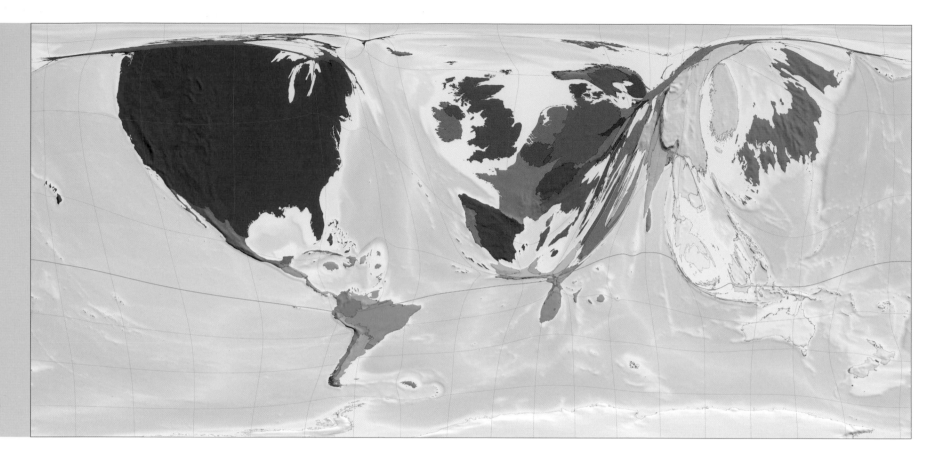

46

MOST AND FEWEST AIRCRAFT PASSENGERS

Rank	Territory	Passengers[a]
1	Antigua and Barbuda	14,260
2	Nauru	12,385
3	Bahamas	6,203
4	Iceland	4,773
5	Qatar	4,455
6	Singapore	3,977
7	Seychelles	3,940
8	Ireland	3,585
9	Malta	3,413
10	Norway	3,374
191	Tanzania	5.3
192	Nigeria	4.2
193	DR Congo	4.0
194	Sierra Leone	4.0
195	North Korea	3.7
196	Burundi	1.8
197	Uganda	1.6
198	Rwanda	1.2
199	Dominican Republic	1.2
200	Lesotho	0.6

[a] passengers per 1,000 people per year by country of airline registration

031 Air Passengers

The size of each territory indicates the total number of flights taken by passengers on aircraft registered there.

There were 1.6 billion air passengers in 2000. In these statistics, every time someone takes a flight they are counted as an aircraft passenger.

40% of all air passengers fly on aeroplanes registered in the United States. This includes both international and domestic flights, and also the many residents of other countries who fly on US aeroplanes.

PASSENGER FLIGHTS
per year, millions

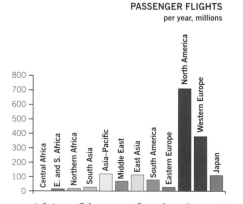

'...most of the world's cities are now within 36 hours of each other...'
Peter Haggett, *Geography: A Global Synthesis*, 2001

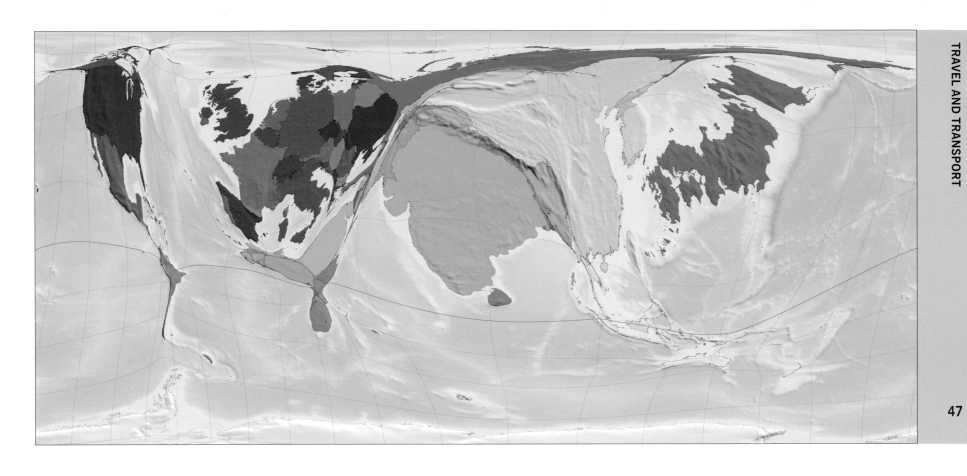

**LONGEST AND SHORTEST DISTANCES
TRAVELLED BY RAIL**

Rank	Territory	Distance[a]
1	Japan	1,876
2	Switzerland	1,783
3	Belarus	1,449
4	France	1,225
5	Russia	1,061
6	Austria	1,039
7	Ukraine	1,034
8	Denmark	1,024
9	Netherlands	887
10	Germany	848
127	Côte d'Ivoire	9.0
128	Mozambique	7.4
129	Tajikistan	6.6
130	Angola	5.6
131	Ghana	4.1
132	Cambodia	3.3
133	DR Congo	3.1
134	Sudan	2.2
135	Philippines	1.6
136	Madagascar	0.6

[a] rail km travelled per person per year (64
countries had no recorded rail system)

032 Rail Travel

The size of each territory indicates the total distance travelled by train passengers
there. Nearly a third of territories on the map have no railway lines.

In 2003 passengers worldwide travelled
a total of 2.2 trillion kilometres by train.
Of this distance, a fifth was covered in
India, another fifth in China and a tenth
in Japan. The annual average distance
travelled by train is 358 kilometres per
person in the world. Within individual
territories Japan has the highest rail
use, with an averageof 1,876 kilometres
travelled by train each year per person,
while at the other end of the scale there are
64 territories on our map (out of a total of
200) that have no railway network at all.

DISTANCES TRAVELLED BY TRAIN
km per person per year

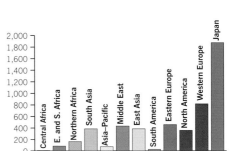

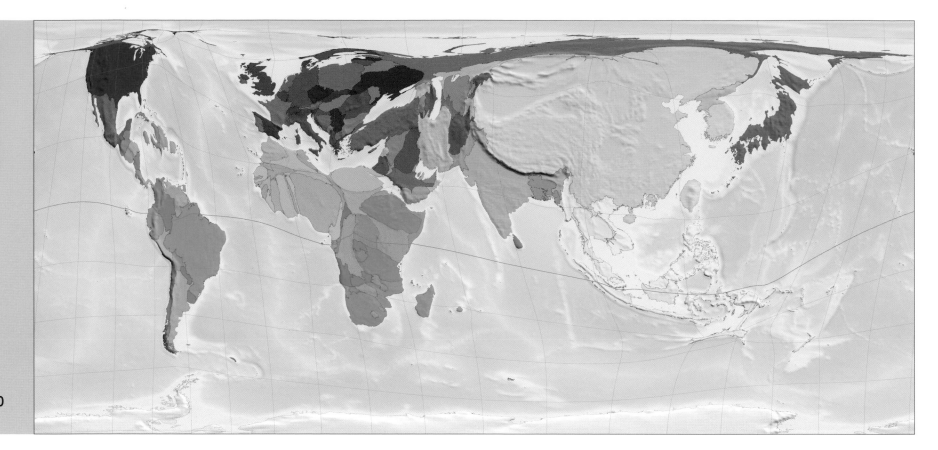

MOST AND LEAST USE OF PUBLIC TRANSPORT

Rank	Territory	%[a]
1	Bosnia Herzegovina	47
2	Armenia	44
3	Russia	43
4	Mongolia	42
5	Moldova	42
6	Bulgaria	41
7	Georgia	39
8	Kazakhstan	39
9	Qatar	37
10	Turkmenistan	35
191	Bangladesh	4.3
192	Sri Lanka	4.1
193	India	3.8
194	Nepal	3.8
195	Bhutan	3.8
196	Maldives	3.8
197	Cuba	3.6
198	Pakistan	3.1
199	Burkina Faso	0.8
200	Cambodia	0[b]

[a] % of the whole population who use public transport to get to work, 1998
[b] 0% reported for Phnom Penh

035 Public Transport

Over a third of all journeys to and from work are made on public transport. The size of each territory in this map indicates the number of people living there who travel to work by this means.

Public transport as defined here includes buses, trams and trains, but not taxis, rickshaws, ferries or animals. 35% of all commuting trips made worldwide take place on public transport. The highest use of public transport for travel to work occurs in Bosnia Herzegovina, where 47% of the population commute by this means. Given that less than half the population of Bosnia actually works, this means that essentially everyone who has a job gets to it by using public transport.

'...with the opening of Beijing's first exclusive Bus Rapid Transit (BRT) line on December 30, car owners may soon be jealous of bus riders...' Yingling Liu, 'Bus Rapid Transit', 2006

PUBLIC TRANSPORT USERS
daily trips to work using public transport, millions

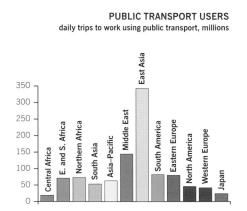

LONGEST AND SHORTEST COMMUTING TIMES

Rank	Territory	Minutes[a]
1	Thailand	71
2	Kenya	54
3	Algeria	53
4	Central African Republic	52
5	Cuba	49
6	Liberia	46
7	Gambia	45
8	Hong Kong (China)	45
9	Nigeria	44
10	South Korea	44
191	Libya	14
192	Panama	13
193	Oman	12
194	Nicaragua	12
195	Bosnia Herzegovina	11
196	Guatemala	11
197	Kuwait	8
198	Peru	8
199	Lebanon	8
200	Malawi	4

[a] total minutes travelling to and from work in 1998 by those in work, divided by total population

036 Commuting Time

The time it takes to get to work depends on how fast as well as how long the trip is.
The size of each territory indicates the total number of person-hours per year spent commuting.

Commuting time is the time it takes to travel from home to work and back – by foot, bus, car, boat, train, bicycle or other means.

In Thailand, which has the longest commuting times in the world, the population spends a combined total of 74 million hours commuting every day, which works out at 71 minutes for every member

of the population. Not everyone works, however, so the average commuting time for people who do is certainly even longer. At the other end of the scale, the quickest journeys to work are in Malawi, where the average member of the population spends just 4 minutes commuting a day.

TIME SPENT TRAVELLING TO WORK
total time travelling to work, divided by total population, in minutes

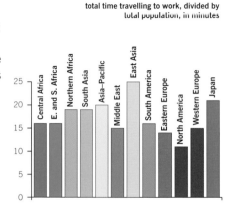

'Currently, the average travel speed in central Bangkok during peak hours is just 7 mph for this lively city of 7.5 million.' Bridges, 2005

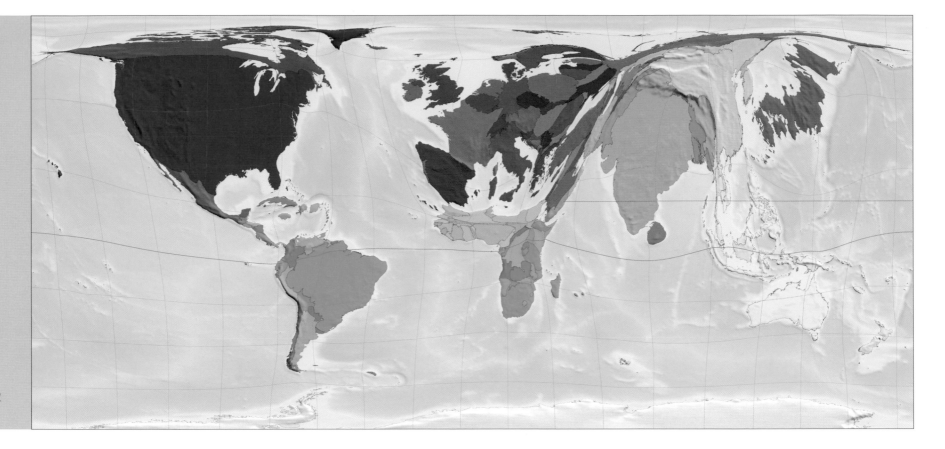

037 Roads

In cities most people can hear road traffic around the clock, whereas in some areas people have never seen a car. The size of each territory indicates the total length of usable roads there.

In 2002 there were 29 million kilometres of road – routes passable by a four-wheeled on-road vehicle, though not necessarily paved – worldwide. If the network of roads were evenly spread out over all territories in a grid system, you would never be more than 4.5 kilometres from a road (provided you stayed on land).

But, like most things in this atlas, roads are not evenly distributed. In cities most people live, work and sleep within a few metres of a road, while some in remote places live so far from roads that they have never seen even a bicycle. Regionally the highest density of roads is found in Japan. The lowest density is in the Middle East and Central Asia, which has less than a quarter the length of road found in North America, despite being the site of some of the world's oldest cities.

'Road transport is…the dominant form of transport in Sub Saharan Africa…It carries 80 to 90 per cent of…passenger and freight transport and provides the only form of access to most rural communities.'

Ian Heggie, *Management and Financing of Roads: An Agenda for Reform*, 1994

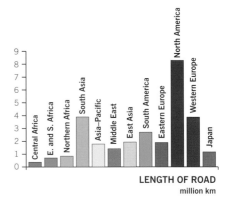

LENGTH OF ROAD
million km

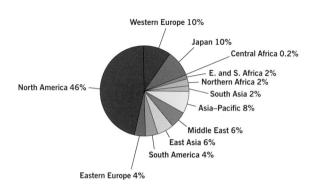

MOST AND FEWEST FREIGHT VEHICLES AND BUSES

Rank	Territory	Freight vehicles[a]
1	United States	299
2	Greenland	219
3	Cyprus	168
4	Suriname	154
5	Japan	153
6	Qatar	129
7	New Zealand	118
8	Malta	111
9	Australia	108
10	Kuwait	106
191	Benin	1.4
192	Mozambique	1.3
193	Somalia	1.3
194	Pakistan	1.3
195	Equatorial Guinea	1.3
196	Afghanistan	1.3
197	Niger	1.2
198	Myanmar	1.0
199	Bangladesh	0.6
200	Ethiopia	0.6

[a] freight vehicles and buses per 1,000 people

038 Freight Vehicles

The size of each territory indicates the number of freight vehicles there, including lorries or trucks, buses and vans.

In 2002 there were 202 million freight vehicles in the world, or 3 for every 100 people. Almost half of all freight vehicles are in North America, most of them in the United States. The United States also has the largest number of freight vehicles per person. The smallest numbers of freight vehicles per person are found in South Asia and Central Africa. Western Europe has 10 times as many freight vehicles per person as Central Africa and Japan has 3 times as many as Western Europe.

FREIGHT VEHICLE DISTRIBUTION

Western Europe 10%
Japan 10%
Central Africa 0.2%
E. and S. Africa 2%
Northern Africa 2%
South Asia 2%
Asia–Pacific 8%
North America 46%
Middle East 6%
East Asia 6%
South America 4%
Eastern Europe 4%

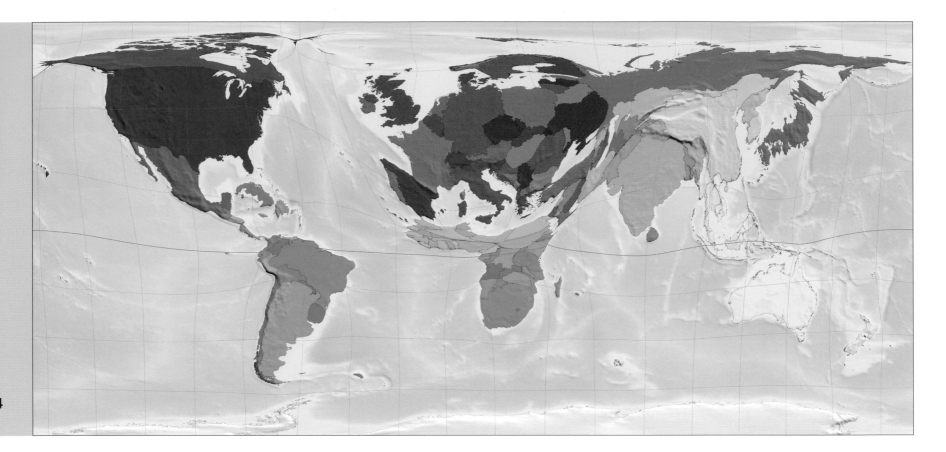

SHORTEST AND LONGEST RAILWAY NETWORKS

Rank	Territory	Metres[a]
1	Czech Republic	123
2	Belgium	107
3	Germany	103
4	Hungary	84
5	Netherlands	83
6	Switzerland	81
7	Slovakia	76
8	United Kingdom	71
9	Austria	69
10	Poland	66
127	Uganda	1.31
128	Mongolia	1.16
129	Paraguay	1.11
130	Mauritania	0.70
131	Ethiopia	0.68
132	Mali	0.60
133	Saudi Arabia	0.50
134	Venezuela	0.49
135	Nepal	0.41
136	Nicaragua	0.05

[a] metres of track per sq km of land
(64 territories recorded no rail network)

039 Railway Lines

The size of each territory indicates total length of usable railway lines there – irrespective of whether they are actually used.

In 2002 there were 1 million kilometres of railway in the world. However, the mere existence of rails is no guarantee that they are used. South America, for example, is home to 9% of the world's railway lines but only 0.5% of all passenger-kilometres travelled and 0.1% of all rail freight. Of the 64 territories without a rail network, many are relatively small islands, while of the 7 territories with the largest land areas, 6 also have the longest distances of railway.

'Japanese railways nationwide pass through some 3800 mountain tunnels totalling 2100 km in length, including the Seikan Tunnel (the world's longest tunnel) completed in 1988.'

Yukinori Koyama, Railway Technical Research Institute of Japan, 1997

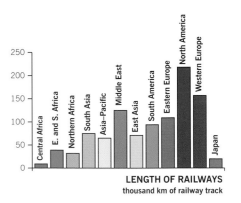

LENGTH OF RAILWAYS
thousand km of railway track

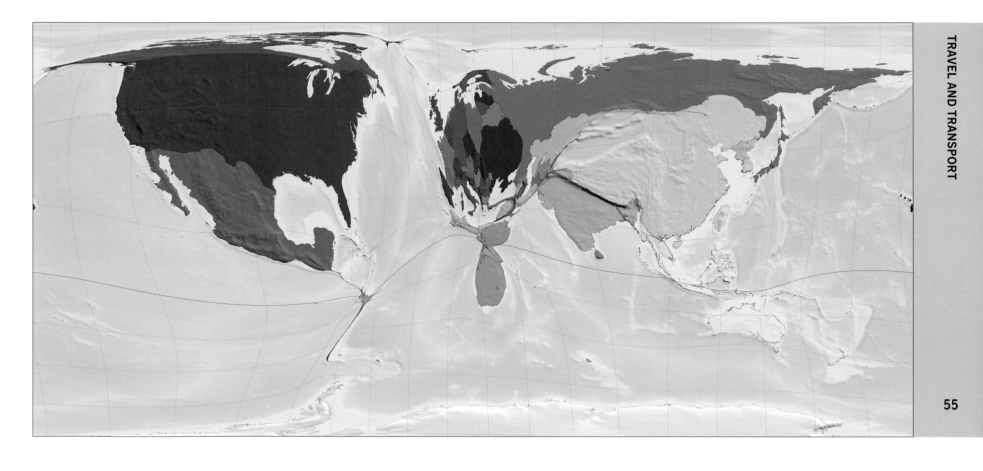

MOST AND LEAST RAILWAY FREIGHT CARRIED

Rank	Territory	Tonne-km[a]
1	Russia	10,480
2	Canada	10,339
3	Kazakhstan	8,586
4	Australia	8,108
5	Mexico	7,822
6	United States	7,561
7	Estonia	7,177
8	Latvia	6,530
9	Lebanon	4,074
10	Ukraine	3,950
127	Mali	15
128	Benin	13
129	Ghana	12
130	Uganda	9
131	DR Congo	8
132	Albania	7
133	Cambodia	7
134	Bangladesh	7
135	Venezuela	1
136	Madagascar	1

[a] tonne-km per person per year
(64 territories recorded no rail network)

040 Rail Freight

The size of each territory indicates the total amount of rail freight carried there, in tonne-kilometres. Nearly a third of all territories have no rail system at all.

8 trillion tonne-kilometres of freight were transported by rail in 2003. A tonne-kilometre is one metric tonne travelling one kilometre; so a territory can transport a large number of tonne-kilometres either by transporting many tonnes of freight or by transporting its freight many kilometres, or both.

The territories where the largest total amounts of freight are moved by rail are the United States, Russia and China. At the other extreme, 64 territories (out of 200) have no rail system and hence carry no rail freight.

'This is the Night Mail crossing the Border, / Bringing the cheque and the postal order, / Letters for the rich, letters for the poor, / The shop at the corner, the girl next door...' W. H. Auden, 'Night Mail', 1936

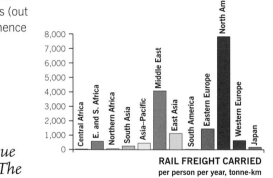

RAIL FREIGHT CARRIED
per person per year, tonne-km

MOST AND LEAST CONTAINER PORT TRAFFIC

Rank	Territory	Containers[a]
1	China	61.62
2	United States	32.64
3	Greenland	23.05
4	Singapore	18.44
5	Japan	14.57
6	South Korea	12.99
7	Germany	10.50
8	Malaysia	10.07
9	Italy	8.47
10	Spain	7.36
144	Dominican Republic	0.48
145	Honduras	0.47
146	Trinidad and Tobago	0.44
147	Mauritius	0.38
148	Yemen	0.38
149	Morocco	0.35
150	Algeria	0.31
151	Uruguay	0.30
152	Lebanon	0.30
153	Poland	0.26

[a] shipping containers loaded per person per year (47 landlocked countries had no container port traffic)

041 Container Ports

More shipping containers are loaded and unloaded off the coasts and rivers of China than in all other countries put together. The size of each territory indicates the numbers of shipping containers loaded and unloaded there.

The rest of the world handles only a third of the container volume that China handles. This implies that the majority of China's shipping must involve goods in transit within the country, rather than between China and other countries, since international trade results in containers being loaded and unloaded in other territories as well. This traffic includes goods being carried to serve the internal Chinese market, transportation of part-finished goods to be finished elsewhere, and goods being loaded and unloaded for transfer between different container ships in the same port.

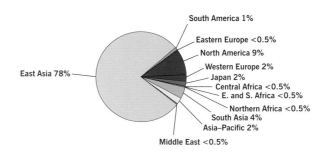

DISTRIBUTION OF SHIPPING CONTAINERS

South America 1%
Eastern Europe <0.5%
North America 9%
Western Europe 2%
Japan 2%
Central Africa <0.5%
E. and S. Africa <0.5%
Northern Africa <0.5%
South Asia 4%
Asia–Pacific 2%
Middle East <0.5%
East Asia 78%

'Mao claimed that China's industrial output could overtake that of the United States and Britain within fifteen years.' Jung Chang, *Wild Swans*, 1991

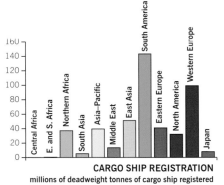

57

042 Cargo Shipping

Most of the world shipping fleet carries cargo or oil. The size of each territory indicates the total carrying capacity of all cargo ships registered there.

The total carrying capacity of all the merchant ships in the world is 470 million tonnes – 57% of the total carrying capacity of the world shipping fleet. The other 43% carries mainly oil, which we count separately from conventional cargo. Only a small percentage by weight of the fleet is made up by passenger and military vessels.

This map and Map 043 overleaf are among the more distinctive images in the atlas, because registration of cargo vessels is concentrated in a small number of countries that offer attractive regulatory or tax regimes to ship-owners. Panama and Liberia have the largest numbers of ship registrations, followed by Malta and Cyprus. Panama registers 85% of all cargo ships in South America and is responsible for 26% of worldwide registrations by weight.

CARGO SHIP REGISTRATION
millions of deadweight tonnes of cargo ship registered

'Big cargo ships can be surprisingly vulnerable. They tend to have just a few crew members to transport and protect goods worth millions of dollars.' Alison Gee, journalist, 2002

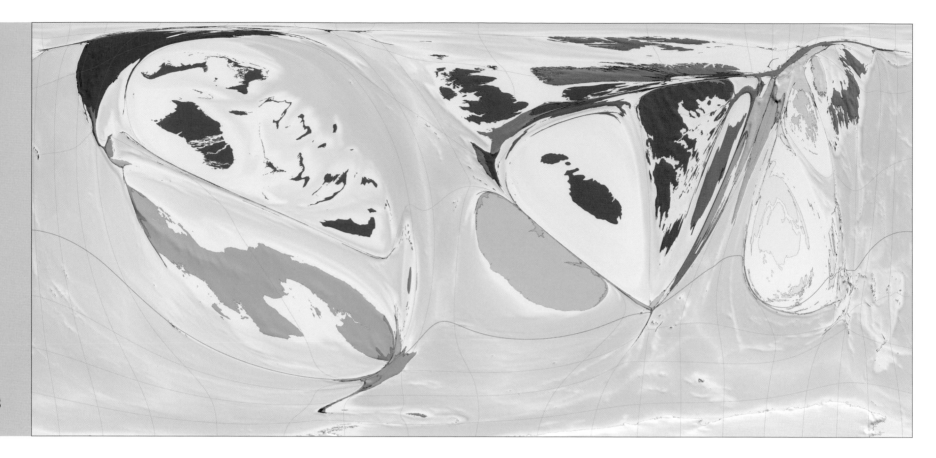

043 Oil Tankers

The size of each territory indicates the total tonnage of oil tankers registered there – not necessarily reflecting actual sailings.

Like Map 042 on the previous page, this map reflects the concentration of ship registration in a small number of countries that offer a favourable regulatory or tax environment. Notice in particular that none of the countries with the largest numbers of registered oil tankers have any significant petroleum imports or exports (see Maps 103 and 104). Luxembourg, which has the tenth largest tonnage of tankers registered per person, is entirely landlocked, with no ports at which any of those tankers could conceivably dock.

The largest tonnage of tankers per person is found in the Bahamas, where there are 93 deadweight tonnes of oil tankers registered for every man, woman and child who lives there.

OIL TANKER REGISTRATION
millions of deadweight tonnes of oil tanker registered

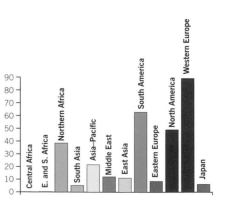

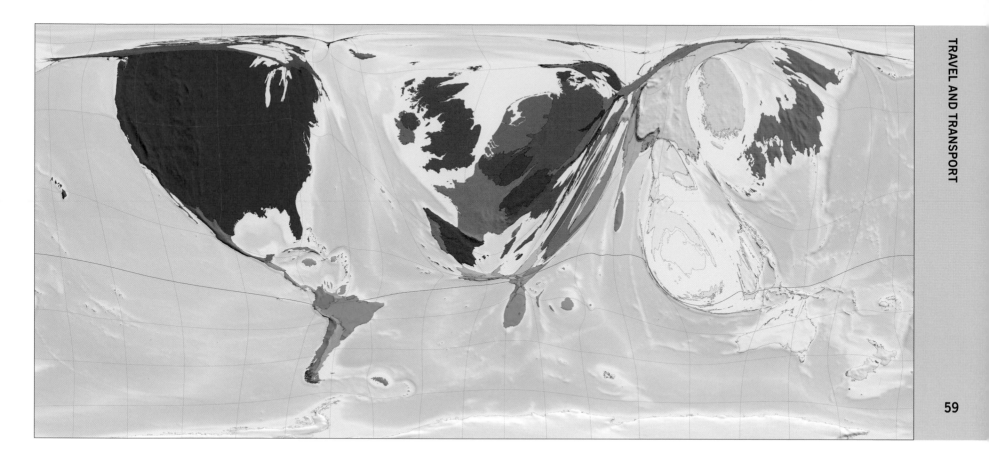

MOST AND LEAST AIRCRAFT FREIGHT

Rank	Territory	Tonne-km[a]
1	Luxembourg	8,933
2	Singapore	3,092
3	Nauru	2,154
4	Iceland	1,637
5	Qatar	1,372
6	Brunei	1,367
7	United Arab Emirates	1,258
8	Seychelles	940
9	New Zealand	791
10	Switzerland	780
191	Tanzania	0.58
192	Haiti	0.49
193	Nigeria	0.47
194	Zambia	0.47
195	Liberia	0.31
196	Lesotho	0.28
197	Rwanda	0.24
198	North Korea	0.22
199	Burundi	0.15
200	Dominican Republic	0.06

[a] tonne-km per person per year

044 Air Freight

The size of each territory indicates the total amount
of freight transported by aircraft registered there.

In 2000, 403 trillion tonne-kilometres of
freight were transported by air. Over half of
the total world air freight is carried by
aircraft registered either in North America
or in Western Europe. Aircraft registered in
Central Africa, on the other hand, carry
only 0.1% of world air freight. A high
tonne-kilometre count can be achieved
either by transporting many tonnes of
freight or by transporting freight many
kilometres, or both.

*'At $400 a day, inspection and storage charges at Jomo Kenyatta
International Airport are the highest in the world. So is the freight
charge of $1.85–$2.2 per [flower] stem.'* Kamau Ngotho, journalist, 2005

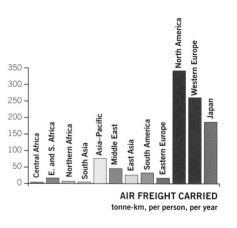

AIR FREIGHT CARRIED
tonne-km, per person, per year

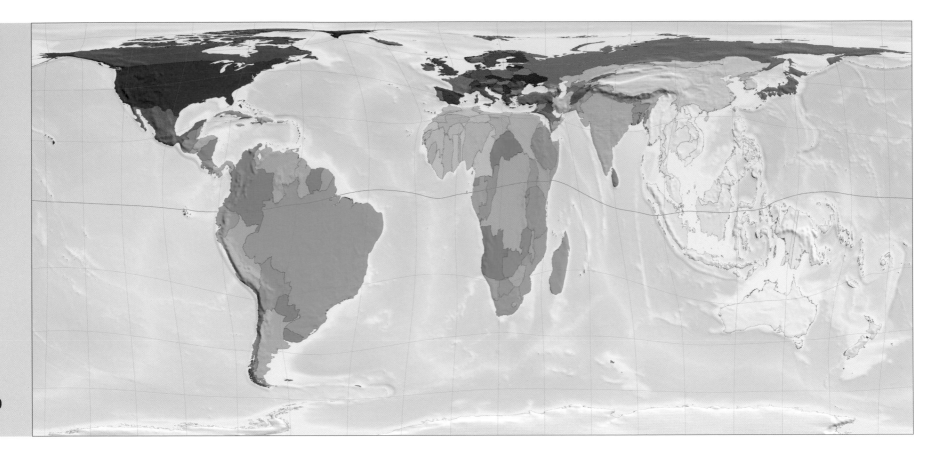

MOST AND LEAST RAINFALL

Rank	Territory	Rainfall[a]
1	Micronesia	357
2	Dominica	344
3	Papua New Guinea	321
4	Solomon Islands	313
5	Tuvalu	303
6	Samoa	300
7	Brunei	298
8	Bangladesh	295
9	Costa Rica	293
10	Malaysia	289
191	Mauritania	9.2
192	Algeria	8.9
193	Oman	8.6
194	Bahrain	8.0
195	United Arab Emirates	7.8
196	Qatar	7.4
197	Saudi Arabia	5.9
198	Libya	5.7
199	Egypt	5.2
200	Western Sahara	4.5

[a] average cm of rainfall per year, 1961–90

045 Rainfall

The size of each territory indicates the volume of rain and other precipitation falling there each year. The monsoons are of particular significance.

The largest volume of rain falls in Brazil, although this is in part simply because of the country's large land area. The territories with the highest rainfall per square kilometre tend to be those that experience monsoons. Malaysia, for instance, which has the tenth highest rainfall, was called by ancient mariners 'the land where the winds meet' – the winds being the monsoons. The word 'monsoon' may originate from the Arabic word *mausim*, meaning seasonal wind patterns that reverse direction. Malaysia and many other territories in the tropics experience two monsoons a year: the southwest monsoon, from May to September, and the northeast monsoon, from November to March. Elsewhere in the world monsoons come and go at different times.

RAINFALL DISTRIBUTION

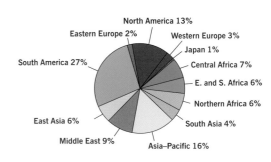

North America 13%
Eastern Europe 2%
Western Europe 3%
Japan 1%
South America 27%
Central Africa 7%
E. and S. Africa 6%
Northern Africa 6%
East Asia 6%
South Asia 4%
Middle East 9%
Asia–Pacific 16%

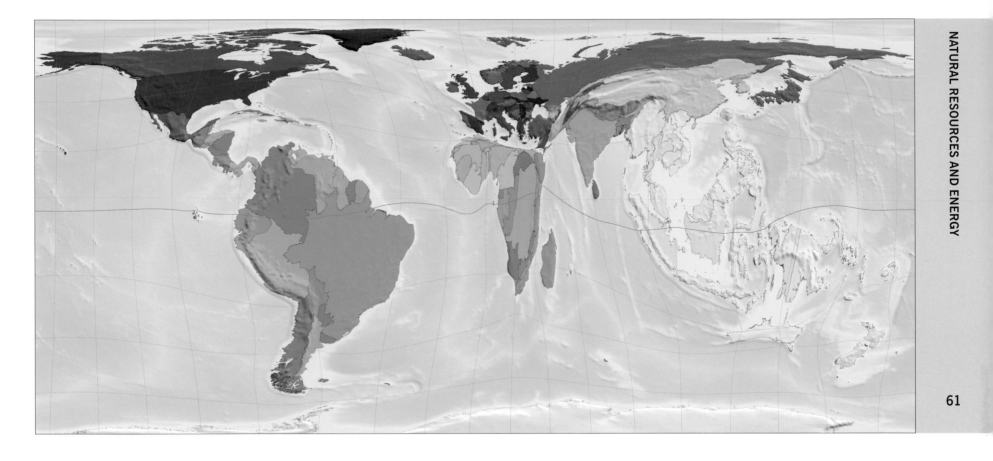

MOST PLENTIFUL AND SCARCEST WATER RESOURCES

Rank	Territory	Water resources[a]
1	Sao Tome and Principe	227
2	Sierra Leone	223
3	Costa Rica	220
4	Liberia	208
5	Colombia	203
6	Bhutan	202
7	Panama	198
8	Taiwan	186
9	Papua New Guinea	177
10	Malaysia	177
191	Oman	0.32
192	Turkmenistan	0.29
193	Niger	0.28
194	Bahamas	0.20
195	Egypt	0.18
196	United Arab Emirates	0.18
197	Saudi Arabia	0.11
198	Mauritania	0.04
199	Libya	0.03
200	Kuwait	0

[a] cu cm water per sq cm land area, 2003 or most recent year for which data were available

046 Water Resources

The size of each territory indicates the annual volume of naturally occurring fresh water available for human use.

This map shows the amount of precipitation that flows into streams, rivers and lakes, along with groundwater recharge occurring within the territory. Worldwide, roughly 43,600 cubic kilometres of fresh water are added to water resources annually, which is about half of the total amount that falls as precipitation. Much of the difference is water lost through evaporation.

Territories with high rainfall, including many in Asia–Pacific and Australasia, and South America, typically also have abundant water resources and so appear large on the map.

Fresh water produced artificially by removing the salt from sea water (desalination) is not shown on the map. Kuwait, for example, has practically no naturally occurring fresh water and produces most of what it consumes by desalination. As a result Kuwait is virtually invisible on this map.

DISTRIBUTION OF WATER RESOURCES

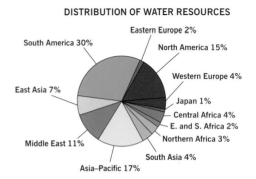

Eastern Europe 2%
South America 30%
North America 15%
Western Europe 4%
East Asia 7%
Japan 1%
Central Africa 4%
E. and S. Africa 2%
Northern Africa 3%
Middle East 11%
South Asia 4%
Asia–Pacific 17%

'The Amazonian basin, where ten of the twenty largest rivers in the world are to be found...represents one fifth of the entire fresh water reserves of the planet.' Brazilian Ministry of External Affairs, 2002

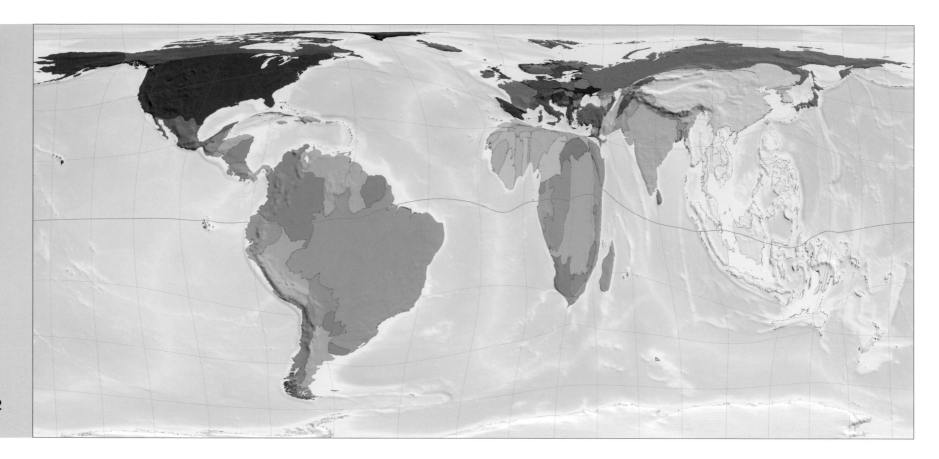

047 Groundwater Recharge

The size of each territory is proportional to the annual volume of groundwater recharge, the natural replenishment of the resource as surface water flows into the ground.

Groundwater is water located in the ground in the soil or in permeable rock, rather than in open bodies of water like lakes and reservoirs or underground aquifers. As nearly 70% of all fresh water is groundwater, it is an important water source. Each year a total of 11,400 cubic kilometres of surface water becomes groundwater. In many places, however, this is not enough to replenish groundwater supplies, because even larger volumes are being withdrawn. Among the regions, South America has the highest groundwater recharge, Japan the lowest.

GROUNDWATER RECHARGE

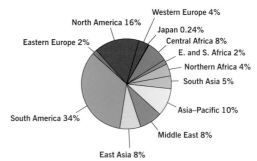

North America 16%
Western Europe 4%
Japan 0.24%
Eastern Europe 2%
Central Africa 8%
E. and S. Africa 2%
Northern Africa 4%
South Asia 5%
Asia–Pacific 10%
Middle East 8%
East Asia 8%
South America 34%

'Groundwater is a vast and slow moving resource, whose volume greatly exceeds that of other available fresh water sources.'

University of New South Wales Groundwater Centre, 2006

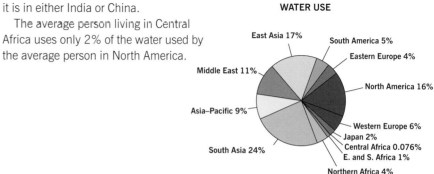

MOST AND LEAST WATER USE

Rank	Territory	Volume[a]
1	Bangladesh	64
2	Bahrain	44
3	Mauritius	31
4	Belgium	27
5	Japan	24
6	Netherlands	24
7	Pakistan	23
8	Maldives	23
9	Viet Nam	23
10	India	22
191	Djibouti	0.036
192	Namibia	0.034
193	Angola	0.029
194	Mongolia	0.029
195	Botswana	0.026
196	Chad	0.020
197	Papua New Guinea	0.017
198	DR Congo	0.017
199	Congo	0.012
200	Central African Republic	0.004

[a] cubic cm water per sq cm land area, 2003
or the most recent year for which data
were available

048 Water Use

Everybody needs water to live, yet people use hugely varying quantities.
The size of each territory in this map is proportional to its annual water use.

The human population of the world uses 4,000 cubic kilometres of water annually for domestic, agricultural and industrial purposes. Only water that is actually consumed is shown here; uses such as energy generation, mining and recreation, which employ water but don't use it up, are excluded.

China, India and the United States use the largest volumes of water. These are also the three territories with the largest populations, so it makes sense that they should have the highest water use. However, water use per person is about three times higher in the United States than

it is in either India or China.

The average person living in Central Africa uses only 2% of the water used by the average person in North America.

WATER USE

East Asia 17%
South America 5%
Eastern Europe 4%
Middle East 11%
North America 16%
Asia–Pacific 9%
Western Europe 6%
Japan 2%
Central Africa 0.076%
E. and S. Africa 1%
South Asia 24%
Northern Africa 4%

'... the right to water emanates from and is indispensable for an adequate standard of living as it is one of the most fundamental conditions for survival.' Céline Dubreuil, report for the World Water Council, 2006

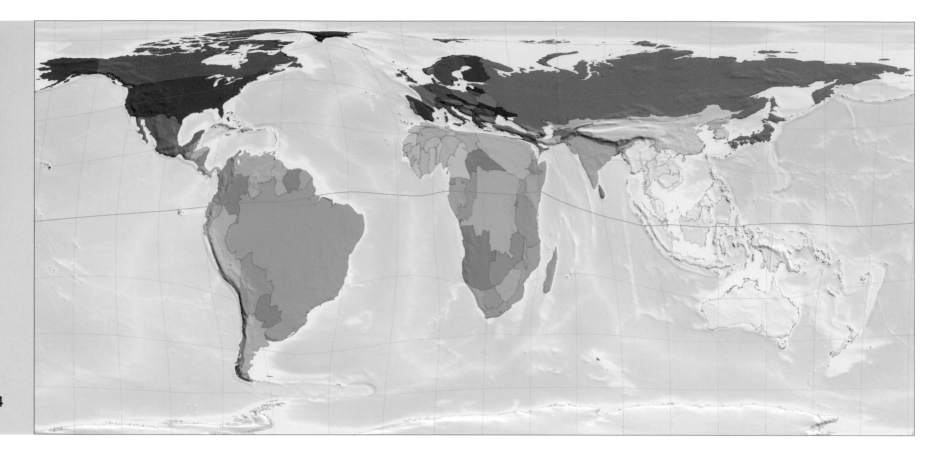

MOST AND LEAST FOREST IN 1990

Rank	Territory	%ᵃ
1	Solomon Islands	92
2	Suriname	90
3	Guyana	88
4	Brunei	86
5	Guinea-Bissau	85
6	Gabon	85
7	Bahamas	84
8	Palau	76
9	Belize	75
10	Finland	72
191	Lesotho	0.46
192	Mauritania	0.41
193	Djibouti	0.26
194	Iceland	0.25
195	Libya	0.18
196	Kuwait	0.17
197	Qatar	0.09
198	Egypt	0.05
199	Oman	0.00
200	Malta	0

ᵃ forest as % of land area in 1990

049 Forests 1990

Large land areas with small populations are conducive to forest growth. The size of each territory in this map indicates the area covered by forests in 1990.

This map reflects the definition of forest used by the World Bank, namely 'land under natural or planted stands of trees'. Russia, Brazil and Canada had the largest areas of forest in 1990. These territories are among the five largest in the world by land area, and also have population densities considerably below the world average, leaving more of their land free for the conservation and cultivation of forest.

Most of the 10 territories with the highest percentage of forest coverage (as opposed to total forest area) in 1990 were located between the Tropics of Cancer and Capricorn. Finland is an anomaly, however,

recording the tenth highest percentage of forested land in 1990: not only is it a long way north of the tropics, part of its territory lies within the Arctic Circle.

Interestingly, many of the territories with the lowest percentages of forest coverage in 1990 are also entirely or partially located within the tropics. Of the bottom 10 territories, Lesotho, Iceland, Kuwait, Malta and Qatar are outside the intertropical zone.

FOREST DISTRIBUTION 1990

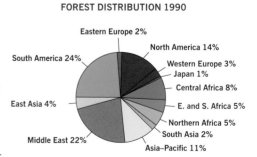

Eastern Europe 2%
North America 14%
South America 24%
Western Europe 3%
Japan 1%
Central Africa 8%
East Asia 4%
E. and S. Africa 5%
Northern Africa 5%
South Asia 2%
Middle East 22%
Asia–Pacific 11%

'Democratic Republic of Congo's forests cover an area of 1.3 million square kilometres, more than twice the size of France.'

The Rainforest Foundation, 2004

050 Forests 2000

The size of each territory indicates the area covered by forests in 2000.
Islands feature at both ends of the spectrum.

In 2000 more than 90% of the land area of the Solomon Islands was covered in forest, making this territory number one in the world for forest cover. In contrast, Malta recorded no forest cover at all in 2000. Of course, it is not the case that Malta has no trees at all; it just doesn't have enough in any one place to count as a forest.

Although there are some marked differences between the pattern of forest cover in 1990 (Map 049) and in 2000, there are also many similarities. The countries with the largest total areas of forest in both years are Russia, Brazil and Canada, and the top and bottom 10 territories in terms of percentage of area covered by forest are almost identical.

'Russian forests act as an environmental shield not only for this country, but for the whole planet.'

Vladimir Putin, president of the Russian Federation, 2006

FOREST DISTRIBUTION 2000

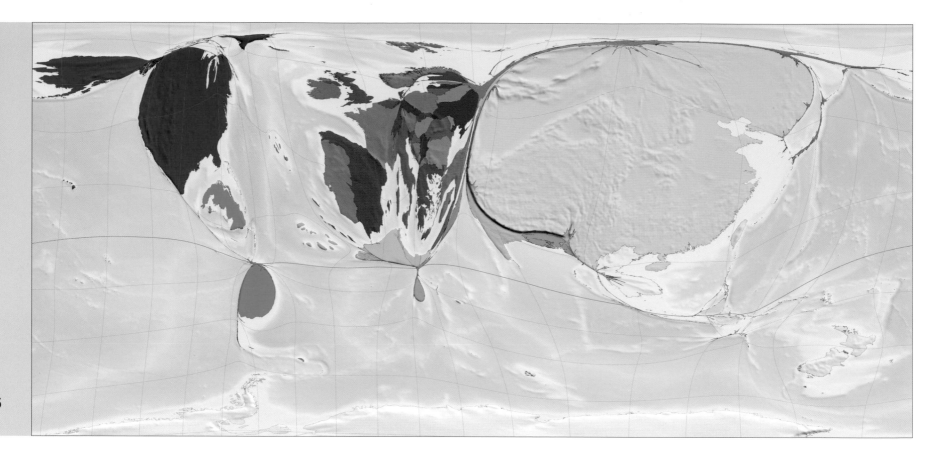

051 Forest Growth

Worldwide there is net forest loss, but some individual territories nonetheless show gains. The size of each territory here is proportional to the net increase in forested area between 1990 and 2000.

The territory with the most forest growth between 1990 and 2000 was China, where 181,000 square kilometres were added during this decade. China is also the territory with the largest population. The United States had the second largest forest growth, but this was just a fraction of China's increase, of 39,000 square kilometres.

It is not surprising that the largest total increase in forested area took place in territories such as these two with large land areas. In terms of forest growth as a percentage of land area, however, smaller territories, such as Cape Verde, Liechtenstein and Portugal, are in the lead.

NET FOREST GROWTH

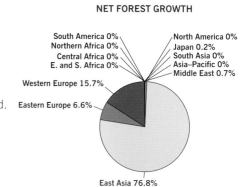

South America 0%
Northern Africa 0%
Central Africa 0%
E. and S. Africa 0%
North America 0%
Japan 0.2%
South Asia 0%
Asia–Pacific 0%
Middle East 0.7%
Western Europe 15.7%
Eastern Europe 6.6%
East Asia 76.8%

'One generation plants the trees; another gets the shade.' Chinese proverb, date unknown

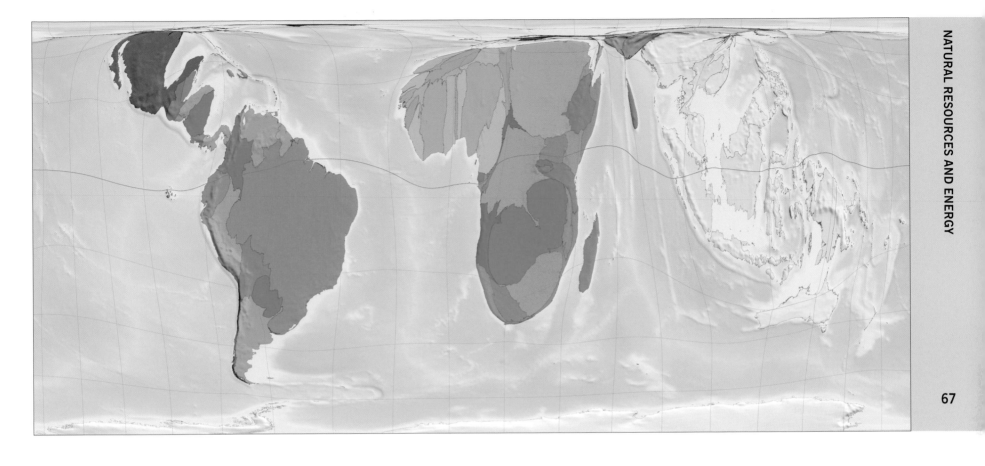

MOST FOREST LOSS

Rank	Territory	%[a]
1	Belize	15.6
2	Zambia	11.4
3	Nicaragua	9.7
4	Samoa	8.8
5	Côte d'Ivoire	8.3
6	Zimbabwe	8.3
7	St Lucia	8.2
8	Liberia	7.9
9	Myanmar	7.9
10	Guinea-Bissau	7.7
11	Malawi	7.5
12	Indonesia	7.2
13	Malaysia	7.2
14	Panama	7.0
15	Benin	6.3
16	Rwanda	6.1
17	Burundi	5.7
18	Nepal	5.5
19	Sri Lanka	5.4
20	Dominica	5.3

[a] net loss of forest area as % of land area, 1990–2000

052 **Forest Loss**

The size of each territory is proportional to the net loss in forested area between 1990 and 2000. A low net loss does not necessarily mean forests are protected: it may be that few remain.

Among the regions, South America and the Asia–Pacific and Australasia had the greatest net loss between 1990 and 2000, with South America accounting for 32% and Asia–Pacific and Australasia 21% of worldwide net losses. Across the globe, territories experiencing net forest loss together lost 1.33 million square kilometres of forest over this decade. Despite this,

South America was still the region with the largest forested area in 2000.

In Africa the area covered by forests shrank by 550,000 square kilometres between 1990 and 2000. 11.4% of the land in Zambia was deforested during this period.

'Indonesia is blessed with some of the most extensive and biologically diverse tropical forests in the world. But the tragedy is that Indonesia has one of the highest rates of tropical forest loss in the world.' E. G. Togu Manurung, *Jakarta Post*, 2006

NET FOREST LOSS

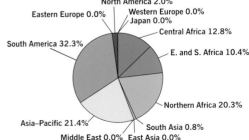

North America 2.0%
Western Europe 0.0%
Japan 0.0%
Eastern Europe 0.0%
Central Africa 12.8%
South America 32.3%
E. and S. Africa 10.4%
Northern Africa 20.3%
Asia–Pacific 21.4%
South Asia 0.8%
Middle East 0.0% East Asia 0.0%

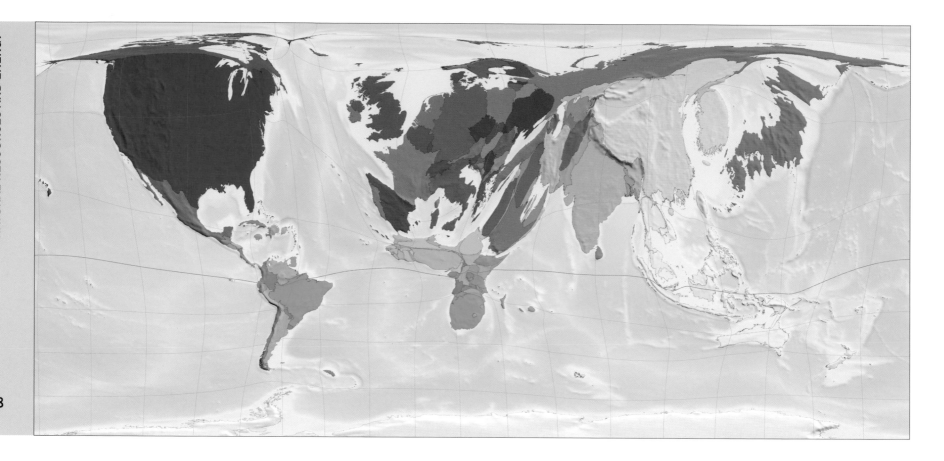

053 Fuel Use

The size of each territory indicates the annual amount of fuel consumed, measured in kilograms of oil needed to produce the same amount of power.

Each year, fuel equivalent to 11,567,000,000,000 kilograms – over 11.5 billion metric tonnes – of oil is used by people worldwide. This fuel includes gas, coal, oil, wood, nuclear fuel and other materials. The amount of power generated by a single kilogram of fuel varies greatly from one type of fuel to another, so for the sake of comparison we measure fuel use in

terms of the equivalent weight of oil needed to produce the same amount of power.

Per person, worldwide fuel consumption averages 1,853 kilograms of oil equivalent per person per year. The territory with the highest fuel use per person is Luxembourg, which uses almost 100 times as much as the territory with the lowest, Bangladesh.

'Currently, vehicles account for about two-thirds of annual fuel use in the United States – twice the consumption rate in Europe.'

Terry Costlow, 'Fuel cell research moving "at light speed"', 2003

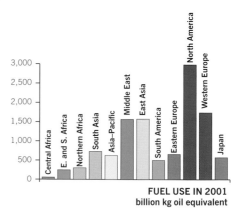

FUEL USE IN 2001
billion kg oil equivalent

HIGHEST INCREASES IN FUEL USE

HIGHEST INCREASES IN FUEL USE

Rank	Territory	Energy[a]
1	Equatorial Guinea	11,394
2	Israel	8,410
3	Singapore	5,431
4	Iceland	5,069
5	South Korea	3,368
6	Seychelles	3,363
7	Trinidad and Tobago	3,250
8	Oman	2,940
9	Mauritius	2,393
10	Saudi Arabia	2,370
11	Latvia	1,972
12	Norway	1,959
13	Cyprus	1,900
14	New Zealand	1,882
15	Botswana	1,769
16	Bahrain	1,679
17	Portugal	1,645
18	Malaysia	1,561
19	Slovakia	1,560
20	Ireland	1,544

[a] additional kg oil equivalent per person used in 2001 compared to 1980

054 Fuel Increase

The size of each territory indicates its increase in fuel use from 1980 to 2001, measured in equivalent kilograms of oil.

The world trend in fuel use during the twentieth century was relentlessly upward. Between 1980 and 2001 the average yearly increase in worldwide fuel consumption was 340 kilograms of oil equivalent per person. Increases were greatest in China, Japan, India and South Korea. However, there was no reported increase in fuel use at all in 58 territories.

The region with the largest increase was East Asia. The region with the smallest was Central Africa.

Per person, Equatorial Guinea saw the largest increase in fuel use between 1980 and 2001. A more than tenfold increase in oil production occurred there following the 1995 discovery of the Zafiro oilfield.

'Higher oil prices…are badly needed to encourage efficient usage. But if that means a heavier burden for the poor in terms of proportion of their income, it will create more problems than it answers…' China Daily, 2006

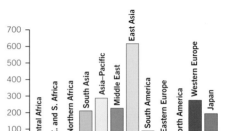

INCREASE IN FUEL USE, 1980–2001
billion kg oil equivalent, per year

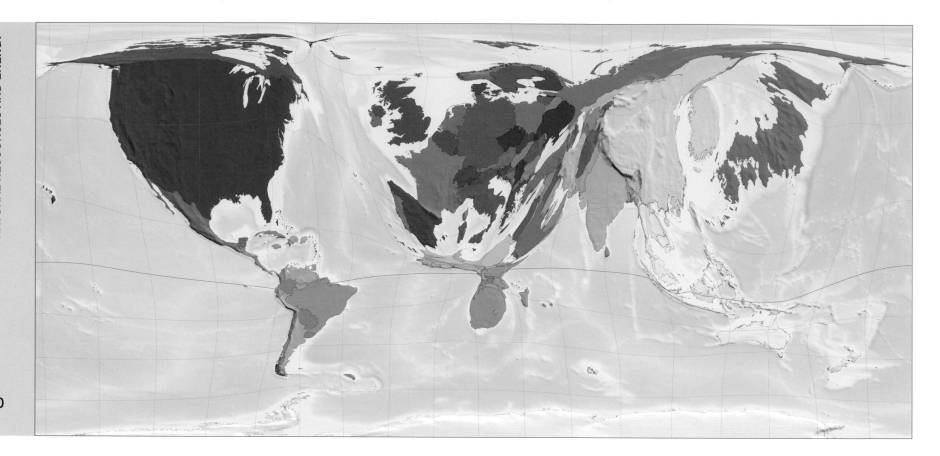

055 Electricity Production

Electricity is generated in a wide variety of ways. The size of each territory here indicates total annual electricity production in kilowatt-hours.

Electricity is generated by burning coal, oil, gas, wood and waste; or by harnessing hydroelectric, solar, geothermal, tidal, wind, wave and nuclear energy. Coal is the biggest source of electricity worldwide, followed by gas, hydroelectric power, nuclear power and oil. In 2002, 2,584 kilowatt-hours of electricity were produced for every person living on Earth. Electricity production is not uniformly distributed among the planet's population, however. On average about 10 kilowatt-hours were generated per person living in Benin and Togo, for instance, while almost 3,000 times as much was generated per person in Norway and Iceland.

'Energy supply and demand plays an increasingly vital role in our national security and the economic output of our nation.'

United States Department of Energy, 2006

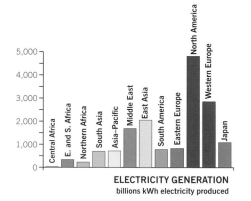

ELECTRICITY GENERATION
billions kWh electricity produced

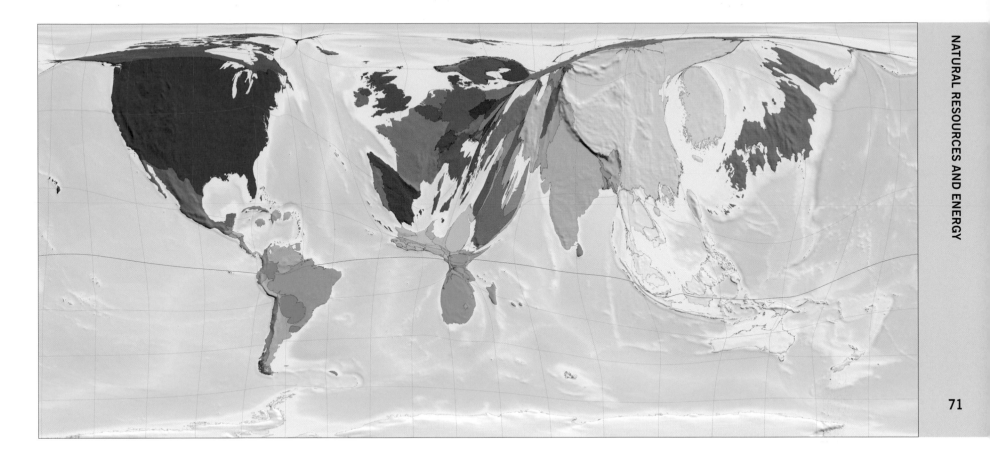

GREATEST INCREASES IN ENERGY PRODUCTION

Rank	Territory	Energy[a]
1	Iceland	16,163
2	Qatar	12,400
3	United Arab Emirates	11,801
4	Kuwait	10,199
5	Norway	7,870
6	Paraguay	7,816
7	Brunei	7,457
8	Sweden	7,335
9	Bahrain	7,313
10	Canada	6,913
11	Finland	6,482
12	Singapore	6,390
13	Australia	6,236
14	South Korea	5,152
15	Israel	4,989
16	Saudi Arabia	4,976
17	United States	4,850
18	France	4,834
19	New Zealand	4,422
20	Greenland	4,172

[a] additional kWh per person in 2001 compared to 1980

056 Increase in Electricity Production

The size of each territory indicates the increase in electricity production between 1980 and 2002.

The total amount of energy produced in the world doubled between 1980 and 2002. The amount produced per person has also increased, although not by so large a factor because the world population has also grown over time. Much of the world's population is now reliant on a steady supply of electricity to maintain its living standards.

Over the period shown here there were increases in the amount of electricity produced in 187 territories. Electricity production fell in 13 territories, more than half of those in Eastern Europe. Electricity production rose most in the United States and China, both of which experienced increases more than twice as large as in any of the other territories.

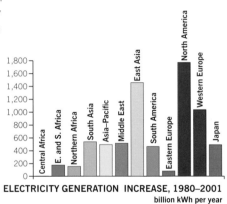

ELECTRICITY GENERATION INCREASE, 1980–2001
billion kWh per year

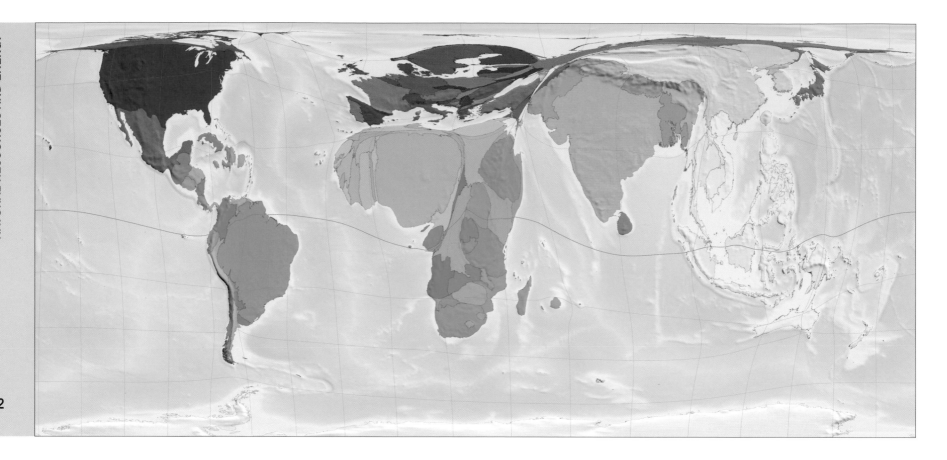

Rank	Territory	Energy[a]
1	Equatorial Guinea	9,068
2	Seychelles	2,532
3	Sweden	2,130
4	Finland	1,761
5	St Kitts–Nevis	1,166
6	Latvia	1,116
7	Botswana	1,059
8	Mauritius	1,053
9	Estonia	868
10	Angola	775
178	Tajikistan	13
179	Mongolia	13
180	Armenia	10
181	Morocco	10
182	Hong Kong (China)	10
183	Singapore	8
184	Yemen	7
185	Saudi Arabia	3
186	Iran	2
187–200	14 territories	0

[a] **number of energy units equivalent to 1 kg
oil used per person living there, 2001**

057 Traditional Fuel Consumption

The size of each territory indicates the quantity of traditional fuels burnt annually,
in terms of the quantity of oil needed to produce the same amount of energy.

Traditional fuels include wood, charcoal, bagasse (sugar-cane waste), and animal and vegetable wastes that can be burned.

Traditional fuels such as wood and charcoal differ from hydrocarbon-based fuels in that they are usually produced locally. Sometimes they are generated as waste material from another process. Sometimes they are obtained for free. Thus it is perhaps not surprising that Central Africa has the highest traditional fuel usage per person, given its relatively poor infrastructure and high levels of poverty. Ironically, Equatorial Guinea, which consumes the largest quantities

of traditional fuel per person, also exports considerable quantities of oil.

The Middle East and Central Asia, where most of the earth's oil originates, uses the traditional fuel equivalent of only 77 kilograms of oil per person per year. And the United States, despite its high consumption of fossil fuels and generation of nuclear power, still consumes, per person, more traditional fuel than Mexico, Argentina, Brazil, Chile, Colombia, Costa Rica, El Salvador or Venezuela. The United States consumes, per person, 4 times as much traditional fuel as China and 2.5 times as much as India.

TRADITIONAL FUEL USE
estimated equivalent in kg oil, per person per year

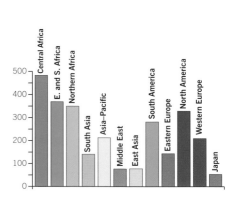

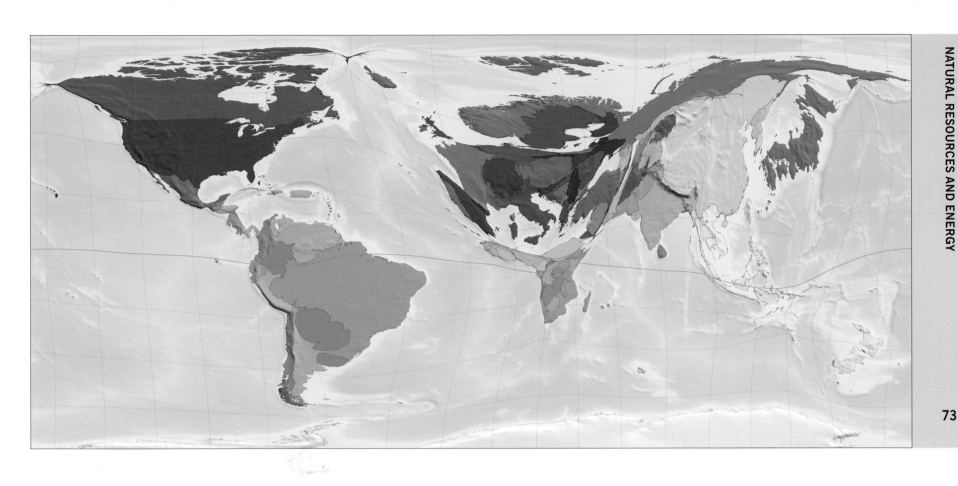

058 Hydroelectric Power

The size of each territory indicates the amount of electricity generated by hydroelectric power plants, which transform the energy of moving water into power.

Large dams and steeply flowing rivers facilitate the generation of hydroelectric power. Sometimes other sources of electricity are used to pump water up into dams so it can be used to generate electricity later, the dam effectively acting as a battery.

Canada, China, Brazil and the United States generate the largest amounts of hydroelectric power, together producing 44% of the world total. Per person, however, Norway and Iceland are the leaders, each producing and using twice as much as third-placed Canada. The territories with the smallest amounts of hydroelectric power generation tend to be those with very little water, low elevations or only small amounts of money to invest, such as Denmark, Turkmenistan and Togo. 15 territories do not use hydroelectric power at all. Most of these are either relatively small islands, or oil producers with little rainfall in territories of the Middle East and Central Asia.

HYDROELECTRIC POWER GENERATION
kWh used, per person per year

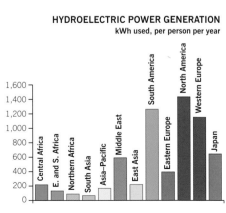

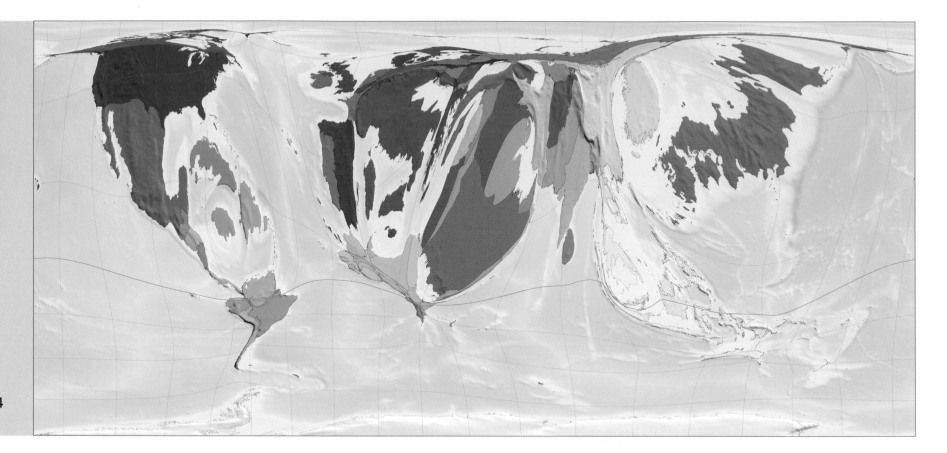

059 Oil Power

The size of each territory indicates the amount of electricity generated from oil. Just 10 territories produce no electricity at all this way.

Each year 183 kilowatt-hours of electricity are generated from oil per person worldwide. A kilowatt-hour is the amount of electricity needed to run a 1 kilowatt electric heater (or any other appliance) for one hour. Alternatively, it is the amount need to run a 100 watt light bulb for ten hours.

Japan, the United States and Saudi Arabia generate the largest total amounts of electricity from oil. Per person, Saudi Arabia generates more than three times as much as Japan, and Japan over three times as much as the United States. Continuing down the rankings, there is also a factor

of three or more between the United States and the United Kingdom, the United Kingdom and China, China and Bangladesh, Bangladesh and Zambia, Zambia and Côte d'Ivoire, and Côte d'Ivoire and Ethiopia. Add it all up, and Saudi Arabia produces more than 13,000 times as much electricity from oil as Ethiopia.

10 territories do not generate any electricity from oil. Of these, 6 are in the Middle East and Central Asia while only 1, South Africa, is in Africa. Nevertheless, as regions, East and Southern Africa and Central Africa generate the lowest total amount of electricity from oil.

MOST AND LEAST OIL-GENERATED POWER

Rank	Territory	Energy[a]
1	Kuwait	12,133
2	Malta	5,130
3	Cyprus	4,731
4	Saudi Arabia	4,082
5	Singapore	3,337
6	Jamaica	2,592
7	Lebanon	2,495
8	Libya	2,210
9	Israel	1,621
10	Italy	1,526
182	Zimbabwe	2.3
183	Trinidad and Tobago	2.3
184	Colombia	2.2
185	Mozambique	1.8
186	Côte d'Ivoire	0.8
187	DR Congo	0.4
188	Ethiopia	0.3
189	Congo	0.3
190	Nepal	0.2
191–200	10 territories	0

[a] kWh generated per person living there, 2002

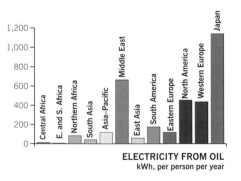

ELECTRICITY FROM OIL
kWh, per person per year

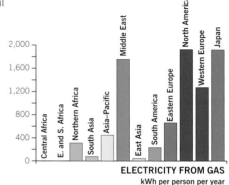

MOST POWER GENERATED FROM NATURAL GAS

Rank	Territory	Energy[a]
1	Qatar	16,838
2	United Arab Emirates	13,388
3	Bahrain	10,397
4	Brunei	8,920
5	Luxembourg	6,485
6	Singapore	4,917
7	Trinidad and Tobago	4,679
8	Netherlands	3,542
9	Kuwait	3,239
10	Oman	3,025
11	Ireland	2,777
12	Russia	2,670
13	New Zealand	2,661
14	United Kingdom	2,574
15	Belarus	2,518
16	United States	2,448
17	Malaysia	2,397
18	Turkmenistan	2,333
19	Finland	2,174
20	Saudi Arabia	2,115

[a] **kWh generated per person living there, 2002**

060 Gas Power

The size of each territory indicates the amount of electricity generated from gas. The United States is way out in the lead.

The United States leads the world in total electricity generated from gas, generating almost twice as much as the next on the list, Russia. Russia is a net exporter of gas and coal, whereas the United States is a net importer.

As regions, Japan and North America generate the most electricity from gas per person, followed by the Middle East and Central Asia. At the other extreme, Central Africa and East and Southern Africa generate barely any. Many territories in Central America also generate no electricity from gas.

'Gazprom, Russia's state-run natural gas monopoly, holds more than one-fourth of the world's natural gas reserves...'

Bernard A. Gelb, specialist in industrial economics, Resources, Science and Industry Division of the Congressional Research Service for the Government of the United States, 2006

ELECTRICITY FROM GAS
kWh per person per year

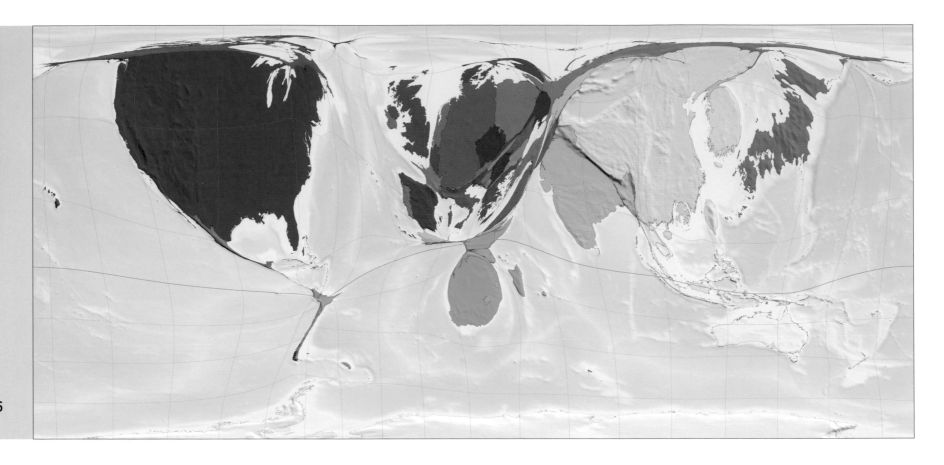

MOST COAL-GENERATED POWER

Rank	Territory	Energy[a]
1	Australia	8,918
2	United States	7,035
3	Estonia	5,960
4	Israel	5,570
5	Greenland	5,163
5	Bahamas	5,163
7	Czech Republic	4,976
8	South Africa	4,537
9	Finland	3,794
10	Canada	3,751
11	Germany	3,538
12	Poland	3,489
13	Denmark	3,381
14	Greece	3,142
15	Hong Kong (China)	3,128
16	South Korea	2,755
17	Slovenia	2,651
18	Kazakhstan	2,631
19	Japan	2,283
20	Ireland	2,280

[a] **kWh generated per person living there, 2002**

061 Coal Power

Territory sizes show the amount of electricity generated from coal.

The United States, China and India generate the highest quantities of electricity from coal, with the United States in the lead by a wide margin. The United States generates 7 times as much power from coal as China, and 18 times as much as India.

As a region, North America generates over twice as much electricity per person from coal as any other region, with the United States making up 93% of the region's total. The United States is also the world's largest producer of nuclear energy, but still generates 2.5 times more in coal-powered stations than in nuclear-powered ones.

71 of the 200 territories generate no electricity from coal at all.

ELECTRICITY FROM COAL
kWh, per person per year

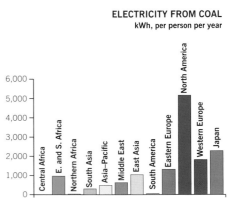

77

MOST NUCLEAR-GENERATED POWER

Rank	Territory	Energy[a]
1	Sweden	7,593
2	France	7,304
3	Belgium	4,598
4	Finland	4,288
5	Lithuania	4,041
6	Switzerland	3,783
7	Slovakia	3,325
8	United States	2,765
9	Slovenia	2,764
10	Bulgaria	2,528
11	South Korea	2,513
12	Canada	2,413
13	Japan	2,314
14	Germany	2,001
15	Czech Republic	1,837
16	Ukraine	1,595
17	Spain	1,537
18	United Kingdom	1,490
19	Hungary	1,409
20	Russia	983

[a] kWh generated per person, 2002

062 Nuclear Power

The size of each territory indicates the amount of electricity generated in nuclear power plants. Only 30 out of 200 territories produce power in this way.

Of the 30 territories that generate electricity using nuclear power plants, 17 are in Europe – including 9 of the top 10 producers per person. In 2002, Sweden produced the most nuclear power per person, followed by France. Major non-European nuclear electricity producers include the United States, Japan, Russia and South Korea. Nuclear power is produced in the Middle East and Central Asia by only Armenia and Russia, and in South America by only Brazil and Argentina (and only in small quantities).

The United States is the country that produces the most nuclear power overall. No nuclear power is generated in any of the territories in Central Africa, Northern Africa, or Asia–Pacific and Australasia.

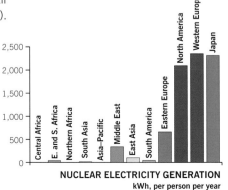

NUCLEAR ELECTRICITY GENERATION
kWh, per person per year

The Trading World

Globalization and Internationalism

Food and Consumables

Minerals, Natural Products and Petrochemicals

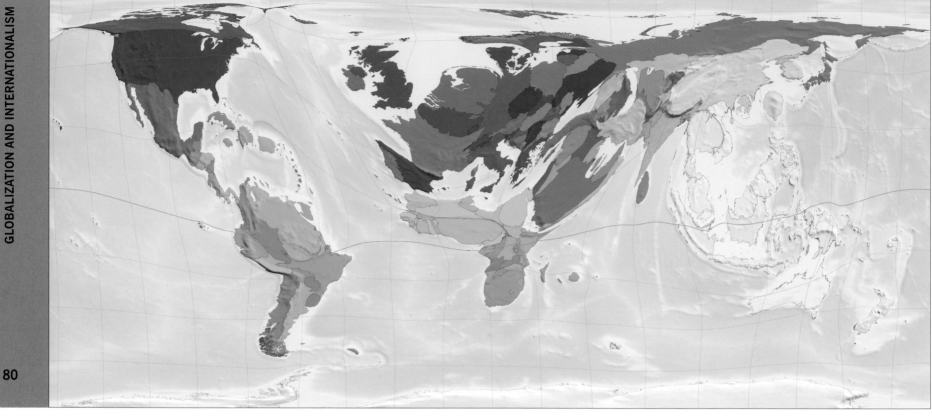

063 Primary Exports in 1990

The size of each territory indicates the dollar value of exports of unprocessed
goods in 1990, adjusted for local purchasing power.

Primary goods are unprocessed goods such
as crude oil, fresh apples or metal ore.
These goods are often exported to territories
that then process them into secondary
(manufactured) goods, and may then in
turn export these. Most territories export
both primary and secondary goods, though
there are exceptions. For example, in
1990, the year shown on this map, Angola
exported primary goods almost exclusively.

The value of primary exports depends
both on volume and on price. The same
prices, however, may be considered high in
one territory and low in another, depending
on the going rate for typical goods and
services in the territories in question.
The figures used to create this map were
adjusted for purchasing power parity (PPP)
to allow for such variations.

PRIMARY EXPORTS
as % of value of goods exported, 1990

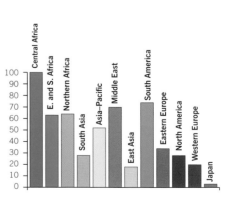

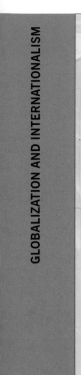

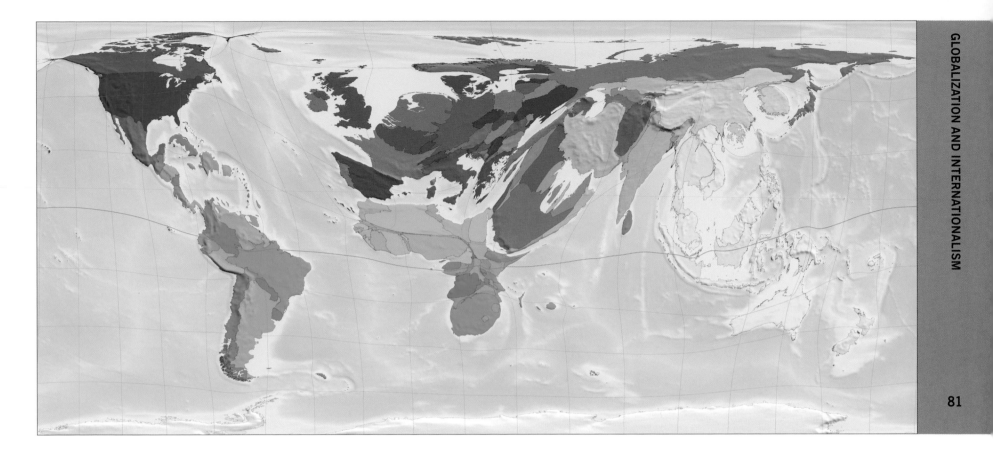

HIGHEST AND LOWEST EXPORTS OF PRIMARY GOODS IN 2002

Rank	Territory	%ᵃ
1	Nigeria	99
2	Rwanda	98
3	Algeria	98
4	Papua New Guinea	98
5	Gabon	98
6	Sudan	97
7	Niger	95
8	Benin	94
9	Cameroon	93
10	Azerbaijan	93
191	Botswana	9
192	South Korea	8
193	Philippines	8
194	Bangladesh	8
195	Ireland	8
196	Switzerland	7
197	Israel	7
198	Hong Kong (China)	5
199	Malta	4
200	Japan	3

ᵃ primary goods as % of total merchandise exports, 2002

064 Primary Exports in 2002

The size of each territory indicates the dollar value of exports of unprocessed goods in 2002, adjusted for local purchasing power.

The regions whose international trade relies most on the export of primary goods are Central Africa, the Middle East and Central Asia, Northern Africa and South America. The regions that rely on primary goods exports least are Japan and East Asia.

No single territory dominates this map. Russia, China, the United States and Iran export the highest total amounts of primary goods in US dollar terms, corresponding to 69%, 10%, 14% and 91% of their total export incomes respectively. Between 1990 and 2002 the value of all exports of primary goods worldwide rose from US$1,755 billion to US$3,293 billion.

PRIMARY EXPORTS
as % of value of goods exported, 2002

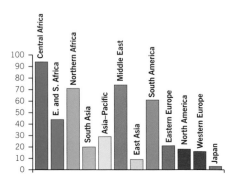

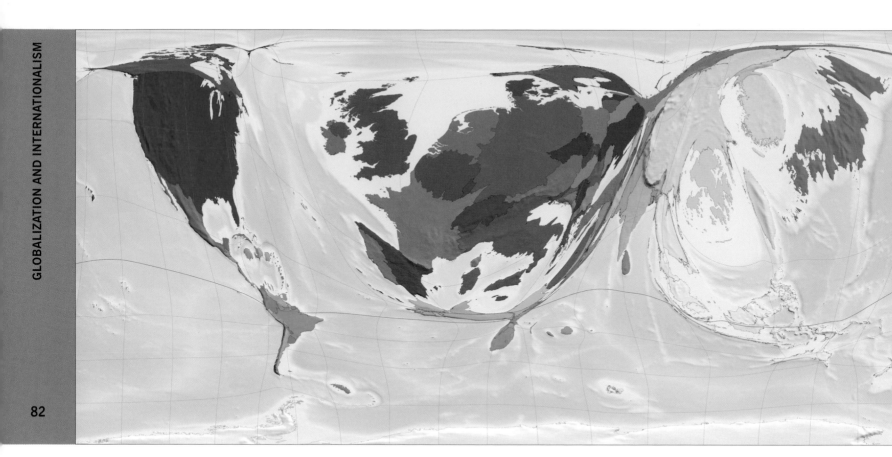

065 Secondary Exports in 1990

The size of each territory indicates the dollar value of secondary exports in 1990, adjusted for local purchasing power.

Secondary exports are manufactured goods, such as clothes, apple pies or cars. In 1990 the combined total earnings from secondary exports of all the territories in the world amounted to US$3,432 billion. US$1,755 billion was earned from primary exports in the same year. Secondary exports constituted more than 90% of the combined export earnings of the five biggest earners.

Western European territories accounted for 38% of the world's secondary exports in 1990. Major Western European exporters included Germany, France, Italy and the United Kingdom. Many of these exports were destined for other European territories.

SECONDARY EXPORTS
as % of value of goods exported, 1990

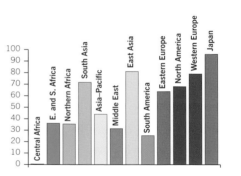

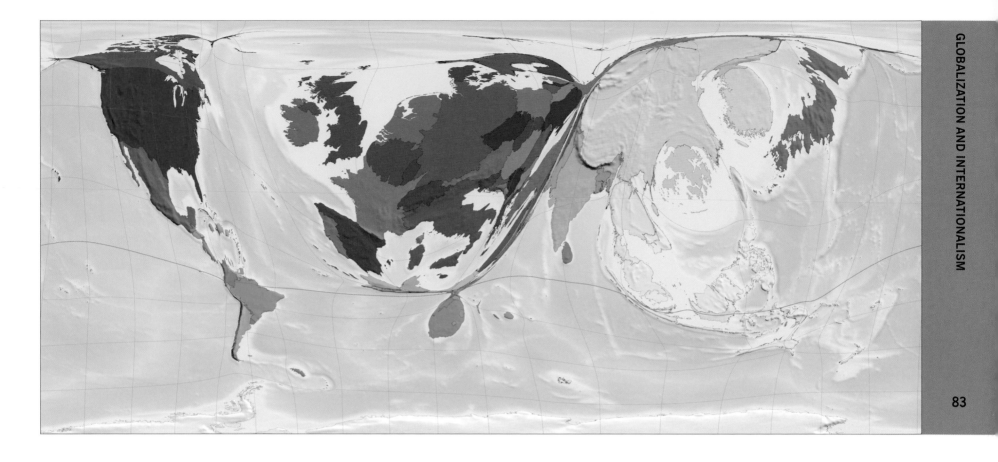

HIGHEST AND LOWEST EXPORTS OF SECONDARY GOODS IN 2002

Rank	Territory	%ᵃ
1	Cape Verde	96
2	Malta	96
3	Hong Kong (China)	95
4	Japan	93
5	Switzerland	93
6	Israel	93
7	South Korea	92
8	Bangladesh	92
9	Botswana	91
10	Taiwan	91
191	Tonga	4
192	Niger	3
193	Rwanda	3
194	Sudan	3
195	Algeria	2
196	Papua New Guinea	2
197	Gabon	2
198	Burundi	1
199	Belize	1
200	Nigeria	1

ᵃ secondary goods as % of total merchandise exports, 2002

066 Secondary Exports in 2002

The size of each territory indicates the dollar value of secondary exports in 2002, adjusted for local purchasing power.

Between 1990 and 2002 the total value of all secondary exports worldwide more than doubled. By 2002 secondary exports made up 71% of all exports in dollar terms. Alongside this growth in volume, the geographical distribution of export earnings changed over this interval. Western Europe's share of the world total

fell, for example, while China's increased, though both remain major exporters of secondary goods.

On the islands of Cape Verde and Malta 96% of all export earnings in 2002 were from secondary exports. At the other extreme, just 1% of Nigeria's export earnings were from secondary exports.

'China is aiming to lift the value of its vehicle and auto parts exports to 120 billion US dollars, or 10 percent of the world's total vehicle trading volume in the next 10 years...' Wei Jianguo, Chinese vice-minister of commerce, 2007

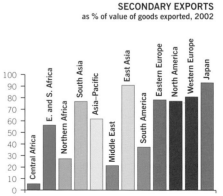

SECONDARY EXPORTS
as % of value of goods exported, 2002

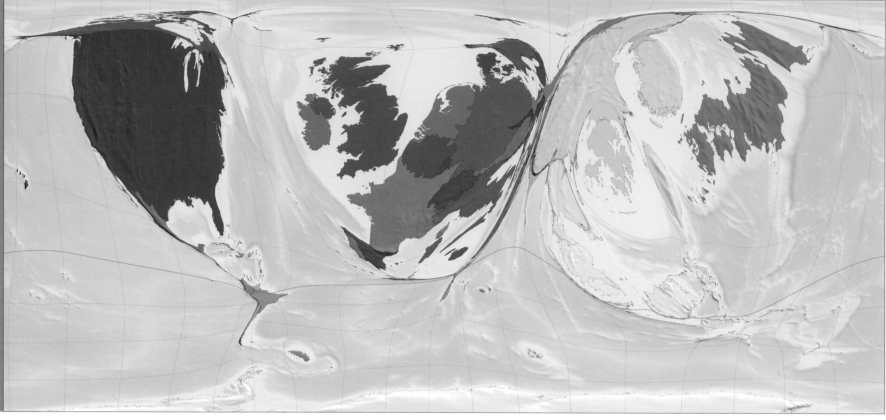

[a] value of high-tech exports as % of total value of secondary goods exported

067 High-Tech Exports in 1990

The size of each territory indicates the combined dollar value of all high-tech exports for the year 1990, adjusted for local purchasing power.

Exports of high-tech goods – usually electronic items, such as computers, mobile telephones, cameras and audio-visual devices – made up 16% of all exports of manufactured goods in 1990.

In that year 72% of all high-tech exports came from just 8 territories: the United States, Japan, China, the United Kingdom, France, Germany, Hong Kong (which is part of China) and South Korea.

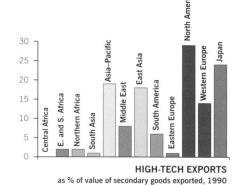

HIGH-TECH EXPORTS
as % of value of secondary goods exported, 1990

'Since its establishment in 1991, the new- and hi-tech industrial belt in the Pearl River Delta region has posted an average annual growth of over 40%.' Monina Wong, researcher for Labour Action China, 2005

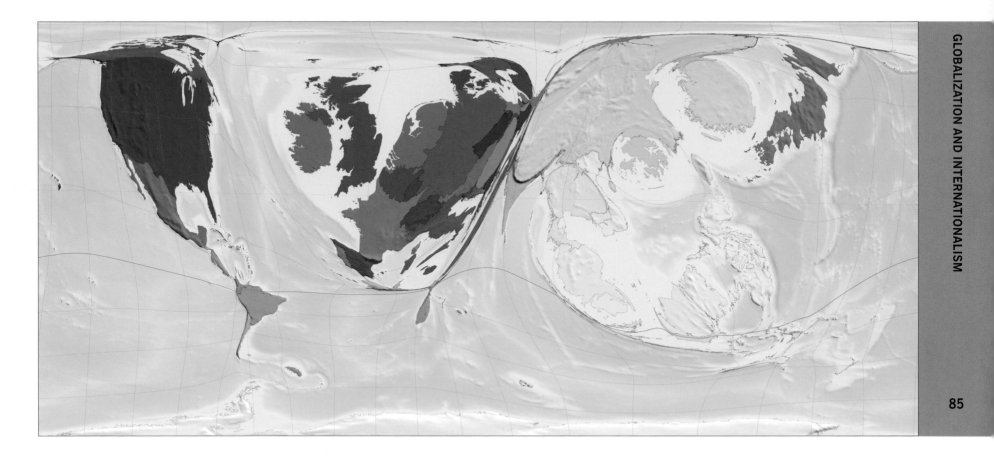

HIGHEST EXPORTS OF HIGH-TECH GOODS IN 2002

Rank	Territory	%ᵃ
1	Philippines	65
2	Malta	62
3	Singapore	60
4	Malaysia	58
5	Tajikistan	42
6	Ireland	41
7	Georgia	38
8	Costa Rica	37
9	South Korea	32
9	United States	32
11	Thailand	31
11	United Kingdom	31
13	Cuba	29
14	Netherlands	28
15	Hungary	25
16	Finland	24
16	Japan	24
18	China	23
19	Denmark	22
19	Norway	22

ᵃ value of high-tech exports as % of total value of secondary goods exported

068 High-Tech Exports in 2002

The size of each territory indicates the combined dollar value of all high-tech exports for the year 2002, adjusted for local purchasing power.

High-tech exports have increased consistently since 1990 and by 2002 accounted for 21% of all exports of manufactured goods. Over that period worldwide earnings from the export of high-tech goods quadrupled, reaching US$2 trillion in 2002. Although the major exporters of high-tech goods remained the same, a significant portion of the increase came from growth on the part of smaller players. Malaysia and Singapore, for example, each reported a 20% increase in high-tech exports, and Malta, Finland and Indonesia reported increases of roughly 15%. Overall, high-tech exports made up 41% of all exports of manufactured goods from Asia–Pacific and Australasian territories, with individual figures ranging from 65% in the Philippines to 0% in Samoa and Tonga.

HIGH-TECH EXPORTS
as % of value of secondary goods exported, 2002

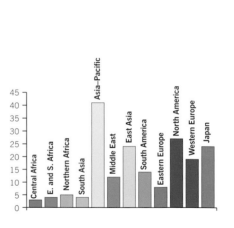

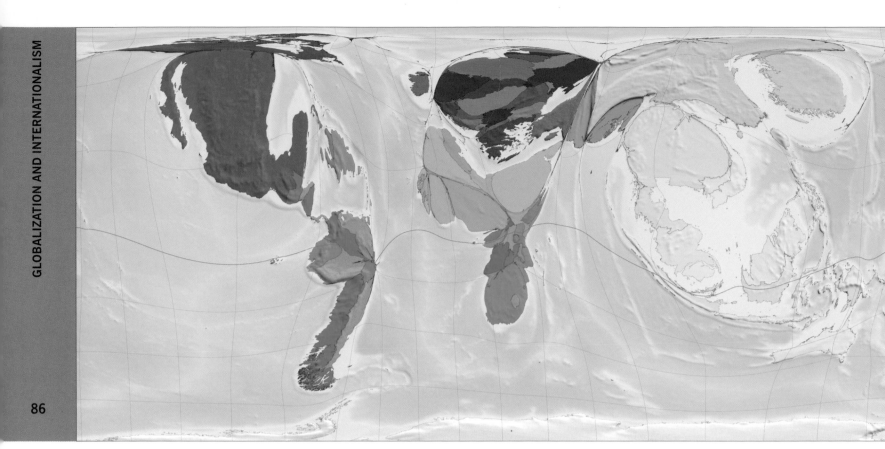

069 Decline in Terms of Trade

The size of each territory indicates the decline in terms of trade between 1980 and 2001.

A territory's terms of trade decline when the market value of exported commodities declines or the cost of imported ones rises. Sometimes this is a result of changes in the types of goods imported or exported. This map shows the amount of money a territory effectively loses each year as a result of changes in terms of trade.

126 territories experienced declining terms of trade between 1980 and 2001.

Those with the largest declines were Burundi, Mexico, Chile and Peru. In Mexico the terms of trade fell by 33% over this period. Another way of putting this is that the decline in terms of trade had the same economic effect as if prices had stayed the same but the country had experienced a decline in productivity to the value of US$176 billion per year.

'A tractor which cost five tons of Tanzanian tea in 1973 cost double that 10 years later. The less developed countries were (and still are) running just to stand still.' Graham Young, *New Internationalist*, 1990

EFFECT OF CHANGING TERMS OF TRADE
annual financial effect of changing terms of trade
1980–2001, PPP US$ billion

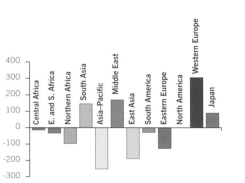

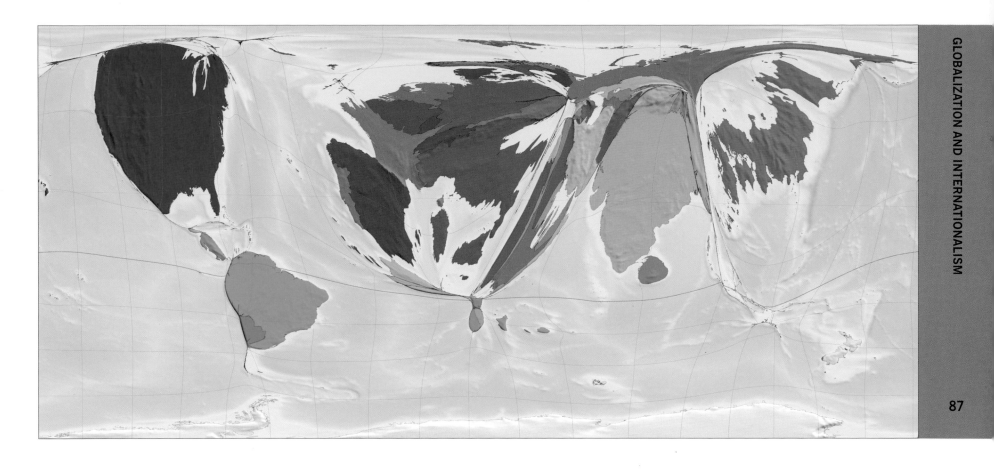

070 Improvement in Terms of Trade

The size of each territory indicates the gains in terms of trade between 1980 and 2001.

A territory's terms of trade improve if the price it receives for its exports increases or the cost of its imports declines. This map shows the effective financial gain resulting from such changes between 1980 and 2001. Territories that experienced a decline in terms of trade during this interval have size zero on this map and so do not appear.

The United States gained the most from improvements in terms of trade over the period: by 2001 US terms of trade had improved by the equivalent of US$202 billion extra per year. India experienced a higher percentage increase in terms of trade, but the total financial gain it represented was lower: an extra US$161 billion per year by 2001.

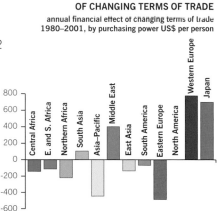

REGIONAL EFFECT PER PERSON OF CHANGING TERMS OF TRADE

annual financial effect of changing terms of trade 1980–2001, by purchasing power US$ per person

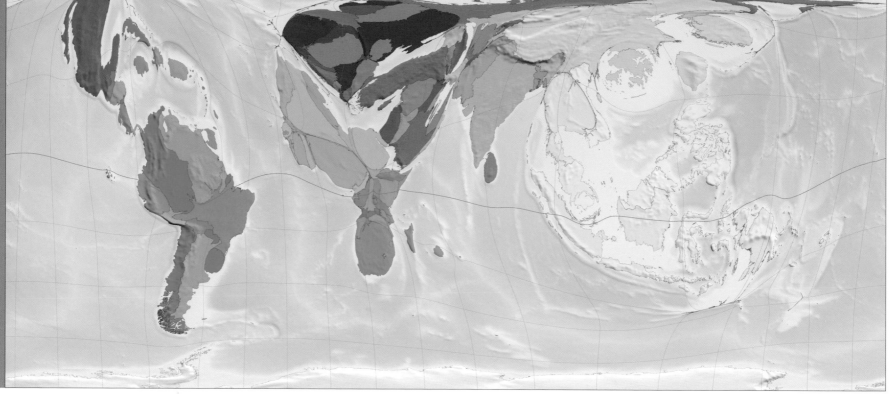

HIGHEST TOTAL DEBT SERVICE IN 1990

Rank	Territory	%ᵃ
1	Uganda	81
2	Algeria	63
3	Madagascar	46
4	Burundi	43
5	Colombia	41
6	Uruguay	41
7	Ethiopia	39
8	Bolivia	39
9	Papua New Guinea	37
10	Argentina	37
11	Ghana	37
12	Côte d'Ivoire	35
12	Kenya	35
14	Congo	35
15	Honduras	35
16	Hungary	34
16	Sao Tome and Principe	34
18	Indonesia	33
19	Tanzania	33
20	Ecuador	33

ᵃ **debt payments as % of earnings from goods and services exports, 1990**

071 Debt Service in 1990

The size of each territory indicates the payments made to service international debts in 1990, adjusted for local purchasing power.

'Debt service' here refers to the payments that must be made to maintain or redeem debts, including interest payments and payments on capital (paying back the original loan). This map shows debt service on publicly guaranteed long-term loans between the governments of different countries and between the IMF and governments, adjusted for local purchasing power. Territories have other forms of debt also, such as debt on bond issues, which are not included on this map.

In 1990 the debt payments of East and Southern Africa, Northern Africa, South Asia and South America were equivalent to more than a quarter of the export earnings of those regions. These statistics inspired the formation in the 1990s of the Jubilee 2000 Campaign, which called for the cancellation of unpayable and unfair debts.

EARNINGS SPENT ON PUBLIC DEBTS
% of income from exporting goods and services spent on paying for debt in 1990

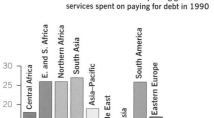
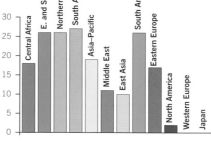

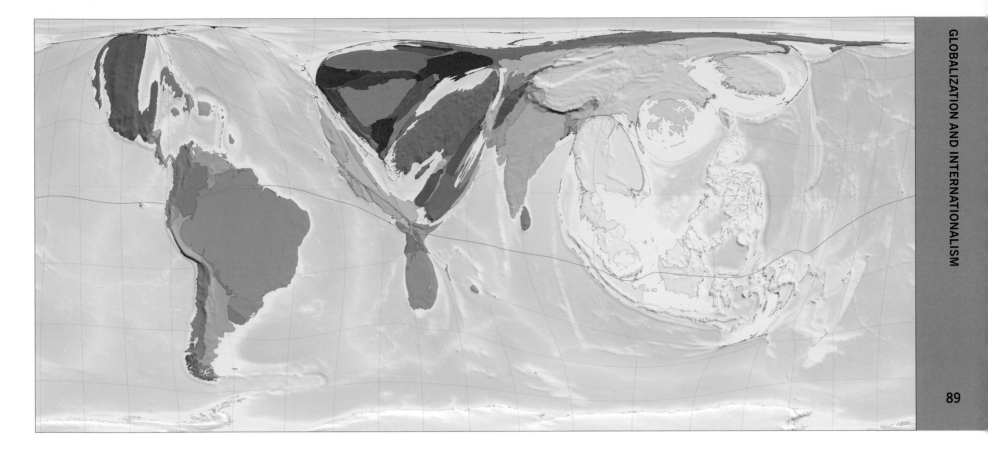

HIGHEST TOTAL DEBT SERVICE IN 2002

Rank	Territory	%ᵃ
1	Brazil	69
2	Burundi	59
3	Lebanon	51
4	Turkey	47
5	Colombia	40
6	Uruguay	40
7	Belize	37
8	Kazakhstan	34
9	Hungary	34
10	Chile	33
11	Peru	33
12	Sao Tome and Principe	32
13	Ecuador	29
14	Bolivia	28
15	Zambia	27
16	Croatia	26
16	Venezuela	26
18	Kyrgyzstan	25
19	Indonesia	25
20	Uzbekistan	24

ᵃ debt payments as % of earnings from goods and services exports, 2002

072 Debt Service in 2002

The size of each territory indicates the payments made to service international debts in 2001, adjusted for local purchasing power.

In 2002 total payments of interest and capital on money borrowed from the IMF and other governments came to US$1,158 billion worldwide, after adjusting for local purchasing power. This was 8.9% of the value of all worldwide exports of goods and services that year. Brazil used the largest proportion of its income (69%) to pay off its debts, leaving only 31% to spend on services within the territory and to bring in imports from abroad. The largest total payments on debts in 2002 were made by Brazil, China, Thailand and India.

Only 30 out of 200 territories were not in debt to the IMF in 2002, all of them in Western Europe and North America except for Australia, New Zealand and Zimbabwe.

'Brazilian society was built on the work, the sweat and the blood of Africans...[Brazil] is in debt to Africa.'
Luiz Inácio Lula da Silva, president of Brazil, 2004

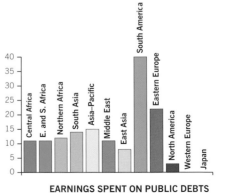

EARNINGS SPENT ON PUBLIC DEBTS
% of income from exporting goods and services spent on paying for debt in 2002

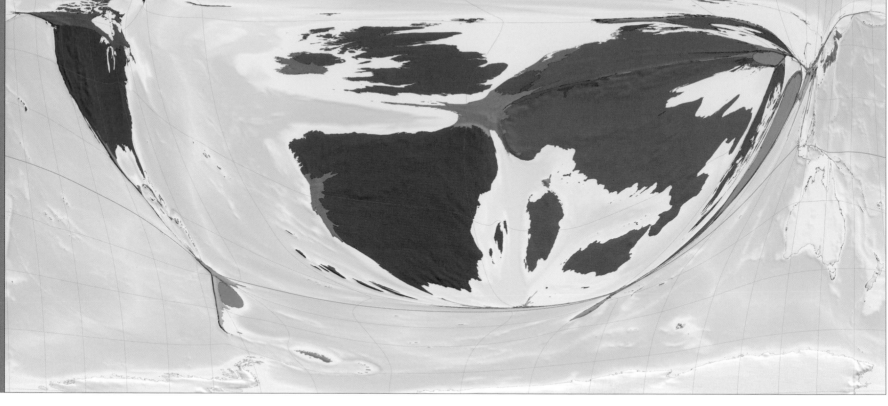

MOST DEMONSTRATORS AGAINST 2003 WAR IN IRAQ

Rank	Territory	Demonstrators[a]
1	Spain	105
2	Italy	80
3	Australia	34
4	Norway	30
5	Ireland	26
6	United Kingdom	26
7	Luxembourg	23
8	Greece	21
9	Uruguay	21
10	Iceland	15
11	Syria	14
12	Germany	14
13	Sweden	12
14	Portugal	11
15	Gaza and West Bank	10
16	Belgium	10
16	Finland	9
18	Canada	8
19	Bahrain	7
20	New Zealand	7

[a] number of demonstrators per 1,000 people in the population, 14–16 Feb. 2003

073 Demonstrations against the 2003 War in Iraq

The size of each territory indicates the number of people who demonstrated on 14, 15 and 16 February 2003 against war in Iraq.

This map shows the distribution of the 15.9 million people worldwide who protested against the invasion of Iraq in 2003 by troops from over 25 countries, including the United States, the United Kingdom, Italy, Spain and Poland.

Protests against the war were recorded in 96 of the 200 mapped territories. Probably there were smaller, unrecorded protests in others. The largest recorded protests were in Italy, Spain, the United Kingdom, Germany and the United States, which together accounted for 80% of the protesters. In Rome, 3 million people marched in the largest popular demonstration ever recorded in human history.

PROTESTERS AGAINST WAR IN IRAQ
number per 1,000 people, in demonstrations 14–16 Feb. 2003

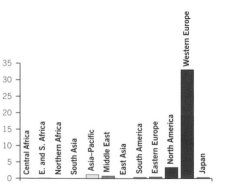

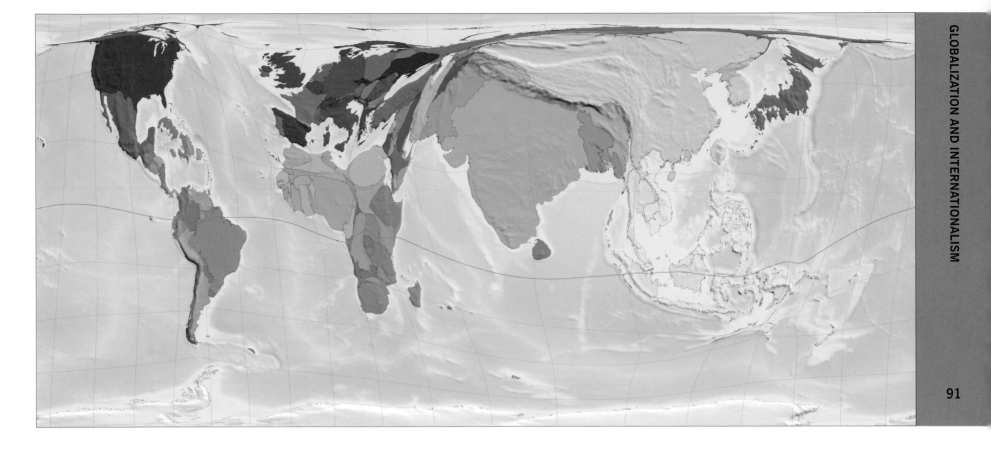

LONGEST AND SHORTEST TIMES AS SIGNATORIES TO THE GENEVA CONVENTION (OR EQUIVALENT)

Rank	Territory	Years[a]
1	Switzerland	55.8
2	Monaco	55.5
3	Liechtenstein	55.3
4	Chile	55.2
5	Austria	55.1
6	Estonia	54.9
7	Germany	54.7
8	Hungary	54.6
9	Kyrgyzstan	54.6
10	Latvia	54.5
191	Nauru	11.4
192	Micronesia	10.3
193	Niue	10.0
194	Palau	9.5
195	Lithuania	9.2
196	Eritrea	5.4
197	Cook Islands	4.6
198	Serbia and Montenegro	4.2
199	Timor-Leste	2.7
200	Marshall Islands	1.6

[a] years as signatories to the Geneva Convention, or equivalent, measured up to 1 Jan. 2006

074 International Justice

The size of each territory indicates the number of years (up to 2006) it has been a signatory of the 1949 Geneva Convention, multiplied by its population.

The Geneva Conventions are a set of four treaties that require of their signatories the protection and respect of individuals who do not (and sometimes cannot) take part in hostilities between and within nations, including the sick and wounded, prisoners of war and civilians. The first Geneva Convention was ratified in 1864. This map shows signatories to the more recent 1949 convention, which is the one most people mean when they refer to 'the Geneva Convention'.

The convention of 1949 is a central element of international humanitarian law. Geneva being a Swiss city, it is perhaps

unsurprising that the first nation to ratify the convention was Switzerland, in March 1950. Other signatories that year included Monaco, Liechtenstein, Chile and India. By 2004 all 200 territories mapped here (or the territories claiming sovereignty over them) were signatories.

Territories can appear large on this map either if they have a large population or if they were early signatories to the convention.

YEARS SIGNED UP TO GENEVA CONVENTION
average years spent as signatories to 1949 Geneva Convention, up to 1 Jan. 2006

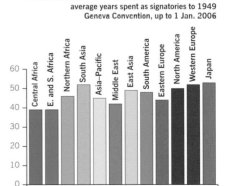

'He who commits injustice is ever made more wretched than he who suffers it.' Plato, Greek philosopher, c.400 BC

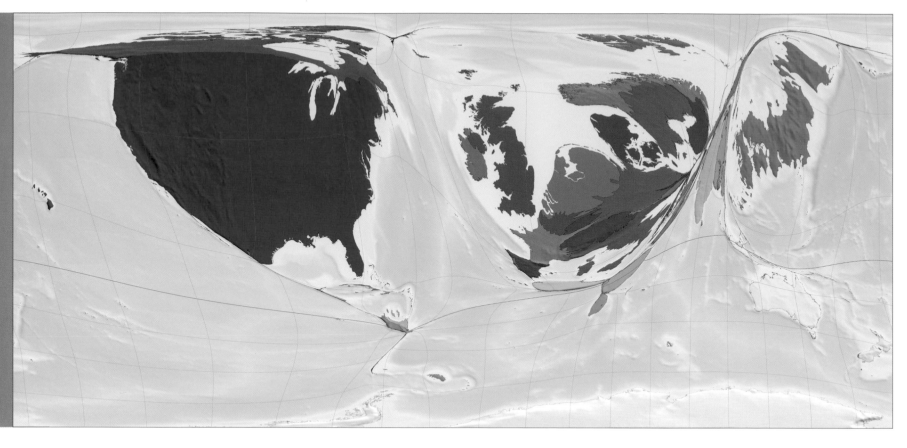

Rank	Territory	US$ᵃ
1	Luxembourg	27
2	Norway	21
3	Vatican City	10
4	Denmark	10
5	Sweden	9
6	Netherlands	7
7	Canada	5
8	Finland	5
9	Ireland	5
10	Switzerland	5
11	United States	4
12	Australia	3
13	New Zealand	3
14	United Kingdom	2
15	Iceland	2
16	Liechtenstein	1
16	Belgium	1
18	Japan	1
19	Monaco	1
20	Germany	1

ᵃ food aid contributions in 2005, US$ per
person per year (no contribution recorded
from 131 territories)

075 International Food Aid

The size of each territory indicates the amount in US dollars
of governmental contributions to international food aid.

Wars, droughts, economic collapse and
other disasters can disrupt access to basic
necessities, including food. This map
shows the amount in dollars given by each
territory in 2005 to provide food for people
whose normal food supplies have failed.
Food aid of this type is a temporary
measure intended to deal only with
immediate food shortages.

In 2005 governments contributed about
US$2.5 billion to food aid programmes.
Half of this came from the United States
and a third from territories in Western
Europe.

A further US$0.5 billion was contributed
by international organizations, individuals
and charities.

*'When I give food to the poor, they call me a saint. When
I ask why the poor are poor, they call me a communist.'*

Dom Helder Camara, Brazilian priest, 2004

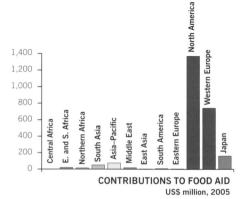

CONTRIBUTIONS TO FOOD AID
US$ million, 2005

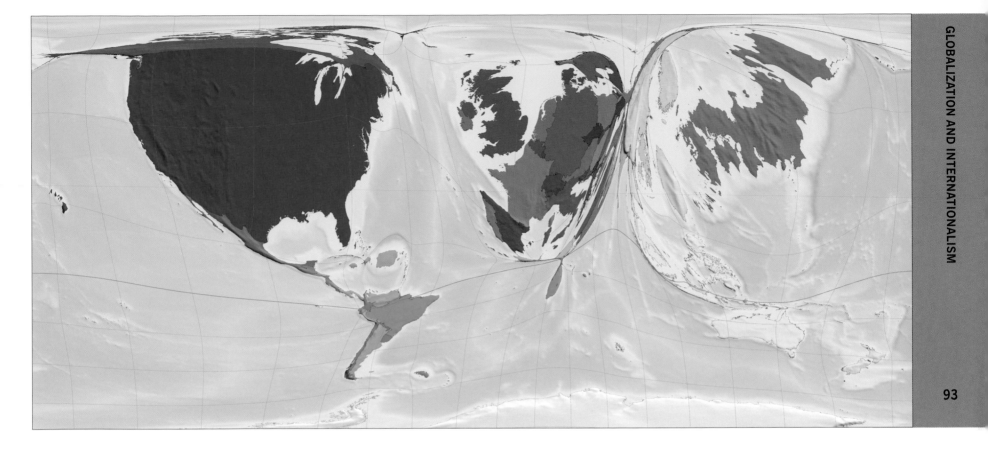

MOST MCDONALD'S RESTAURANTS

Rank	Territory	Restaurants[a]
1	United States	47
2	Canada	44
3	Andorra	43
4	New Zealand	38
5	Australia	37
6	San Marino	37
7	Singapore	31
8	Hong Kong (China)	30
9	Japan	30
10	Liechtenstein	30
11	Monaco	29
12	Puerto Rico	29
13	Sweden	27
14	United Kingdom	21
15	Malta	20
16	Austria	20
16	Switzerland	20
18	Cyprus	19
19	Ireland	18
20	France	18

[a] **McDonald's restaurants per million people, 2004 (102 countries did not have this chain of restaurants in 2004)**

076 International Fast Food

The size of each territory indicates the number of McDonald's fast food restaurants.

This map shows the distribution of one major brand of fast food restaurant, McDonald's. In 2004, the year shown on this map, there were 30,496 McDonald's outlets worldwide. Of these, 45% were located within the United States, which is why it appears so large. The next largest numbers of outlets were in Japan, Canada and Germany.

The world average number of restaurants for this one brand alone is 5 per 1 million people. In the United States there are 47 per million people, in Argentina and Chile a 10th of that, and in Indonesia, China and Georgia a 100th. In all the territories of Africa combined there were only 150 outlets, most of them in South Africa.

FAST FOOD RESTAURANTS
thousands of McDonald's outlets, 2004

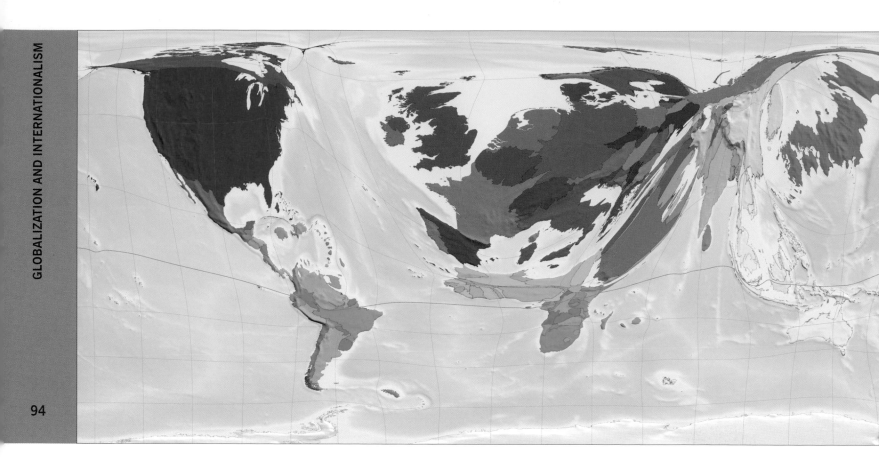

077 Votes in the International Monetary Fund

The size of each territory indicates its voting power in the IMF, which varies according to its contribution to the Fund's resources.

The IMF is an international organization that provides loans to countries and aims to promote monetary cooperation and to foster economic growth and high levels of employment. In 2006 the IMF had 184 member countries. Not all those countries, however, have equal voting power in the decisions of the IMF; members have different numbers of votes according to their financial contributions to IMF funds.

This map shows votes per territory in 2006. The largest numbers of votes were held by the United States, which had as many as the next three combined, namely Japan, Germany and the United Kingdom. Central Africa, which does not have the financial resources to make significant contributions to IMF funds, had less than 1% of all votes; East and Southern Africa had just under 2%.

'Go tell the International Monetary Fund that privatisation is a big disaster in Zambia…' Joyce Nonde, quoted in *New Statesman*, 2004

VOTES IN THE IMF
votes per 1,000 people, 2006

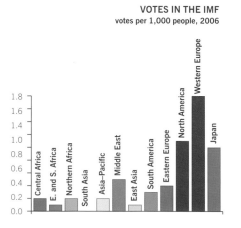

HITS ON THE WORLDMAPPER WEBSITE

Rank	Territory	Hits[a]
1	Luxembourg	76
2	Finland	61
3	Liechtenstein	54
4	United States	50
5	Australia	43
6	Switzerland	41
7	Iceland	32
8	United Kingdom	30
9	Canada	28
10	Belgium	27
150	Angola	0.005
151	Sao Tome and Principe	0.005
152	Syria	0.005
153	Madagascar	0.004
154	Bangladesh	0.004
155	Cameroon	0.004
156	Zambia	0.003
157	Rwanda	0.002
158	Mozambique	0.001
159–200	42 countries	0.000

[a] **hits on http://www.worldmapper.org per 1,000 people, Jan.–Oct. 2006**

078 Who's Looking at Us?

The size of each territory indicates the distribution of hits on the authors' Worldmapper website, which aims to inform people about the social, economic and geographical state of the world today, using maps.

Many of the maps that appear in this book also appear on the website www.worldmapper.org, maintained by the authors. This map shows the geographical distribution of the people visiting the website up to October 2006. More precisely, it shows the number of 'hits' – viewings of pages on the site – so that people who view more than one page, or who make repeated visits to the site, are counted more than once.

The largest numbers of hits on the website come from the United States, the United Kingdom, Germany, Canada, Australia and South Korea. The distribution is certainly affected by the fact that the website is, for the moment at least, in English only. By 2007, when we were preparing this book, over 1 million different people had viewed the website. The success of the website has formed part of the inspiration for this atlas.

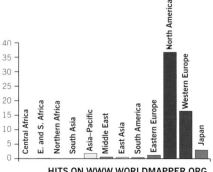

HITS ON WWW.WORLDMAPPER.ORG
hits on the worldmapper website during its first 10 months, in 2006, per 1,000 people

'...it is of the greatest importance that the peoples of the earth learn to understand each other as individuals across distances and frontiers.'

Bertil Lindblad, director of the Stockholm Observatory at Saltsjöbaden, 1938

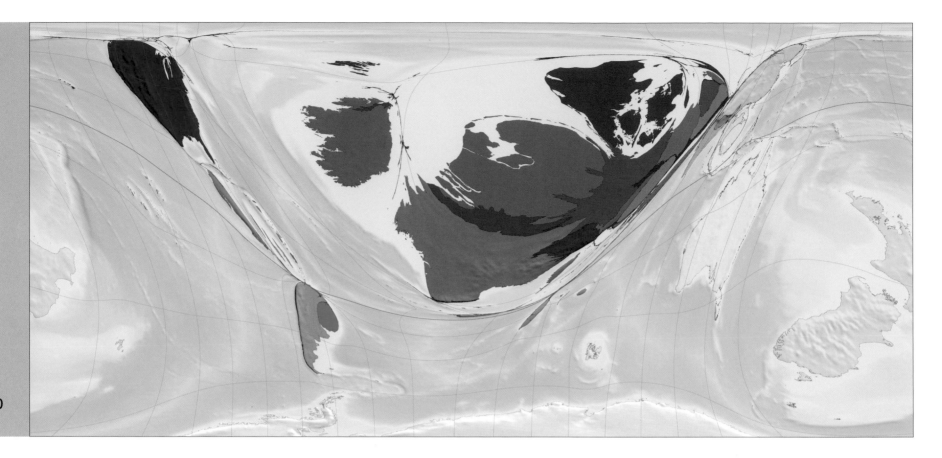

MOST NET DAIRY EXPORTS

Rank	Territory	US$ᵃ
1	New Zealand	640.6
2	Ireland	320.0
3	Denmark	282.1
4	Netherlands	170.3
5	Australia	59.5
6	France	38.1
7	Belgium	36.0
8	Lithuania	29.4
9	Switzerland	28.9
10	Uruguay	28.7
11	Costa Rica	24.9
12	Estonia	17.7
13	Hungary	14.4
14	Belarus	10.8
15	Argentina	9.0
16	Austria	9.0
17	Germany	8.6
18	Poland	5.1
19	Finland	4.6
20	United States	3.7

ᵃ annual US$ worth of net dairy exports per person

083 Dairy Exports

The size of each territory indicates the dollar value of its net exports of dairy produce, including milk, butter, cheese and eggs.

Dollar values depend on the local values of goods and the exchange rates between the dollar and other currencies, as well as the volume of goods moved.

Dairy products constitute 0.8% of world trade. Less than a quarter of our 200 territories have net dairy exports.

As with many of the trade maps, Europe is large and Africa is small here, their relative sizes resulting from the combination of the quantity of goods moved and the dollar value of these goods.

New Zealand is the largest per capita net exporter in dollar terms, with dairy exports valued at US$641 a year. New Zealand exports dairy produce to 140 other territories, and is home to 3.5 million dairy cows.

NET DAIRY EXPORTS
annual earnings, US$ billion

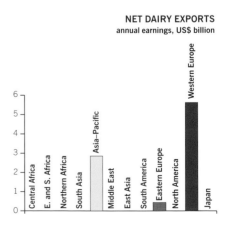

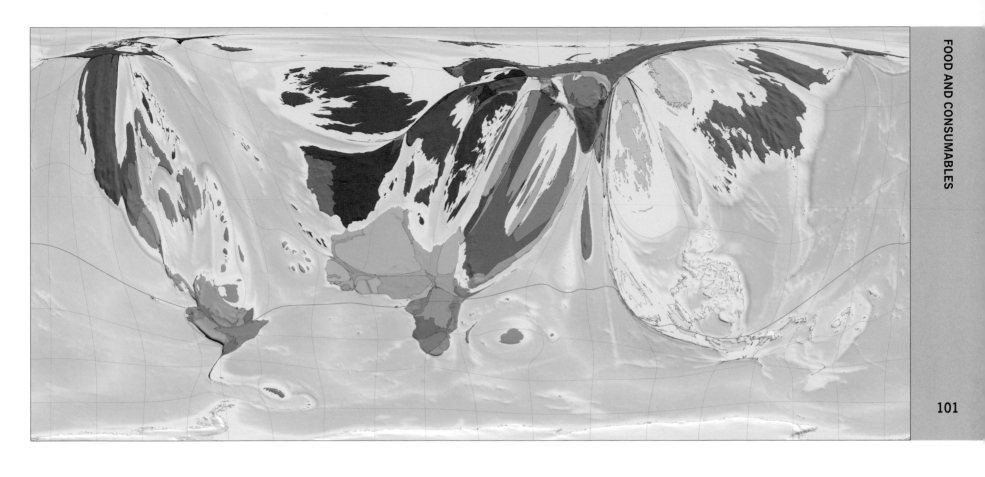

084 Dairy Imports

The size of each territory indicates the dollar value of its net imports of dairy produce, including milk, butter, cheese and eggs.

Net dairy imports exceed exports in every region except for the Asia–Pacific and Australasia, Eastern Europe and Western Europe. The Middle East and Central Asia, Northern Africa and Japan are the largest regional net importers.

NET DAIRY IMPORTS
annual spending, US$ billion

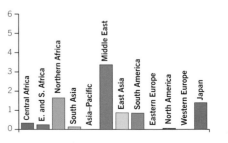

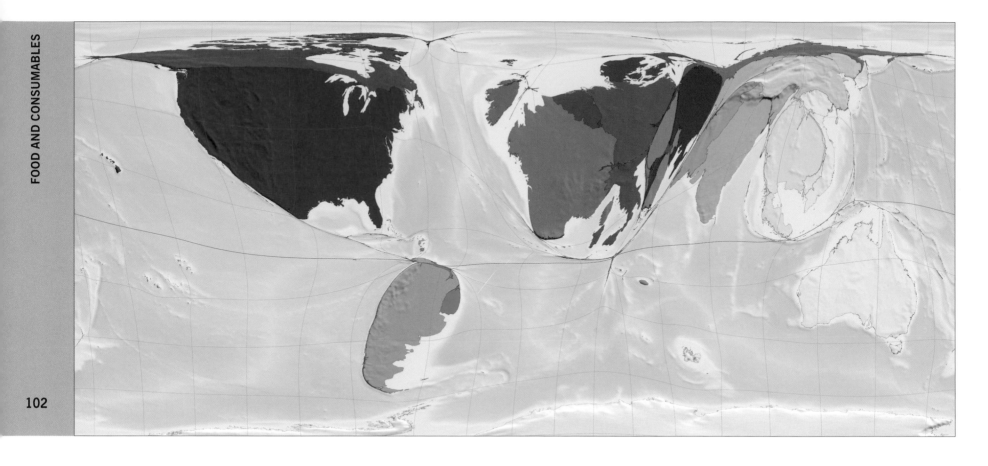

MOST NET CEREAL EXPORTS

Rank	Territory	US$[a]
1	Australia	157.9
2	Ireland	84.4
3	Canada	73.1
4	Denmark	69.5
5	France	60.8
6	Argentina	60.6
7	Uruguay	39.6
8	Belgium	37.3
9	Guyana	35.1
10	Hungary	32.6
11	United States	30.7
12	Thailand	26.5
13	Kazakhstan	24.7
14	Bulgaria	20.5
15	Ukraine	20.2
16	Germany	19.4
17	Italy	9.1
18	Serbia and Montenegro	8.7
19	Moldova	8.0
20	Lithuania	7.5

[a] annual US$ worth of net cereal exports per person

085 Cereal Exports

The size of each territory indicates the dollar value of its net exports of cereals, which provide the main carbohydrate component of the human diet in most parts of the world.

Dollar values depend on the local values of goods and the exchange rates between the dollar and other currencies, as well as the volume of goods moved. Cereals include wheat, rice, barley, maize (called corn in North America and Australia) and flour.

The United States, France and Australia are the three largest net exporters of cereals. No region dominates cereal exports: at least one territory in each region is a net exporter, the exports often going to feed people in other parts of the same region.

'Lao food is traditionally eaten with sticky rice, with the fingers. In the countryside, people will all eat family style, sitting on the floor, sharing a few dishes.' Visit Laos, Laos tourist website, 2006

NET CEREAL EXPORTS
annual earning, US$ billion

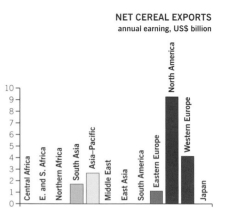

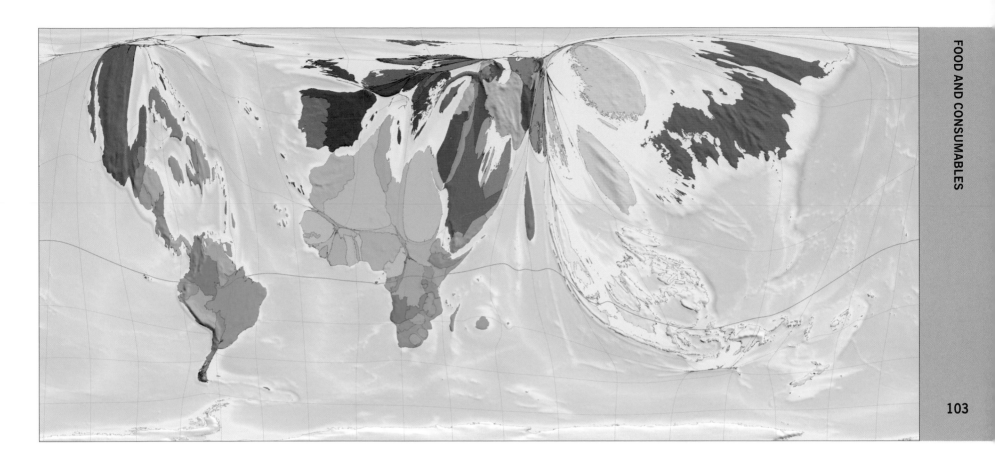

ᵃ **annual US$ worth of net cereal imports per person**

086 Cereal Imports

The size of each territory indicates the dollar value of its net imports of cereals. Only a few countries produce more cereals than they consume.

4 out of every 5 territories are net importers of cereals in terms of dollar value. Africa, the Middle East and Central Asia, East Asia, South America and Japan, as regions, are all net importers of cereals. African territories account for a higher share of world net imports of cereals than of world net imports of fruit, vegetables, meat, fish, groceries or alcohol and tobacco.

NET CEREAL IMPORTS
annual spending, US$ billion

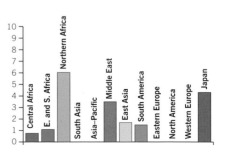

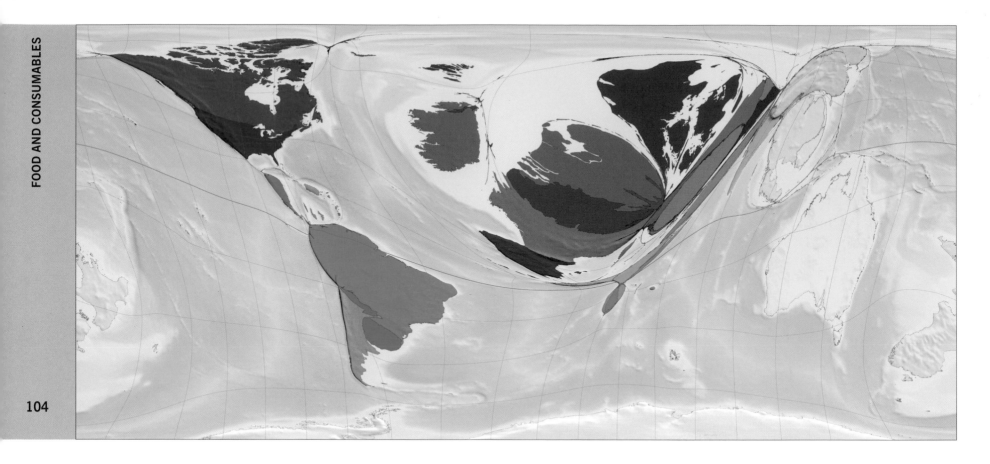

087 Meat Exports

The size of each territory indicates the dollar value of its net exports of meat, including chilled, frozen, dried, salted and smoked meat, as well as living animals destined for slaughter as food.

Denmark (famous for its bacon) and New Zealand (famous for its lamb) each earn over US$500 per person every year in meat exports. The highest net meat-exporting regions are the Asia–Pacific and Australasia, and South America. Within South America, net meat exports from Brazil are more than 5 times as large as those from any other South American territory.

NET MEAT EXPORTS
annual earnings, US$ billion

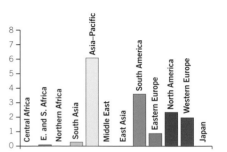

'Between 1990 and 2001 the percentage of Europe's processed meat imports that came from Brazil rose from 40 to 74 percent.'

Centre for International Forestry Research, 2004

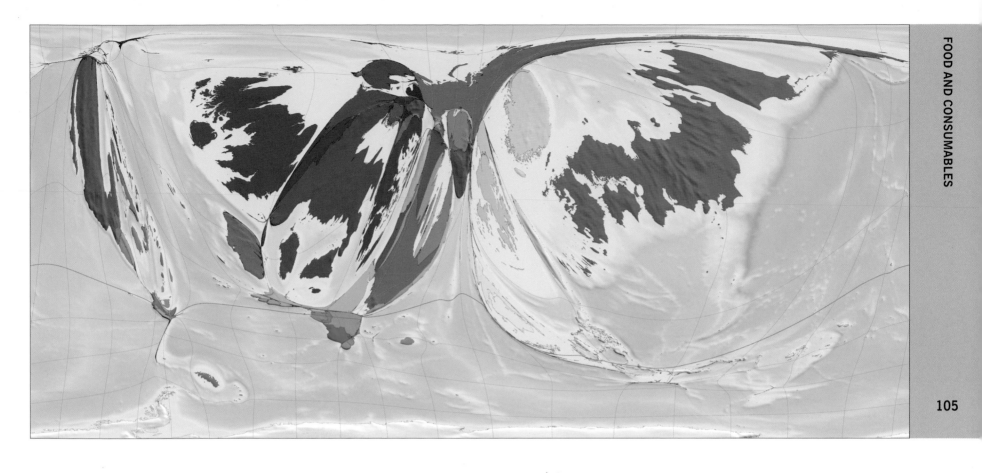

MOST AND LEAST NET MEAT IMPORTS

Rank	Territory	US$[a]
1	Andorra	405
2	Tuvalu	248
3	Greenland	242
4	Bahamas	196
5	Luxembourg	192
6	Cook Islands	175
7	Qatar	158
8	Niue	151
9	Hong Kong (China)	148
10	St Lucia	143
11	United Arab Emirates	124
12	Bahrain	111
13	St Kitts–Nevis	106
14	Singapore	102
15	Tonga	95
16	Grenada	84
17	St Vincent and the Grenadines	82
18	Kuwait	77
19	Malta	77
20	Greece	71

[a] **annual US$ worth of net meat imports per person**

088 Meat Imports

The size of each territory indicates the dollar value of its net imports of meat, including chilled, frozen, dried, salted and smoked meat, as well as living animals destined for slaughter as food.

Japan has the highest net imports of meat per year, measured in terms of US dollars, accounting for a quarter of the world total. The only territory on the map that is considered a region in its own right, Japan also has the highest net imports of any region.

The highest net imports of meat per person are found in Andorra, a country of just 70,000 people. Andorra imports US$405 of meat annually for every member of its population.

NET MEAT IMPORTS
annual spending, US$ billion

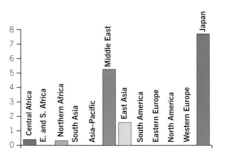

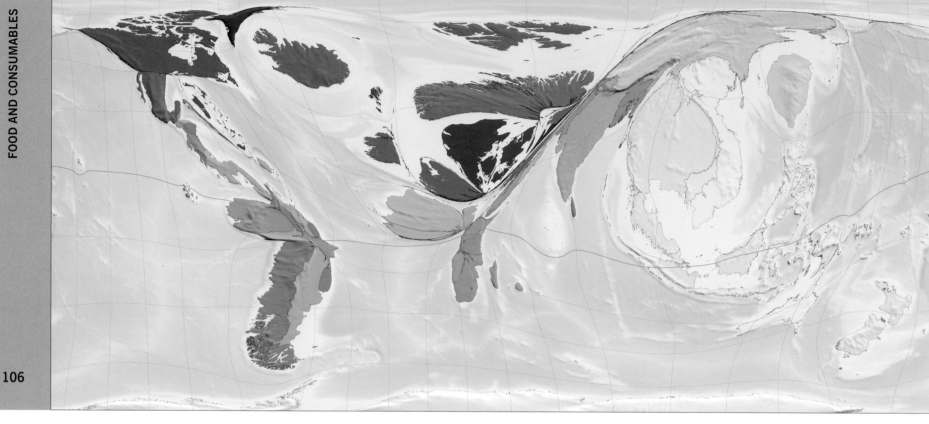

089 Fish Exports

The size of each territory indicates the dollar value of its net exports of fish. Not all island territories have substantial fish exports.

The Asia–Pacific and Australasia region, and South America, have the highest regional net fish exports in terms of dollar value. Among individual territories, Thailand, China and Norway have the highest exports; together they provide one-third of worldwide net fish exports.

Island territories, which often have active fishing industries, tend to have high net fish exports per person, although there are exceptions. While many islands in the Asia–Pacific and Australasia region, for example, are strong net exporters, many Central American islands are net importers. Similarly, Japan, which has a large fishing industry, is nonetheless a net importer of fish and so does not appear on this map.

'There is no firm evidence to show that fish exports are detrimental to food security in the exporting country as generally the products exported are different from those consumed locally.'

Helga Josupeit, fishery industry officer, Fish Utilization and Marketing Service, Rome, 2003

MOST NET FISH EXPORTS

Rank	Territory	US$[a]
1	Greenland	4,681
2	Iceland	3,751
3	Norway	677
4	Bahamas	362
5	Denmark	264
6	Maldives	174
7	New Zealand	169
8	Kiribati	101
9	Panama	100
10	Chile	97
11	Suriname	85
12	Mauritania	75
13	Ireland	73
14	Guyana	66
15	Canada	61
16	Estonia	57
17	Namibia	55
18	Thailand	54
19	Solomon Islands	53
20	Ecuador	52

[a] annual US$ worth of net fish exports per person

NET FISH EXPORTS
annual earnings, US$ billion

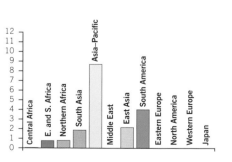

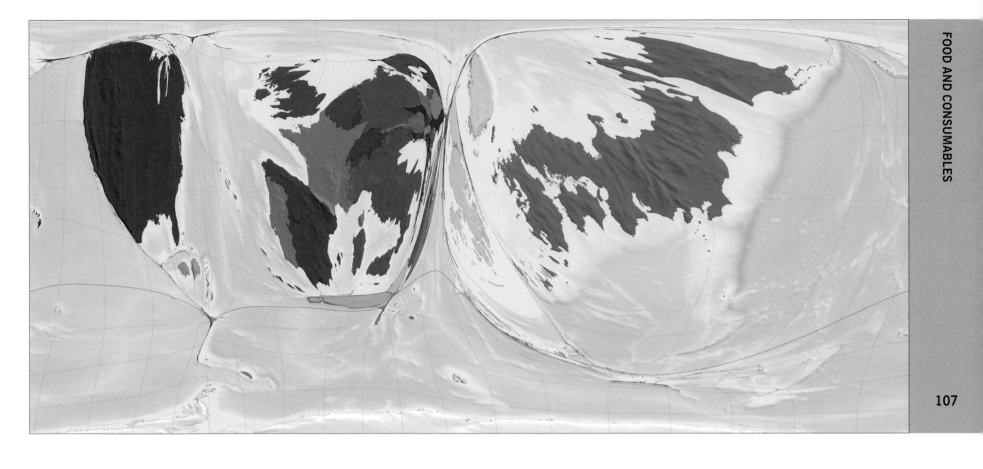

090 Fish Imports

The size of each territory indicates the dollar value of its net imports of fish. As in most trade, the wealthy tend to be the biggest players.

One might imagine that the leading net importers of fish would be landlocked territories, but in fact only 3 of the 10 territories with the highest net fish imports per person are landlocked. Of course, lack of an ocean coast does not prevent a country from catching fish: fish can be caught in fresh-water rivers and lakes, as well as in the sea.

With most trade, the largest importers in dollar terms tend to be rich countries, and fish imports are no exception. The biggest net importer of fish (in total, not per person) is Japan, followed by Western European territories and the United States. Imports to these three constitute 89% of all net fish imports worldwide.

NET FISH IMPORTS
annual spending, US$ billion

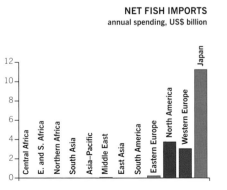

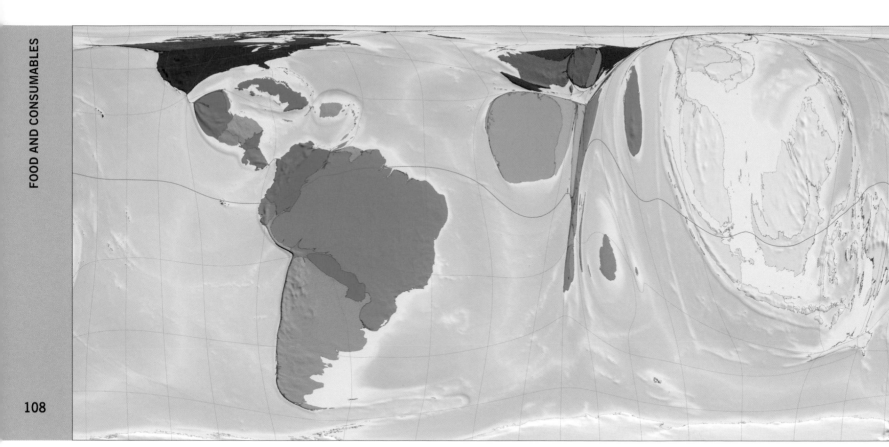

091 Grocery Exports

Almost half of all grocery exports are oils and fats from vegetables and animals. The size of each territory on this map indicates the dollar value of its net exports of groceries.

Groceries include sugar, honey, edible oils, cocoa, chocolate, tea, maté (a herbal infusion) and spices.

Territories in South America and in the Asia–Pacific and Australasia region together account for three-quarters of net grocery exports. Net exporters at the regional level are South America, Asia–Pacific and Australasia, North America, East and Southern Africa, and Northern Africa. Most territories in Asia and Europe are not visible on this map because they are net importers of groceries. Mauritius, which is a major producer of cane sugar and as a result ranks 19th in the world for net exports of groceries, is number one in terms of exports per person, on account of its relatively small population.

'"The food of the gods", as cocoa was called 500 years ago when the Spanish came upon it in South America, remains a precious commodity.' International Cocoa Organization, 2006

NET GROCERY EXPORTS
annual earnings, US$ billion

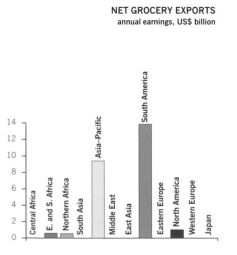

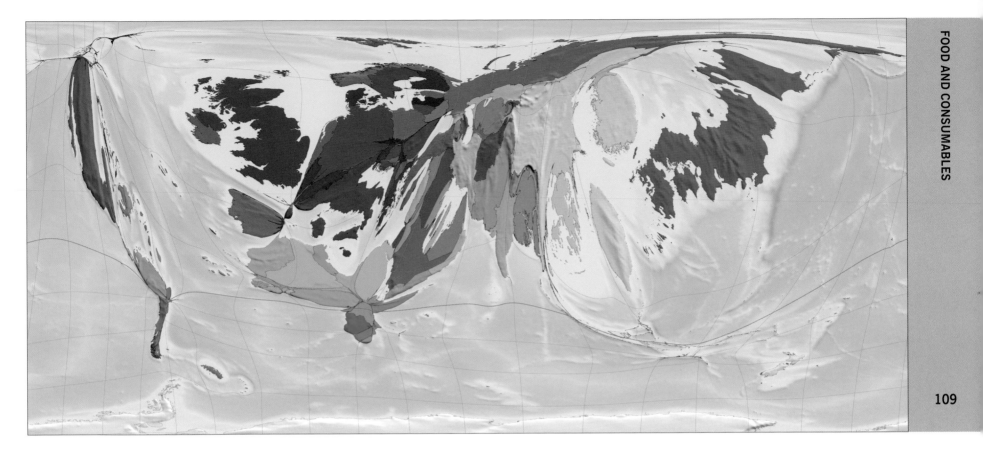

ᵃ **annual US$ worth of net grocery imports per person**

092 Grocery Imports

The size of each territory indicates the dollar value of its net imports of groceries. Over half of all territories import more than they export.

62% of territories are net importers of groceries. As the map shows, territories in more northerly latitudes tend to be net importers, while those in southerly latitudes, particularly the Asia–Pacific and Australasia region and South America, tend to be net exporters. Southern Africa has neither large net imports nor large exports.

Japan is the largest net importer of groceries, accounting for 1.5 times the net imports of the next largest, China: a remarkable observation given that China has ten times the population of Japan. (As a result, Japan's net imports of groceries per person are about 15 times those of China in dollar terms.)

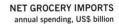

NET GROCERY IMPORTS
annual spending, US$ billion

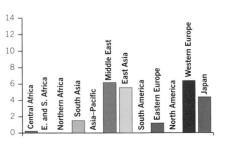

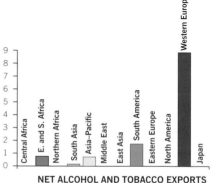

093 **Alcohol and Tobacco Exports**

The size of each territory indicates the dollar value of its net exports of alcohol and tobacco. Western Europe is by far the largest net exporter, accounting for more than two-thirds of the total.

Alcohol and tobacco account for 0.9% of all international trade in terms of dollar value. Within Western Europe, France and the Netherlands are the leading exporters, both ranking in the top 10 of exports per person. After Western Europe, South America is the next largest regional exporter, although only one South American territory, Chile, makes it into the top 10 exporters per person.

'A custom loathsome to the eye, hateful to the nose, harmful to the brain, dangerous to the lungs, and the black stinking fume thereof...'
King James I of England and VI of Scotland, 1604

NET ALCOHOL AND TOBACCO EXPORTS
annual earnings, US$ billion

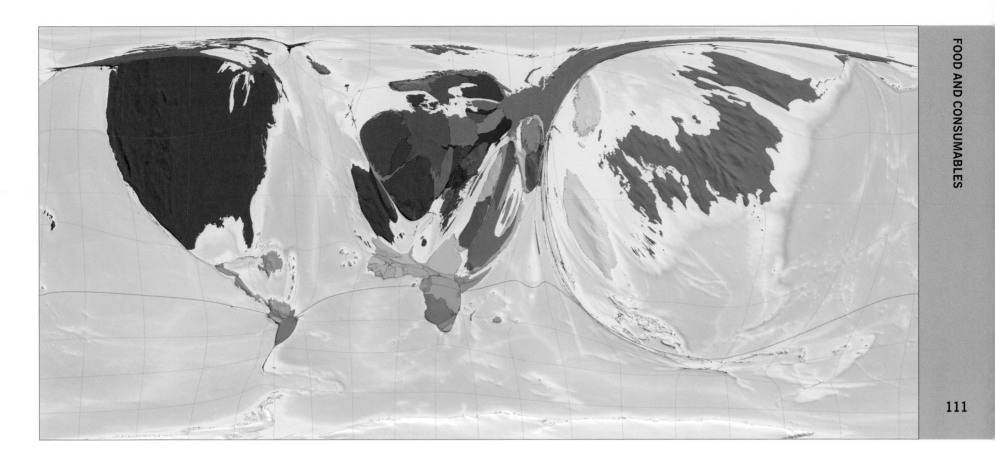

MOST NET IMPORTS OF ALCOHOL AND TOBACCO

Rank	Territory	US$ᵃ
1	Andorra	1,172
2	Luxembourg	302
3	Greenland	213
4	Brunei	123
5	Iceland	117
6	Bahrain	92
7	Switzerland	77
8	Tuvalu	75
9	Norway	58
10	Belgium	48
11	Malta	43
12	Estonia	42
13	St Kitts–Nevis	41
14	Qatar	40
15	Lebanon	36
16	Maldives	36
17	Antigua and Barbuda	35
18	Finland	35
19	Taiwan	33
20	Grenada	29

ᵃ annual US$ worth of net alcohol and tobacco imports per person

094 Alcohol and Tobacco Imports

The size of each territory indicates the dollar value of its net imports of alcohol and tobacco. Western Europe ranks high in imports as well as in exports.

The United States and Japan have the highest net alcohol and tobacco imports in terms of dollar value. The territories with the highest net imports per person, however, are mostly in Western Europe, where wealth and tradition combine to make both alcohol and tobacco popular.

NET ALCOHOL AND TOBACCO IMPORTS
annual spending, US$ billion

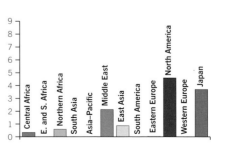

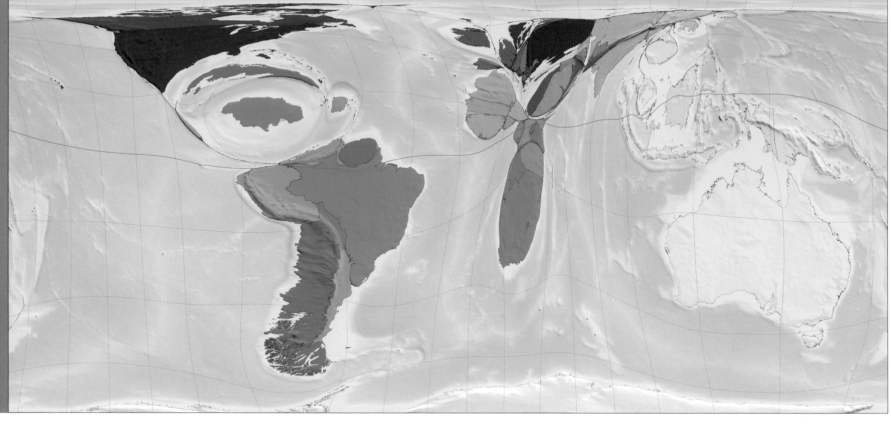

095 Ore Exports

The size of each territory indicates the dollar value of its net exports of ores – minerals in their unprocessed state – along with sand, stone, gravel and scrap metals.

Mineral ores are the raw material for a variety of products, including fertilizer and steel, and can be refined to produce precious metals. 1.2% of worldwide earnings from exports come from ores. Territories in South America and in the Asia–Pacific and Australasia have the largest total net ore exports in dollar terms, and the top 5 territories for net ore exports per person are also in these regions.

NET ORE EXPORTS
annual earnings, US$ billion

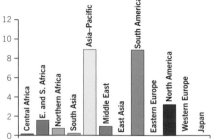

'As global metals prices rise, a new Latin American mining rush is underway...' Kelly Hearn, 'South America's mining wars heat up', 2005

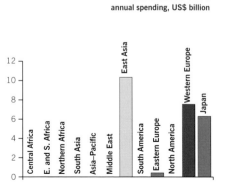

MOST NET ORE IMPORTS

Rank	Territory	US$ª
1	Luxembourg	809
2	Iceland	244
3	Norway	193
4	Finland	147
5	Belgium	85
6	South Korea	72
7	Qatar	71
8	Taiwan	67
9	Slovenia	60
10	Japan	49
11	Tuvalu	46
12	Andorra	46
13	New Zealand	36
14	Italy	34
15	Bahrain	33
16	Spain	28
17	Slovakia	27
18	St Kitts–Nevis	25
19	Greenland	23
20	Maldives	22

ª annual US$ worth of net ore
imports per person

096 Ore Imports

The size of each territory indicates the dollar value of its net
imports of ores, sand, stone, gravel and scrap metals.

East Asia, Western Europe and Japan are
the main ore-importing regions. Over two-
thirds of the territories in these regions are
net ore importers in dollar terms. In all
other regions the territories either import
only small amounts, in dollar terms, or
are net exporters.

In many cases the territories that import
ores are not the final destinations for the
refined product – be it iron or steel, other
metals or finished goods – derived from
the ores. The refined products are often
exported to other territories for manufacture
or consumption. (See Maps 097, 099.)

NET ORE IMPORTS
annual spending, US$ billion

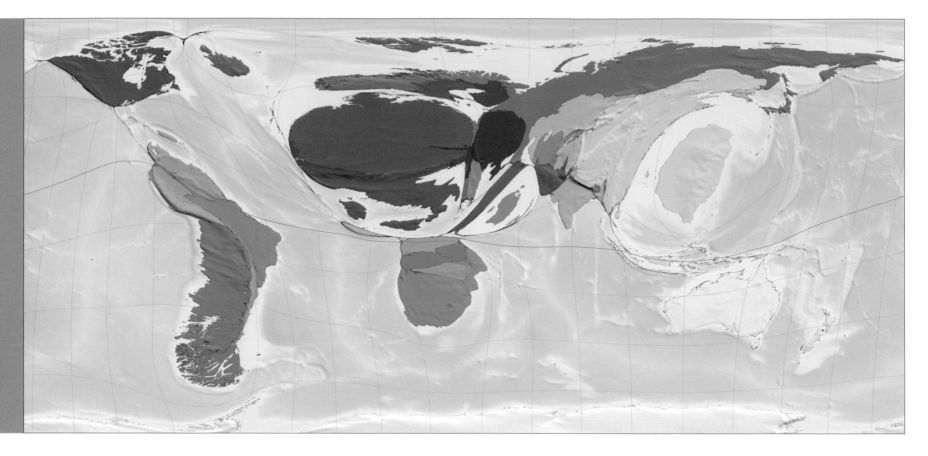

Rank	Territory	US$[a]
1	Iceland	1,222
2	Bahrain	948
3	Norway	429
4	Chile	271
5	Australia	154
6	Taiwan	132
7	Luxembourg	116
8	Canada	83
9	Kazakhstan	82
10	Germany	81
11	Finland	72
12	Tajikistan	64
13	Zambia	55
14	New Zealand	47
15	Sweden	46
16	Russia	45
17	Italy	43
18	South Africa	41
19	Switzerland	38
20	Venezuela	33

[a] annual US$ worth of net metals exports per person

097 Metals Exports

The size of each territory indicates the dollar value of its net exports of metals, including bulk metals and metal items such as tools and cutlery.

This map shows the dollar value of net exports of metals, including copper, nickel, aluminium and zinc, but excluding steel and iron (which are shown separately on Map 099) and precious metals (which are grouped with other valuables on Map 131). The exports shown in the map account for 3.8% of all international trade.

Only 52 (out of 200) territories are net metal exporters. The biggest exporters are Germany, Russia, China and Chile. Although a quarter of all net metals exports come from territories in Western Europe, the region as a whole has relatively low net metal exports. This is because other Western European territories are net metal importers and the imports and exports roughly balance out.

NET METALS EXPORTS
annual earnings, US$ billion

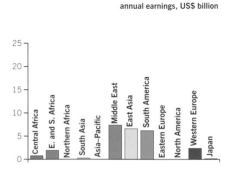

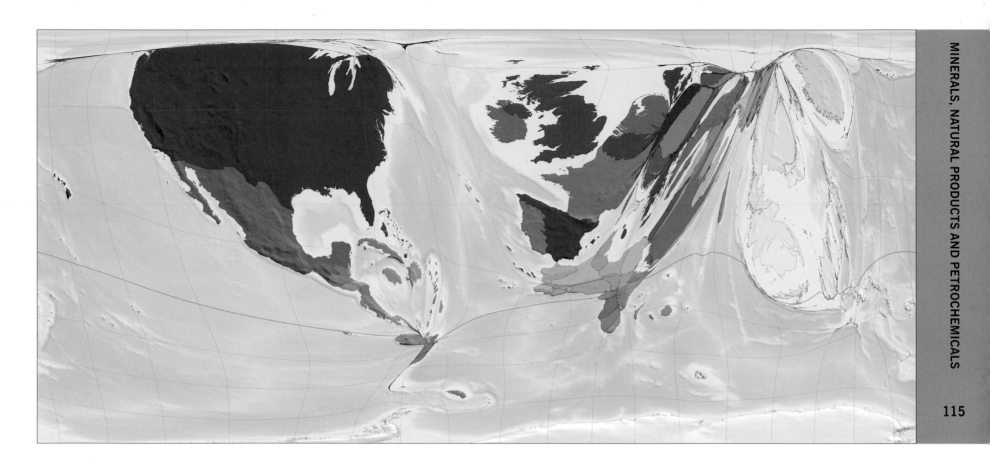

MOST NET METALS IMPORTS

Rank	Territory	US$[a]
1	Andorra	460
2	Greenland	353
3	Qatar	323
4	St Kitts–Nevis	283
5	Singapore	281
6	Bahamas	239
7	Tuvalu	226
8	Seychelles	178
9	Ireland	161
10	Cyprus	156
11	St Lucia	126
12	Malta	123
13	Brunei	119
14	Cook Islands	117
15	Kuwait	109
16	Barbados	108
17	Antigua and Barbuda	99
18	Grenada	96
19	Hungary	83
20	Niue	77

[a] annual US$ worth of net metals imports per person

098 Metals Imports

The size of each territory indicates the dollar value of its net imports of metals and metal items. About three-quarters of our 200 territories are net importers of metals.

Among regions, North America has by far the largest share of net metal imports in US dollar terms, at 82%. North America also contains the two leading territories for net metals imports, the United States and Mexico. In the remaining regions the biggest metal importers are: Angola in Central Africa; Botswana in East and Southern Africa; Algeria in Northern Africa; Bangladesh in South Asia; Thailand in the Asia–Pacific and Australasia; Saudi Arabia in the Middle East and Central Asia; Taiwan in East Asia; Guatemala in South America; Hungary in Eastern Europe; and the United Kingdom in Western Europe. Japan, which is a region in its own right on our map, is a net exporter of metals, so this region has no net importers.

'Mankind has been using copper, lead and tin for thousands of years and yet today more aluminium is produced than all other non-ferrous metals combined.' World Aluminium Institute, 2000

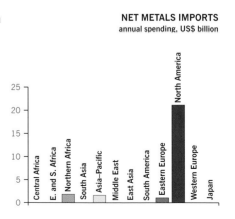

NET METALS IMPORTS
annual spending, US$ billion

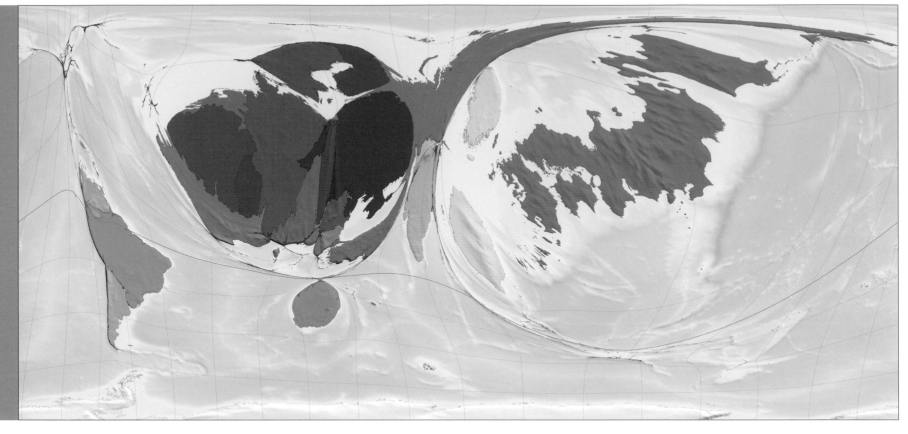

MOST NET STEEL EXPORTS

Rank	Territory	US$ᵃ
1	Luxembourg	2,339.0
2	Belgium	331.6
3	Finland	215.3
4	Sweden	192.2
5	Austria	147.2
6	Slovakia	120.5
7	Japan	101.5
8	Ukraine	94.1
9	Taiwan	73.0
10	Germany	51.0
11	Macedonia, FYR of	39.8
12	Russia	33.0
13	France	30.3
14	Netherlands	25.8
15	South Korea	22.9
16	South Africa	22.5
17	Czech Republic	22.1
18	Bulgaria	21.6
19	Bosnia and Herzegovina	19.7
20	Romania	17.4

ᵃ annual US$ worth of net steel exports per person

099 Steel and Iron Exports

The size of each territory indicates the dollar value of its net exports of steel and iron, key commodities in both industry and consumer manufacturing.

Earnings from steel exports represent 2.2% of all export earnings worldwide. The exports mapped here include steel and iron, but not iron ore (which is shown on Map 095). Steel and iron are used to make pipes, wires, building materials and railway tracks, among other things, as well as consumer goods such as cars.

Only 32 out of 200 territories are net steel and iron exporters. None of these 32 territories are in Northern Africa, Central Africa, North America or the Asia–Pacific and Australasia.

The 5 biggest net exporters of steel and iron per person in terms of dollar value are all in Western Europe, as are the 5 biggest net importers of ores per person.

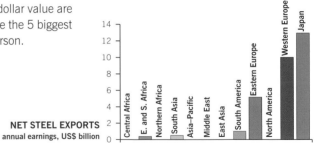

NET STEEL EXPORTS
annual earnings, US$ billion

MOST NET STEEL IMPORTS

ª annual US$ worth of net steel imports per person

100 Steel and Iron Imports

The size of each territory indicates the dollar value of its net imports of steel and iron.

Of the 200 territories shown on the map, 162 are net steel importers. The leading steel- and iron-importing territories are China and the United States, followed by Thailand, Mexico and Malaysia.

NET STEEL IMPORTS
annual spending, US$ billion

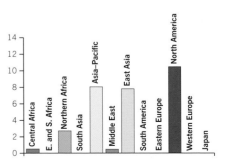

'Indian steel products are currently being exported to Pakistan via Dubai. North Pakistan gets imported steel and steel melting scrap via Karachi, over 1,500 km away...' Anand S. T. Das, *Chandigarh Newsline*, 2006

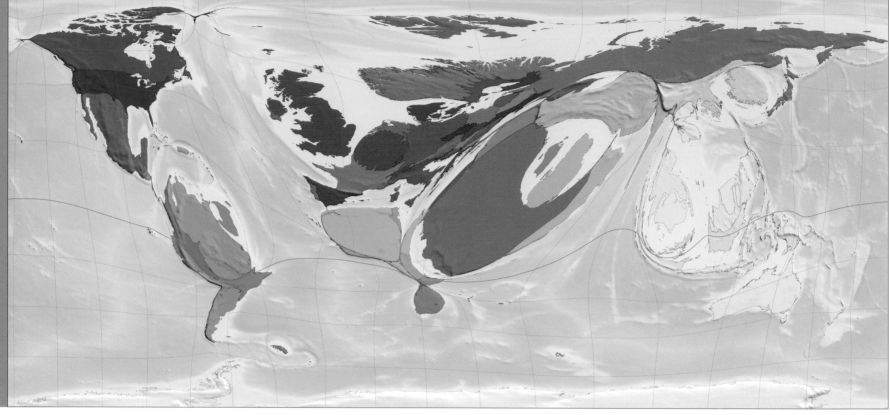

Rank	Territory	US$ᵃ
1	Qatar	16,420
2	Brunei	11,861
3	Norway	8,041
4	Bahrain	5,734
5	Oman	3,083
6	Saudi Arabia	2,735
7	Singapore	2,296
8	Trinidad and Tobago	1,793
9	Canada	1,023
10	Belgium	952
11	Venezuela	780
12	Australia	722
13	Denmark	659
14	Algeria	578
15	Gaza and West Bank	417
16	Russia	388
17	Iran	367
18	United Kingdom	367
19	Malaysia	333
20	Lithuania	296

ᵃ US$ value of gross fuel exports
per person, 2002

101 Fuel Exports

The size of each territory indicates the dollar value of its gross exports of oil, gas, coal, nuclear and traditional fuels. Just over half the territories export some fuel.

Of the 200 territories mapped, only 123 record any fuel exports, though this is partly because of poor data recording in some places. Also, many of those that do export fuel also import it.

Territories in the Middle East and Central Asia have the biggest gross income from exports of fuel; the two biggest fuel-exporting territories are Saudi Arabia and Russia. The third biggest is Norway. These major exporters earn more from oil and gas than they do from coal and traditional fuels.

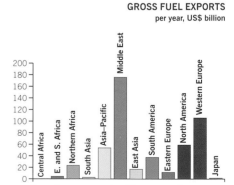

GROSS FUEL EXPORTS
per year, US$ billion

'...the achievement of high human development indicators in some countries, such as Norway, has been enabled by services funded in part through natural resource exploitation.' Melissa Dell, 'The devil's excrement', 2004

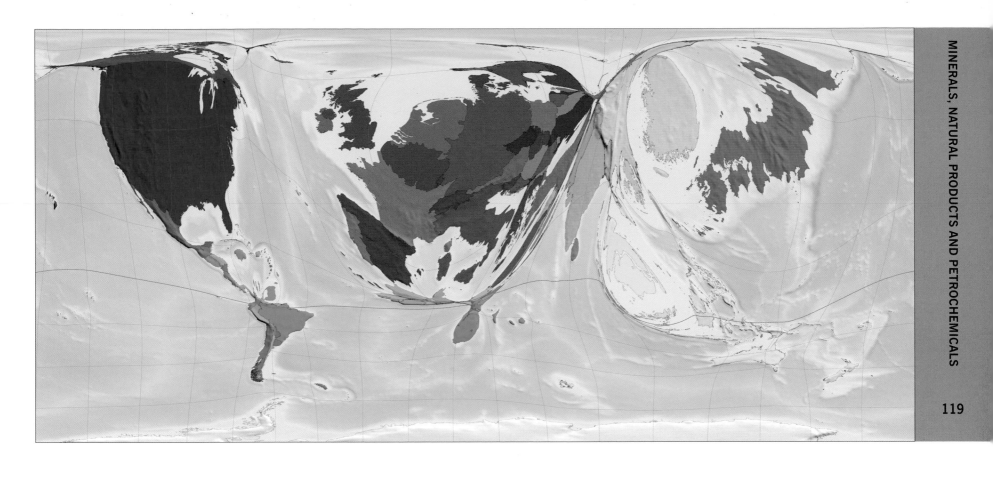

102 Fuel Imports

The size of each territory indicates the dollar value of its gross fuel imports.
The biggest importers are all wealthy territories and regions.

The regions with the highest US dollar values of gross fuel imports are North America and Western Europe. The region with the lowest is Central Africa, where 6 of the 10 territories report no fuel imports at all.

Fuel imports per person are highest in Singapore and Bahrain, both small and relatively wealthy island territories. Singapore is a long-established trading port. Bahrain is a group of islands in the Arabian Gulf that has traditionally been an oil exporter, but because of declining oil reserves now also imports crude oil, refines it, then exports it again.

Among the 12 regions, Japan imports the most fuel per person, with Western Europe coming a close second. Japan imports 338 times as much fuel per resident as the Central Africa region.

'Japan's city gas companies import natural gas...from seven countries including Alaska, Malaysia, Indonesia, Australia, Brunei, Qatar, and Oman.' Kunio Anzai, chairman of the Japan Gas Association, 2004

GROSS FUEL IMPORTS
per year, US$ billion

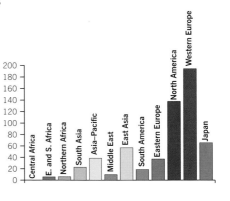

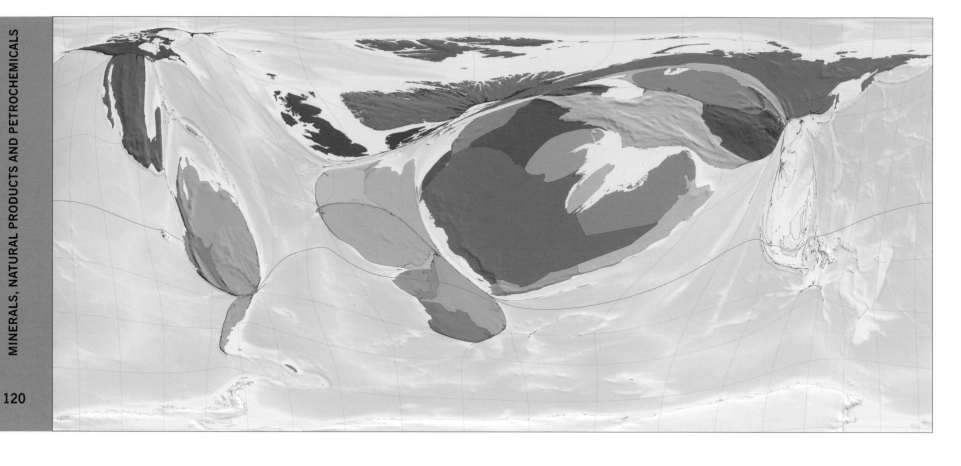

Rank	Territory	US$[a]
1	United Arab Emirates	5,964
2	Norway	5,706
3	Qatar	4,802
4	Brunei	4,370
5	Kuwait	3,634
6	Oman	2,690
7	Saudi Arabia	2,267
8	Gabon	1,705
9	Libya	1,550
10	Venezuela	727
11	Angola	367
12	Iran	351
13	Congo	330
14	Denmark	314
15	Kazakhstan	294
16	Syria	232
17	Algeria	197
18	Russia	187
19	Azerbaijan	178
20	DR Congo	173

[a] annual US$ worth of net crude petroleum exports per person

103 Crude Petroleum Exports

The size of each territory indicates the dollar value of its net exports of crude petroleum, showing the continued dominance of the Middle East and Central Asia.

Exports of crude petroleum account for 5.3% of total worldwide net export income. Territories in the Middle East and Central Asia export 58% of all crude petroleum. Saudi Arabia alone has net exports worth twice as much as those of any other territory in dollar value. The United Arab Emirates has the highest per person net export earnings from crude petroleum. Other important exporters of crude petroleum include Norway, Venezuela, Nigeria and Mexico; there are no net exporters at all in East and Southern Africa, South Asia or Japan.

'The cost of getting oil out of the ground is going up, the amount of water in it is increasing, and there's less and less of the really good oil down there. All of this is forcing the prices up.'

James Brock, energy analyst, 2006

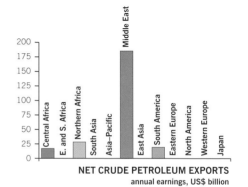

NET CRUDE PETROLEUM EXPORTS
annual earnings, US$ billion

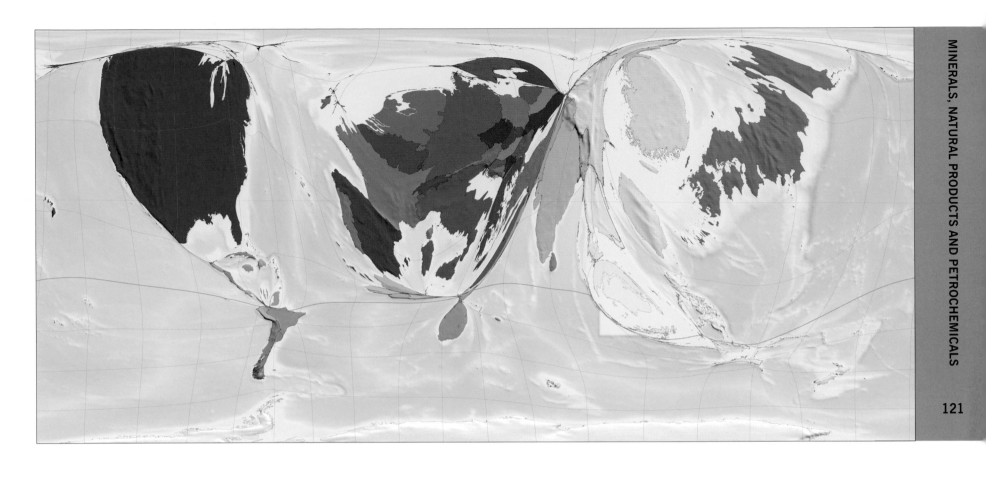

MOST NET CRUDE PETROLEUM IMPORTS

Rank	Territory	US$[a]
1	Singapore	1,808
2	Netherlands	733
3	Belgium	618
4	Finland	430
5	South Korea	417
6	Sweden	392
7	Lithuania	340
8	Taiwan	331
9	Greece	316
10	Japan	294
11	United States	292
12	Trinidad and Tobago	278
13	Israel	269
14	Cyprus	259
15	Spain	255
16	Italy	254
17	France	253
18	Germany	225
19	Portugal	214
20	Czech Republic	203

[a] annual US$ worth of net crude
petroleum imports per person

104 Crude Petroleum Imports

The size of each territory indicates the dollar value of its net imports of crude petroleum.
Wealth and active trade contribute to high consumption and therefore high imports.

Per person, Singapore has higher net
imports of crude petroleum than any other
territory, with around US$1,808 worth of
imports per person per year. Singapore is
a relatively affluent island nation, well
positioned on trade routes, and so can
afford to consume large amounts of oil.
Oil is also required to fuel the Singaporean
shipping industry.

*'Aside from the effects of high oil prices, growth in imports in general
can be interpreted as a sign that domestic demand is robust, another
reason to say that the Japanese economy is on the right track...'*
Koji Kobayashi, senior economist at Mizuho Research Institute, 2006

NET CRUDE PETROLEUM IMPORTS
annual spending, US$ billion

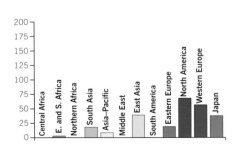

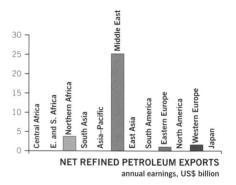

Rank	Territory	US$[a]
1	Bahrain	3,160
2	Kuwait	1,995
3	United Arab Emirates	1,285
4	Qatar	869
5	Trinidad and Tobago	853
6	Netherlands	447
7	Singapore	380
8	Lithuania	278
9	Belgium	230
10	Libya	225
11	Norway	173
12	Finland	146
13	Belarus	141
14	Seychelles	121
15	Canada	105
16	Slovakia	100
17	Turkmenistan	77
18	Russia	76
19	Sweden	71
20	Algeria	67

[a] annual US$ worth of net refined petroleum exports per person

105 Refined Petroleum Exports

The size of each territory indicates the dollar value of its net exports of refined petroleum, which takes several forms and is used for a wide range of domestic and industrial purposes.

Income from the export of refined petroleum constitutes 2.4% of worldwide export earnings. As a region, the Middle East and Central Asia has the highest net refined petroleum exports, measured in US dollars. The same region is also where most petroleum extraction occurs and where the largest known oil reserves are located.

The refinement of petroleum involves several processes: separation of the hydrocarbon compounds within the crude petroleum; conversion of heavier hydrocarbons into lighter, more valuable ones; treatment to remove impurities; and blending to produce the finished products. The products of refinement include fuel oils (for heating and heavy engines), kerosene (also called paraffin, used primarily as aircraft fuel), gasoline (also called petrol, used primarily as fuel in cars) and lubricating oils. All of these contribute to the export earnings shown on this map.

'The rapid economic growth of the 1960s was driven by expansion of energy-consuming industries based on lavish use of cheap Middle East oil.' Yoko Kitazawa, 'The Japanese economy and South-East Asia', 1990

NET REFINED PETROLEUM EXPORTS
annual earnings, US$ billion

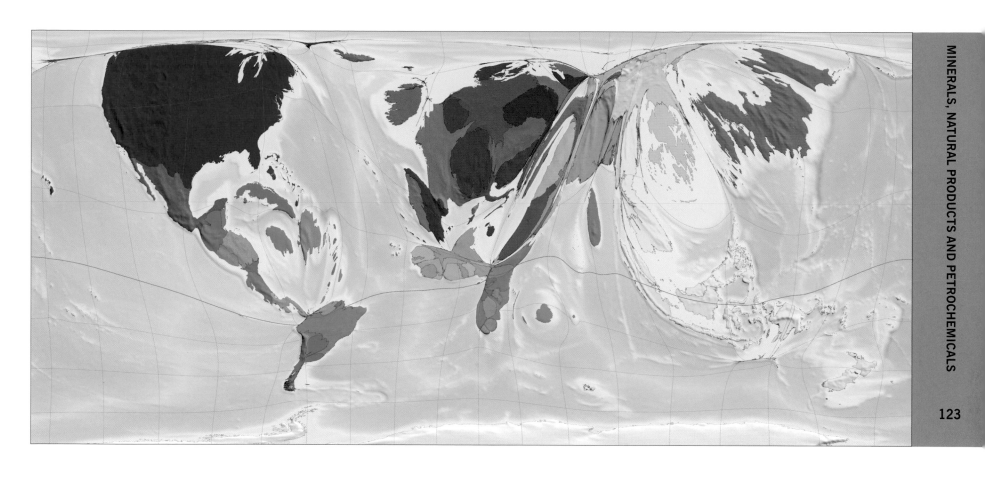

MOST NET REFINED PETROLEUM IMPORTS

Rank	Territory	US$ᵃ
1	Luxembourg	1,547.4
2	Andorra	776.9
3	Greenland	632.7
4	Iceland	538.5
5	Bahamas	389.4
6	Hong Kong (China)	372.5
7	Antigua and Barbuda	301.9
8	Malta	288.6
9	Slovenia	277.4
10	Switzerland	252.6
11	Lebanon	245.7
12	St Kitts–Nevis	243.3
13	St Lucia	234.2
14	Ireland	219.6
15	Cyprus	209.7
16	Belize	180.3
17	Grenada	179.9
18	Mauritius	164.3
19	Cook Islands	154.4
20	Maldives	154.3

ᵃ annual US$ worth of net imports of refined petroleum per person

106 Refined Petroleum Imports

The size of each territory indicates the dollar value of its net imports of refined petroleum.
Over two-thirds of territories are net importers of refined petroleum products.

Refined petroleum products have a wide range of uses, including as fuel for vehicles of various types, heating and the generation of electricity. The biggest net importers are the United States, Japan and Hong Kong (which is part of China). Of these, the United States is easily the leader, with almost 3 times the net imports of the first runner-up, Japan.

As a region, North America is the biggest net importer of refined petroleum products, followed by Western Europe and the Asia–Pacific and Australasia. Asia–Pacific and Australasian imports are dominated by Indonesia and Viet Nam, both of which have imports more than four times those of any other territory in the region.

NET REFINED PETROLEUM IMPORTS
annual spending, US$ billion

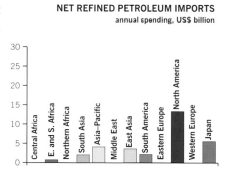

'In fact China, with a fifth of the world's population, consumes only 4% of the world's daily oil output. It imports about three million barrels a day. A lot to be sure, but far below American consumption.'

Rupert Wingfield-Hayes, 'Satisfying China's demand for energy', 2006

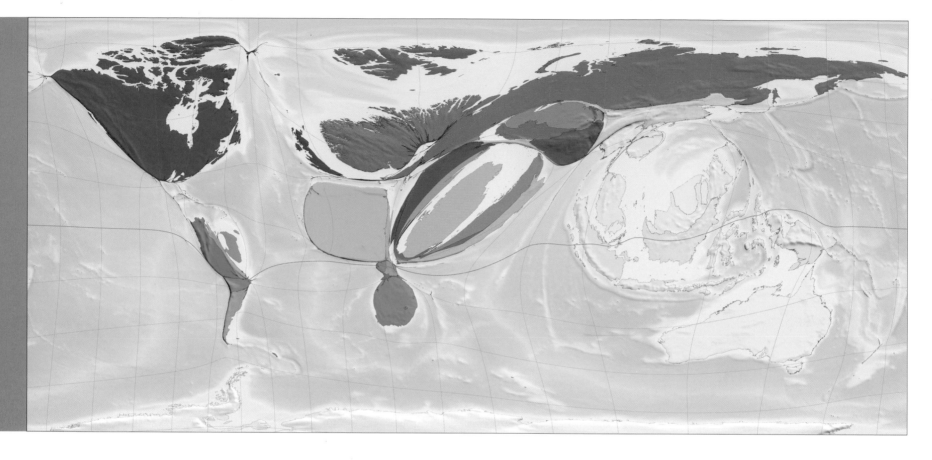

MOST NET GAS AND COAL EXPORTS

Rank	Territory	US$ [a]
1	Qatar	6,225
2	Brunei	3,943
3	Norway	1,872
4	United Arab Emirates	628
5	Canada	447
6	Australia	445
7	Oman	406
8	Kuwait	323
9	Trinidad and Tobago	300
10	Turkmenistan	194
11	Algeria	190
12	Malaysia	118
13	Russia	116
14	Luxembourg	63
14	Singapore	56
14	South Africa	39
14	Indonesia	34
18	Libya	33
19	Bolivia	31
20	Congo	24

[a] annual US$ worth of net gas and coal exports per person

109 Gas and Coal Exports

The size of each territory indicates the dollar value of its net exports of gas and coal, which together make up 2% of all world exports.

The Middle East and Central Asia, and the Asia–Pacific and Australasia, are the main net exporting regions. Territories in these regions account for up to 60% of the world's net gas and coal exports in US dollar terms. Of the 53 net exporting territories, 15 are in the Asia–Pacific and Australasia, and 13 are in the Middle East and Central Asia.

In some regions there is only one net exporting territory: in South Asia it is Bhutan; in Eastern Europe it is Poland; and in North America it is Canada. Japan has no net gas and coal exports and relies on imports.

NET GAS AND COAL EXPORTS
annual earnings, US$ billion

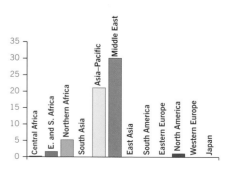

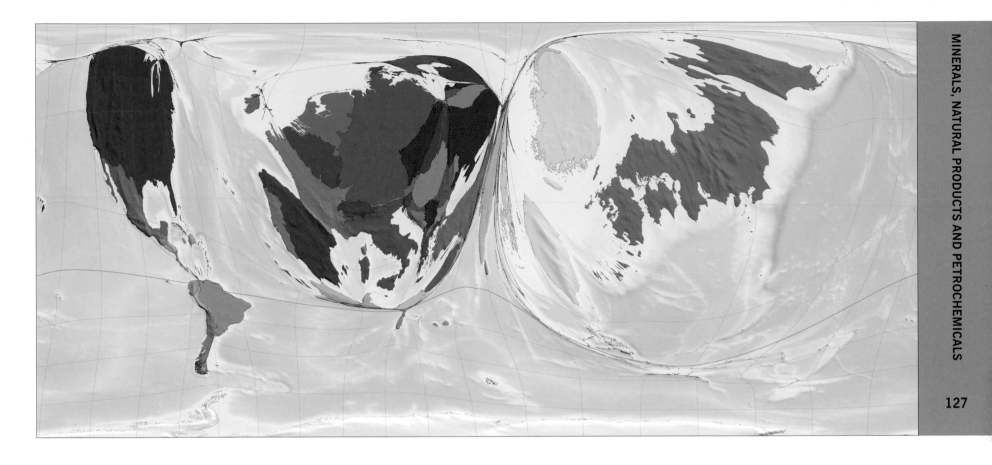

MOST NET GAS AND COAL IMPORTS

Rank	Territory	US$[a]
1	Belgium	199
2	Finland	170
3	Slovakia	164
4	Japan	157
5	Netherlands	156
6	South Korea	150
7	Hong Kong (China)	146
8	Czech Republic	138
9	Taiwan	136
10	Hungary	134
11	Germany	116
12	Spain	86
13	Portugal	75
14	Israel	71
15	Sweden	67
16	Switzerland	67
17	Belarus	65
18	Ukraine	63
19	Ireland	59
20	Latvia	54

[a] annual US$ worth of net gas and coal imports per person

110 Gas and Coal Imports

The size of each territory indicates the dollar value of its net imports of gas and coal, used primarily to produce heat and electricity.

Gas and coal are significant sources of energy in some (but not all) territories. Other power sources include wood, petroleum, nuclear power and hydroelectric generation. Imports of gas and coal are affected by the demand for power and the availability of other sources of power. Japan is the largest net importer of gas and coal overall, but Belgium has the highest imports per person. If Belgium's imports were divided equally among its population, the value of its net imports of gas and coal would be US$199 per year for each of the 10.3 million people living there.

NET GAS AND COAL IMPORTS
annual spending, US$ billion

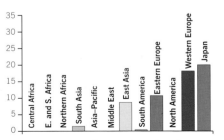

'Eighty percent of Russia's gas exports pass through Ukraine, a crucial weak point in what is acknowledged to be a powerful lever of Russian foreign policy.' Steven Eke, BBC journalist, 2005

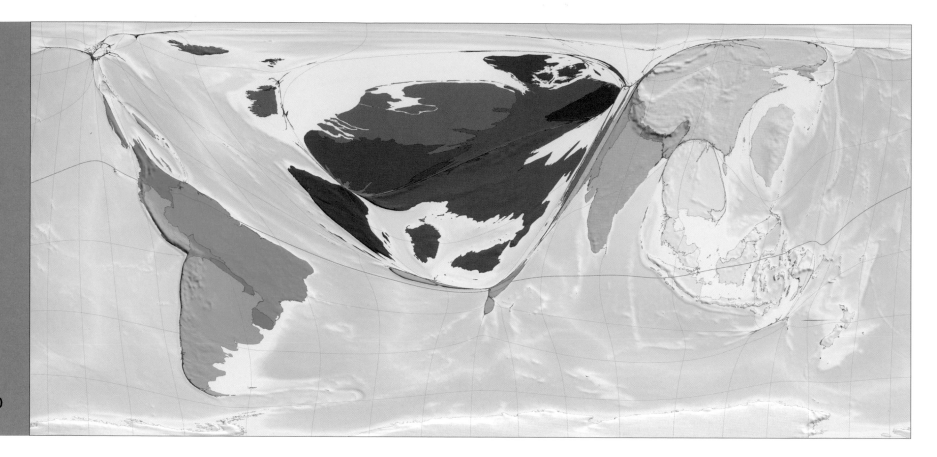

113 Natural Products Exports

The size of each territory indicates the dollar value of its net exports of natural products, including rubber, animal feed and leather.

Pets and zoo animals (but not animals exported for food) also fall into this category, though they make up only a small fraction of the total dollar value.

Earnings from the export of natural products constitute 6.7% of earnings from all exports in US dollar terms. Western Europe tops the list for total earnings, both among individual territories and collectively as a region, although this is partly the result of the high prices commanded by European exports, rather than export volume. A number of South American territories also have large net earnings from natural products but do so primarily by exporting greater volume at lower cost. Argentina and Brazil are particularly noticeable, together accounting for 80% of net natural product exports from South American territories.

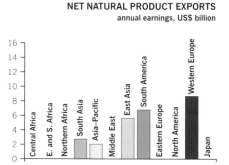

NET NATURAL PRODUCT EXPORTS
annual earnings, US$ billion

'Malaysia has a long history of internationally valued exports, being known from the early centuries AD as a source of...exotics such as birds' feathers, edible birds' nests, aromatic woods, tree resins etc.'

John Drabble, economic historian, University of Sydney, 2004

114 Natural Products Imports

Three-quarters of the 200 territories on the map are net importers of natural products. The size of each territory in this map indicates the dollar value of these net imports.

Topping the list of net importers is tiny, landlocked, mountainous Andorra, which imports (net) over US$600 worth of natural products per inhabitant per year. The high figure reflects in part Andorra's relative wealth, but also the fact that the territory has almost no exports of natural products. Its immediate neighbour, France, on the other hand – also a wealthy country – makes net exports of natural products worth only 27 US cents per inhabitant, because those exports, although substantial, are almost perfectly balanced by its imports. Contrast this with the Central African Republic, which also has low net imports per inhabitant, but in this case by virtue of the fact that it imports and exports very little in the way of natural goods at all.

'Saudi Arabia has tightened restrictions on trading in wildlife animals in a renewed attempt to ensure protection of rare animal species.' M. Ghazanfar Ali Khan, *Arab News*, 2004

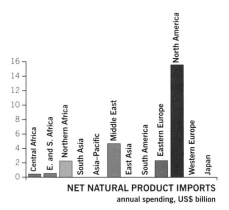

NET NATURAL PRODUCT IMPORTS
annual spending, US$ billion

The Economic World

Manufactured Goods and Services

Wealth and Poverty

Employment and Productivity

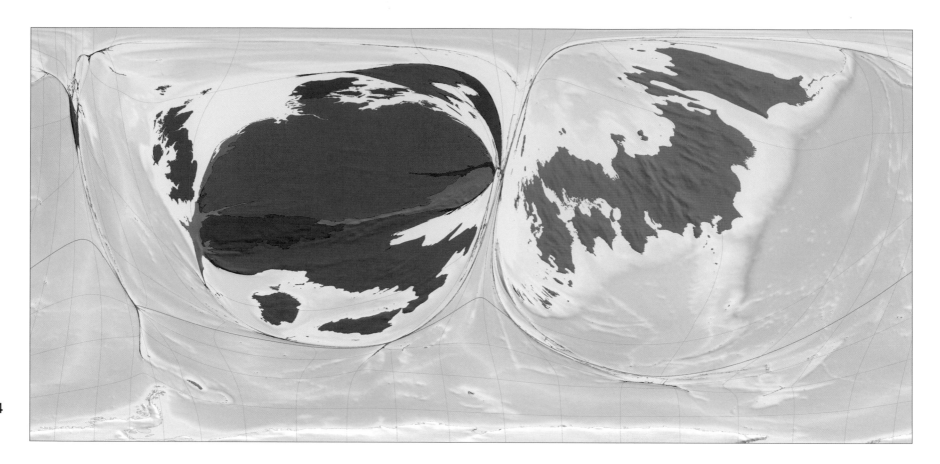

HIGHEST NET EXPORTS OF MACHINERY

Rank	Territory	US$[a]
1	Switzerland	1,075
2	Germany	659
3	Denmark	506
4	Italy	502
5	Sweden	486
6	Austria	485
7	Finland	441
8	Japan	423
9	United Kingdom	84
10	Netherlands	47
11	Czech Republic	25
12	Taiwan	24
13	France	22
14	United States	2
15	Hungary	1

[a] **annual US$ value of net machinery exports per person (the data source records only 15 net exporters; Liechtenstein, Monaco, San Marino and Vatican City are estimated to be net exporters, but these estimates are based on patterns in neighbouring countries and so are not included here)**

115 Exports of Machinery

The size of each territory indicates the dollar value of its net exports of machinery: engines, turbines and pumps as well as machines for making food or binding books.

Only 19 territories, most of them in Western Europe and Japan, are net exporters of machinery in dollar terms, while in 7 of the 12 world regions no territories are net machine exporters. Nonetheless, earnings from machinery exports account for 10.5% of all earnings from international exports worldwide. Of the net exporting territories, Switzerland earns the most per person, at just over US$1,000 per person per year.

'There in the flickering light of the lamp was the machine sure enough, squat, ugly, and askew; a thing of brass, ebony, ivory, and translucent glimmering quartz.' H. G. Wells, *The Time Machine*, 1898

NET MACHINERY EXPORTS
annual earnings, US$ billion

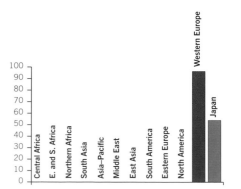

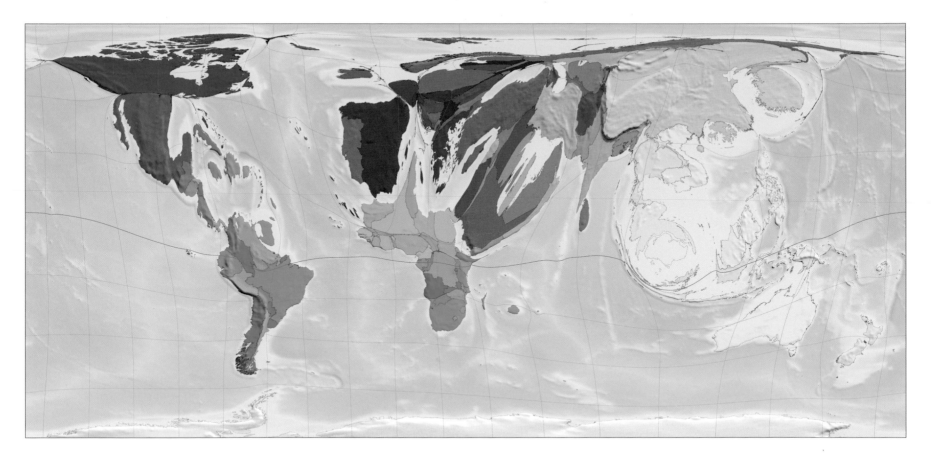

HIGHEST NET IMPORTS OF MACHINERY

Rank	Territory	US$ᵃ
1	Qatar	1,317
2	United Arab Emirates	1,026
3	Singapore	883
4	Andorra	861
5	Greenland	799
6	Cyprus	606
7	Brunei	592
8	Bahrain	535
9	Trinidad and Tobago	478
10	Kuwait	448
11	Canada	395
12	Iceland	362
13	Bahamas	350
14	St Kitts–Nevis	345
15	Australia	339
16	Norway	332
17	Malta	315
18	Estonia	303
19	New Zealand	303
20	Oman	274

ᵃ annual US$ value of net machinery
imports per person

116 Imports of Machinery

90% of the territories in the world import more machinery than they export. The size
of each territory in this map indicates the dollar value of its net imports of machinery.

The Middle Eastern territories of Qatar and
the United Arab Emirates spend the largest
amount per person, at more than
US$1,000 a year each. The term
'machinery' covers a wide range of goods,
but the category accounting for the largest
part of net import spending is non-electrical
machinery parts, which covers spare parts
for existing machinery as well as parts that
are assembled into complete machines
upon arrival at their destination.

NET MACHINERY IMPORTS
annual spending, US$ billion

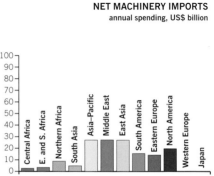

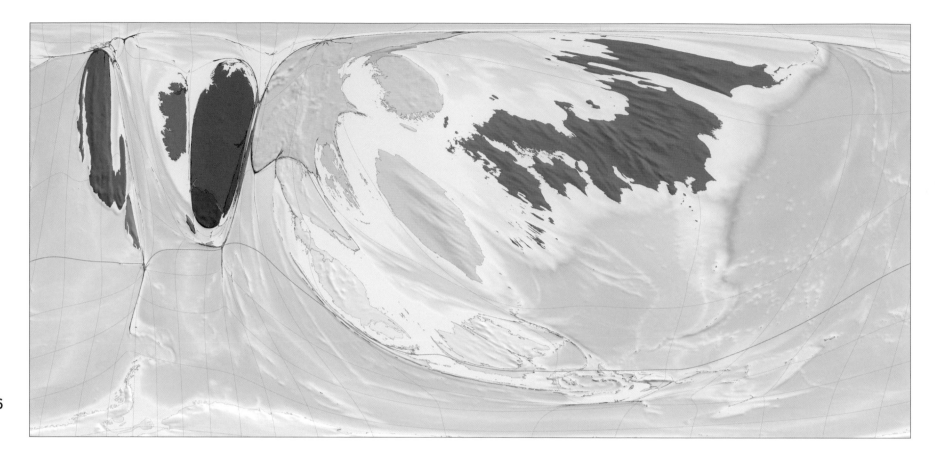

HIGHEST NET EXPORTS OF ELECTRONICS

Rank	Territory	US$ᵃ
1	Ireland	716
2	Taiwan	535
3	Malaysia	404
4	Hong Kong (China)	342
5	Japan	326
6	Singapore	279
7	Slovenia	270
8	Switzerland	219
9	St Kitts–Nevis	203
10	Samoa	187
11	Costa Rica	151
12	Germany	130
13	South Korea	116
14	Thailand	78
15	Mexico	75
16	North Korea	27
17	Hungary	26
18	Indonesia	18
19	China	14
20	Portugal	12

ᵃ annual US$ value of net electronics
exports per person

117 Exports of Electronics

The size of each territory indicates the dollar value of its net exports
of electronics, from domestic sound systems to medical technology.

This category includes consumer
electronics such as televisions, radios and
stereo equipment, photographic equipment
and medical electronics.

Worldwide, electronics account for
10.2% of all export earnings. Asian
territories, particularly Japan, China,
Taiwan and Malaysia, are the leading
sources of electronics. These territories and
others like them are sometimes referred to
as 'tiger economies', a label that derives
from their high growth in the recent past,
rapid industrialization and export-driven
trade strategies. Exports from Asia meet
85% of the demand for electronics in the
net importing territories.

*'The ability of nations to compete globally is not a function of size, as
Korea, Taiwan, Singapore, Hong Kong, and Malaysia demonstrate.'*
Michael J. Kelly, School of Engineering and Technology, California State University in Los Angeles, 1997

NET ELECTRONICS EXPORTS
annual earnings, US$ billion

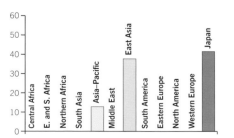

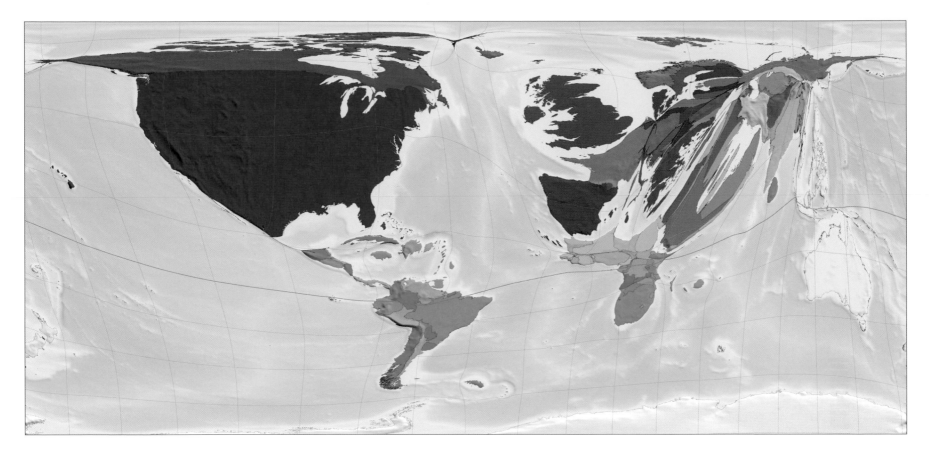

HIGHEST NET IMPORTS OF ELECTRONICS

Rank	Territory	US$[a]
1	Andorra	1,680
2	Qatar	658
3	Iceland	649
4	United Arab Emirates	611
5	Bahrain	403
6	Norway	400
7	Canada	399
8	Brunei	396
9	Estonia	306
10	Kuwait	287
11	Cyprus	279
12	Australia	269
13	Tuvalu	243
14	Bahamas	241
15	Barbados	203
16	New Zealand	197
17	Cook Islands	196
18	Antigua and Barbuda	188
19	Niue	175
20	Greece	173

[a] annual US$ value of net electronics imports per person

118 Imports of Electronics

The size of each territory in this map indicates the dollar value of its net imports of electronics. The spread of reliable electricity supplies facilitates this trade.

While many people use electronic items as part of their daily lives, others have very little contact with them, particularly in places where electricity supply is absent or unreliable. The availability of reliable electricity supplies is growing, however, especially in China, and with it the widespread use of electronics.

The United States is the world's biggest net importer of electronics. Together, its inhabitants spent US$47 billion on imported electronics in 2002, and North American territories account for almost half of the world's net electronics imports in dollar terms. Northern African territories, on the other hand, with a total population size similar to that of North America, account for just 5% of electronics imports.

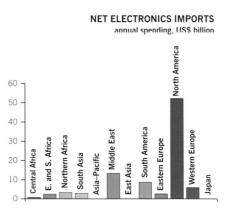

NET ELECTRONICS IMPORTS
annual spending, US$ billion

119 Computer Exports

Computers are now pervasive in the richer territories of the world. The size of each territory in this map indicates the dollar value of its net computer exports.

Computers play an important role in many aspects of our lives, helping us in our work, controlling machinery such as cars, trains and aeroplanes, and allowing us to communicate with friends and colleagues on the other side of the world.

10.6% of all net export earnings derive from computer exports. Per person, Singapore has the highest earnings from computer exports, with 2.5 times as much as Ireland, the second biggest per-person earner. Among the regions, the Asia–Pacific and Australasia, East Asia and Japan have the biggest net exports of computers. Western Europe, although it contains a number of territories with substantial net exports, is only barely a net exporter overall, the exports being almost completely counterbalanced by net imports into its other territories.

'Singapore is to be transformed into an intelligent island, where IT permeates every aspect of the society – at home, work, and play.'

Chun Wei Choo, 'Singapore's vision of an intelligent island', 1997

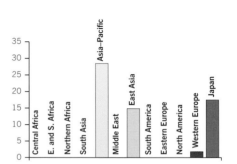

NET COMPUTER EXPORTS
annual earnings, US$ billion

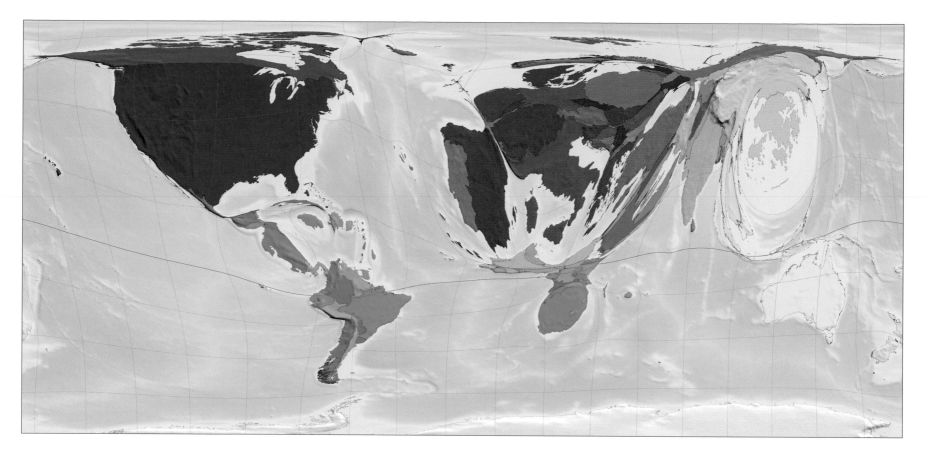

HIGHEST NET IMPORTS OF COMPUTERS

Rank	Territory	US$ᵃ
1	Andorra	722
2	Hong Kong (China)	616
3	Switzerland	476
4	Luxembourg	437
5	Iceland	373
6	Costa Rica	291
7	Norway	270
8	Australia	246
9	St Lucia	231
10	Denmark	203
11	Canada	201
12	New Zealand	187
13	Antigua and Barbuda	186
14	United Arab Emirates	183
15	Barbados	179
16	Cyprus	178
17	Portugal	177
18	Greenland	171
19	Qatar	163
20	Bahamas	156

ᵃ annual US$ value of net computer imports per person

120 Computer Imports

The size of each territory indicates the dollar value of its net computer imports, showing the greatest appetite in the United States.

The first commercial electronic computer was sold only about 60 years ago. Since then, rapid reductions in cost and physical size, as well as continual increases in processing power and memory capacity, have combined to produce what is now a huge annual flow of computers around the world. As it does with imports of so many goods, the United States leads the world in total net imports of computers, with five times as much as any other territory in dollar terms.

'...the short lifetime of today's IT equipment leads to mountains of waste...' Tim Hirsch, BBC environment correspondent, 2004

NET COMPUTER IMPORTS
annual spending, US$ billion

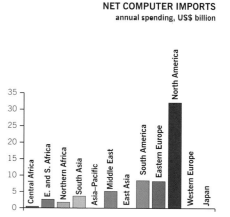

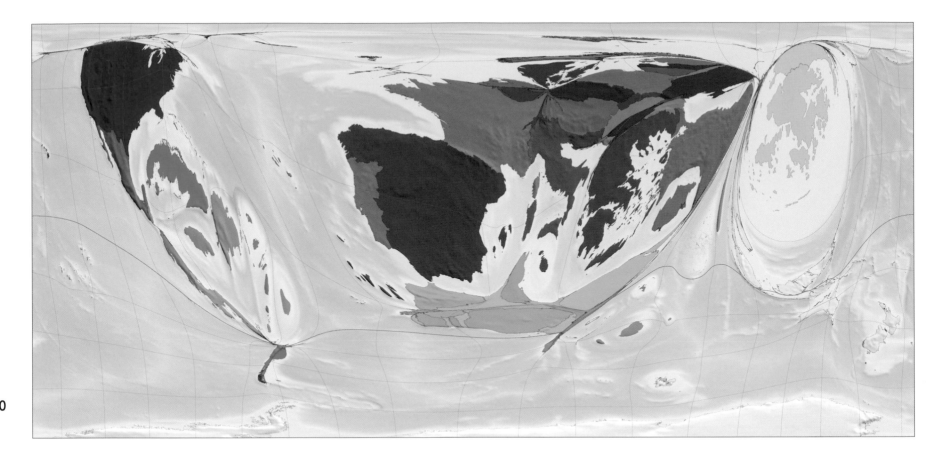

140

121 Transport and Travel Exports

The size of each territory indicates the dollar value of its net exports of transport and travel services: the movement of goods and people by air, sea and land.

It is because of the cost of transport services that the total cost of all imports worldwide adds up to more than the total earnings from exports: the buyer has to pay for both the goods and their transport, whereas the seller receives only the money for the goods. If, however, we consider transport services themselves to be an 'export' and account for them separately, then the totals add up. That's what this map does, showing the total earnings coming into each territory in payment for transport services offered by individuals or organizations in that territory.

When transport services are sold to tourists we call them travel services, but they are essentially the same thing. You may like to compare this map with Map 027, which shows net income (profits) from tourism. The two maps differ because tourism profits include money made from sources other than transport services, such as hotels and restaurants, and do not include money made from non-tourist transport, such as cargo shipping.

NET TRANSPORT AND TRAVEL EXPORTS
annual earnings, US$ billion

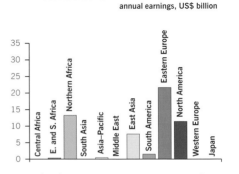

'Ranging from architecture to voice-mail telecommunications and to space transport, services are the largest and most dynamic component of both developed and developing country economies.' WTO, 2006

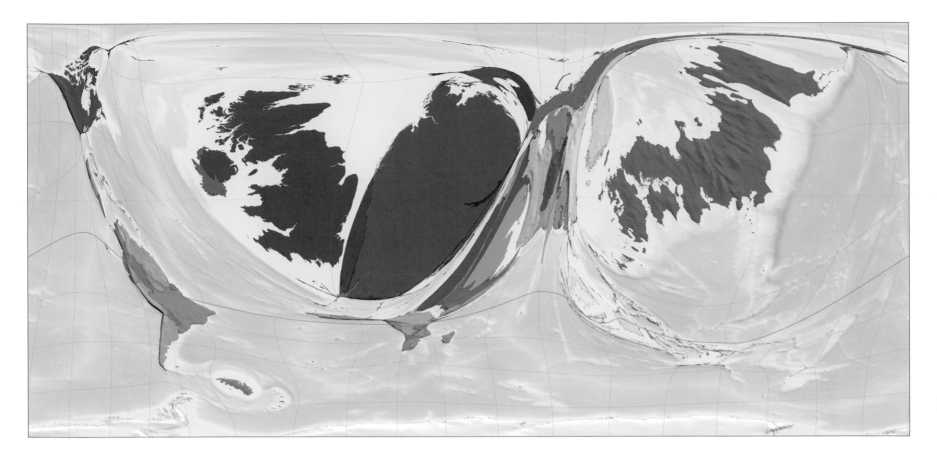

Transport and Travel Imports

The size of each territory indicates the dollar value
of net imports of transport and travel services.

Japan and the Middle East and Central
Asia are the regions with the highest
net spending on imported transport and
travel services. This is because, taken as
a whole, people living in these regions
spend more money with travel and
transport organizations based in other
regions than they earn from providing
travel and transport for others.

The territory with the highest net
spending per person is Kuwait. Net
transport and travel imports to Kuwait
amount to three times as much in dollar
terms as those of the second and third
highest importers, the United Kingdom
and Germany respectively.

**HIGHEST NET IMPORTS OF TRANSPORT
AND TRAVEL SERVICES**

Rank	Territory	US$[a]
1	Kuwait	1,467
2	United Kingdom	482
3	Germany	477
4	Israel	370
5	Japan	240
6	Belgium	233
7	Singapore	228
8	Ireland	213
9	Finland	196
10	Croatia	171
11	Suriname	139
12	Sweden	104
13	Canada	93
14	Venezuela	69
15	South Korea	55
16	Congo	35
17	Russia	31
18	Argentina	27
19	Ecuador	24
20	Yemen	21

[a] annual US$ value of net transport
and travel imports per person

NET TRANSPORT AND TRAVEL IMPORTS
annual spending, US$ billion

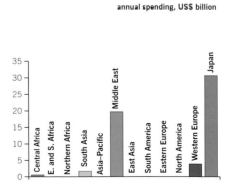

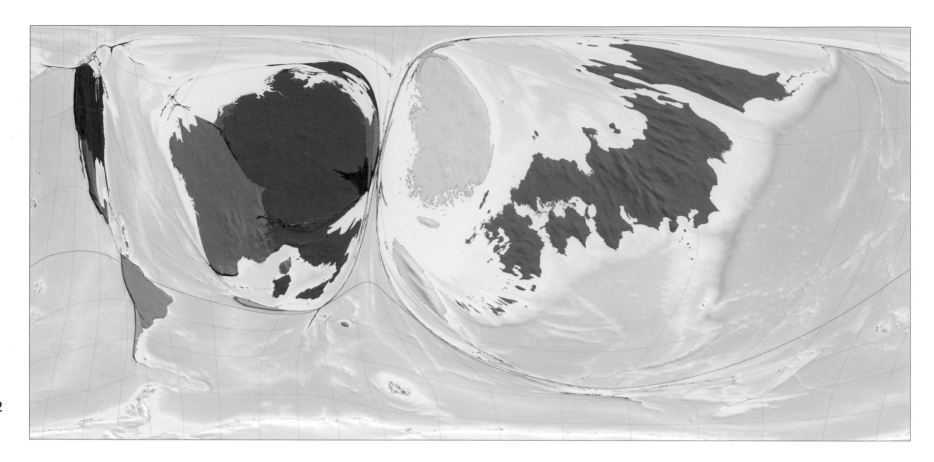

HIGHEST NET EXPORTS OF VEHICLES

Rank	Territory	US$[a]
1	Japan	255
2	South Korea	239
3	Germany	235
4	France	191
5	Taiwan	83
6	Italy	75
7	Sweden	68
8	Czech Republic	41
9	Belarus	35
10	Finland	30
11	Netherlands	14
12	Côte d'Ivoire	13
13	Brazil	12
14	Poland	11
15	Trinidad and Tobago	10
16	United States	9
17	North Korea	9
18	Mexico	9
19	Argentina	7
20	Liberia	5

[a] annual US$ value of net vehicle exports per person

123 Vehicle Exports

Only 27 territories have net vehicle exports in dollar terms. The size of each territory in this map indicates the dollar value of its net exports of vehicles, excluding cars.

Vehicles include lorries (also called trucks), bicycles, motorcycles, mopeds, trains, buses, ships, aeroplanes and car parts, but not complete cars, which are shown separately on Map 125.

Vehicle exports account for 6% of all export earnings worldwide. The major exporters are Japan, Germany, France and South Korea. No territories in South Asia, the Middle East and Central Asia, Central Africa or East and Southern Africa have net vehicle exports. Japan has the highest net vehicle exports, both in total and per person.

'We'll sell a lot more [of the A380 'superjumbo' airliner] than 250. We'll sell 700 or 750. You know it is a plane which will fly for 30 or 40 years.' Noel Forgeard, chief executive of Airbus.

NET VEHICLE EXPORTS
annual earnings, US$ billion

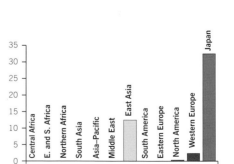

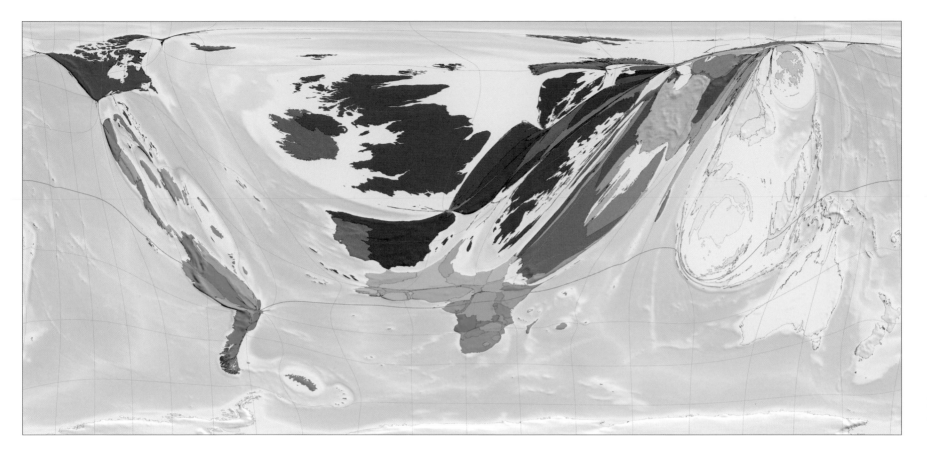

124 Vehicle Imports

The size of each territory indicates the dollar value of its net vehicle imports, excluding cars.

Vehicles on this map do not include cars, which are shown separately on Map 126.

Asia–Pacific and Australasia, and the Middle East and Central Asia, are the biggest vehicle-importing regions. Each spends over US$15 billion net per year, which is more than three times the amount spent by the third in line. Although there are some territories in Western Europe that are big importers of vehicles, the region is a net exporter.

NET VEHICLE IMPORTS
annual spending, US$ billion

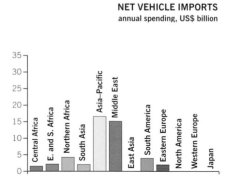

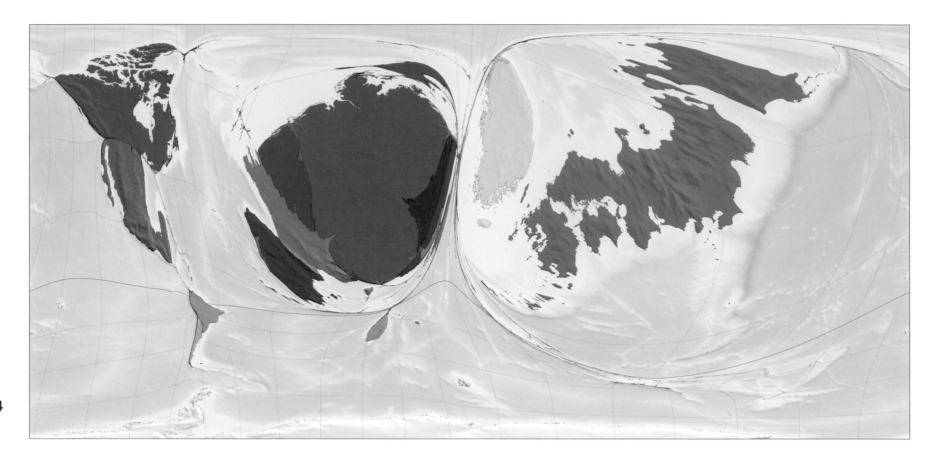

Rank	Territory	US$[a]
1	Belgium	695.0
2	Germany	589.4
3	Canada	480.9
4	Japan	442.9
5	Czech Republic	317.1
6	Slovakia	300.2
7	South Korea	271.4
8	Sweden	197.8
9	Slovenia	182.2
10	Spain	133.0
11	France	122.7
12	Mexico	77.4
13	Hungary	29.2
14	Portugal	19.8
15	Niue	16.6
16	South Africa	14.7
17	Argentina	11.4
18	Thailand	8.3
19	Bosnia Herzegovina	7.7
20	Brazil	7.2

[a] annual value of net car exports per person

125 Car Exports

The size of each territory indicates the dollar value of its net
car exports, showing the dominance of Japan and Germany.

Earnings from car exports constitute 5.3%
of all net income from exports worldwide in
terms of US dollars. Japan and Germany
alone receive 61% of that income between
them. Japanese car brands include Toyota,
Nissan, Suzuki, Subaru, Honda, Mitsubishi
and Mazda. German car brands include
Audi, BMW, Mercedes, Volkswagen and
Porsche. The highest earnings per person
from net car exports are in Germany, Belgium,
Canada and Japan. Because of disparity in
population size, net car exports from Belgium
are greater than those from Germany, and
those from Canada greater than those from
Japan, when measured per person.

Note that this map shows the total value of
cars exported, not the number of cars. Even
if a territory exports more cars than it
imports, it is not necessarily a net exporter in
terms of value; such a country could still be
a net importer if it exports many cheap cars
and imports fewer more expensive ones, so
that the total value of its imports exceeds
that of its exports.

*'The press shop at Wolfsburg processes 1500 metric tonnes
of sheet metal every day...'* Volkswagen, 2006

NET CAR EXPORTS
annual earnings, US$ billion

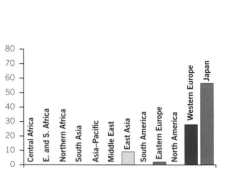

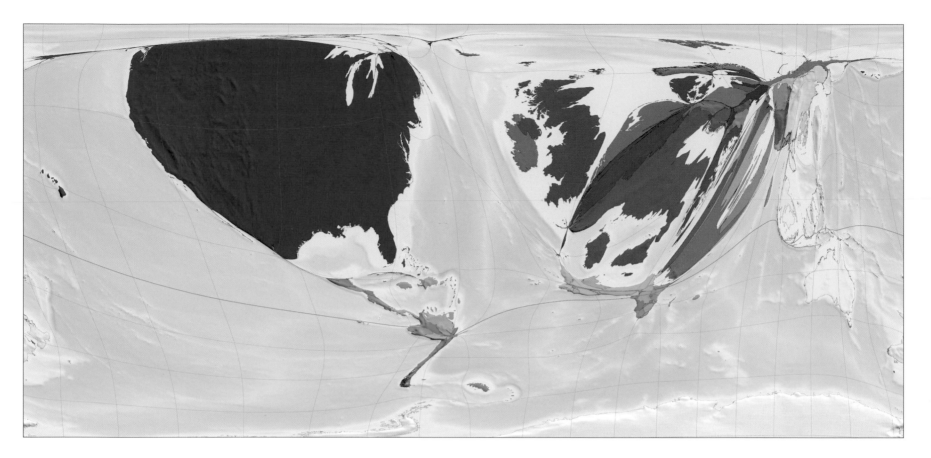

126 Car Imports

The size of each territory indicates the dollar value of its net car imports.
The United States stands head and shoulders above other importers.

The United States alone is responsible for 55% of all net spending on car imports when measured in US dollar terms. Italy and the United Kingdom are in second and third place, respectively, but each spends less than a sixth of what the US does.

Per person, the highest net imports of cars are to countries in Western Europe and in the Middle East and Central Asia, with Luxembourg taking first place by this measure.

Western European territories spend a total of US$43 billion annually on net imports of cars, but the region as a whole is a net exporter. As a region, North America is the biggest net importer, with gross imports of US$94 billion per year.

NET CAR IMPORTS
annual spending, US$ billion

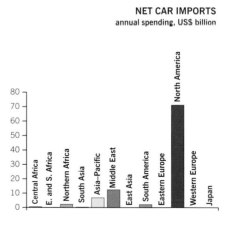

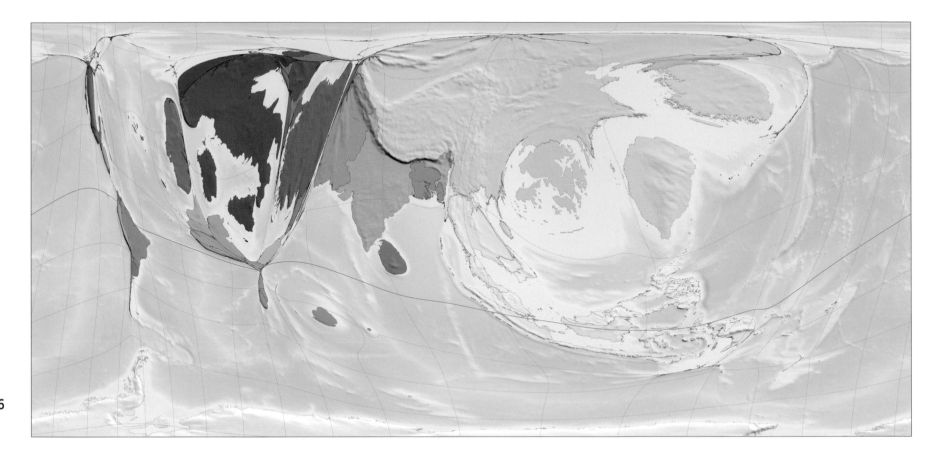

ᵃ annual US$ value of net clothing exports per person

127 Clothing Exports

The size of each territory indicates the dollar value of its net exports of clothes, a sector that accounts for 7% of all earnings from international trade.

Exports of clothing include footwear, bags and fabrics as well as garments. China leads the rankings for net clothing exports, exporting more in US dollar terms than any other territory, while East Asia, which includes China, exports over five times as much as any other region.

It is common for cloth to be exported from one territory to another, where it is sewn into a garment, and then exported for sale in a third. In the process it will usually gain in value because finished items are worth more than raw materials and because they command higher prices in the more affluent territories to which they are exported.

'...Bangladesh...relies on garments for more than three-quarters of its exports.' Roland Buerk, BBC journalist, 2005

NET CLOTHING EXPORTS
annual earnings, US$ billion

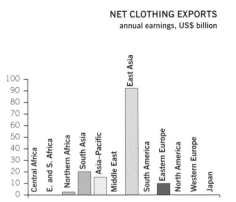

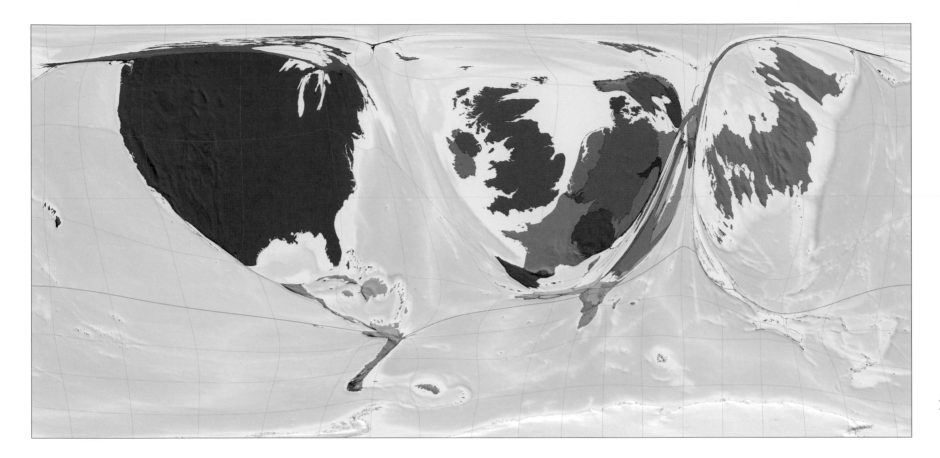

HIGHEST VALUE OF NET CLOTHING IMPORTS

Rank	Territory	US$[a]
1	Niue	3,102
2	Andorra	2,071
3	United Arab Emirates	523
4	Switzerland	435
5	Norway	410
6	Iceland	365
7	Greenland	359
8	Cyprus	346
9	Ireland	318
10	United Kingdom	279
11	Luxembourg	277
12	United States	276
13	Kuwait	260
14	Austria	232
15	Bahamas	227
16	Sweden	207
17	Finland	172
18	Japan	155
19	Netherlands	151
20	Barbados	150

[a] annual US$ value of net clothing imports per person

128 Clothing Imports

Despite the wide variation in clothing styles around the world, there is a large international trade in clothes. The size of each territory in this map indicates the dollar value of its net imports of clothing.

The United States, Western Europe and Japan in particular import substantial amounts of clothing. (People in these regions also throw more clothes away.)

Most clothes are made in territories where wages for labour are relatively low, which lowers prices for buyers in the importing territories and tends to bias purchasing decisions towards imported goods. The high spending per person on clothing imports in, for instance, many Western European territories is a result not only of those territories' greater wealth, but also of their tendency to spend what money they have for clothing purchases on imports.

'...we are trying to balance a number of interests here because obviously people want less expensive goods...'
Tony Blair, British prime minister, 2005

NET CLOTHING IMPORTS
annual spending, US$ billion

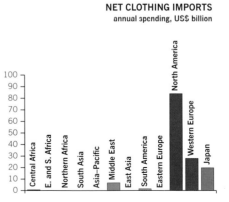

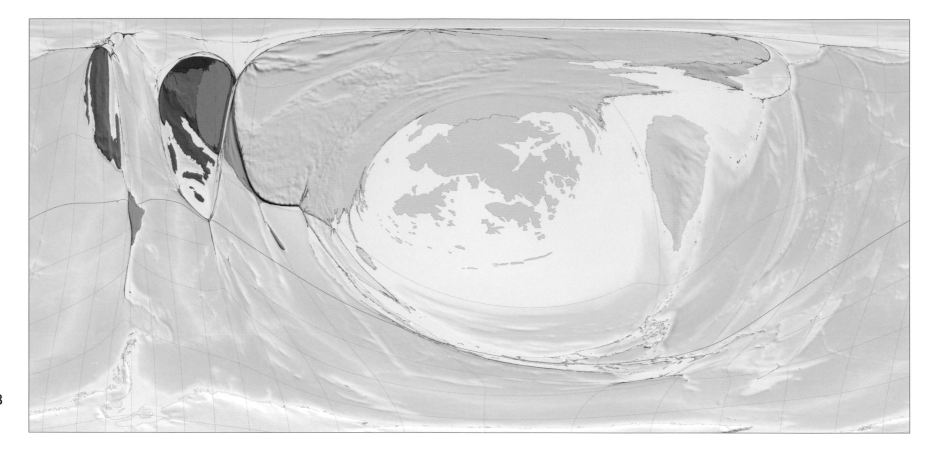

HIGHEST NET TOY EXPORTS

Rank	Territory	US$[a]
1	Hong Kong (China)	799.8
2	Malta	91.3
3	Taiwan	74.7
4	Hungary	61.9
5	Austria	38.0
6	Slovenia	15.0
7	North Korea	14.1
8	Czech Republic	11.1
9	China	9.5
10	Italy	8.4
11	Thailand	7.5
12	Malaysia	6.3
13	Mexico	6.0
14	Bosnia Herzegovina	2.3
15	Romania	2.2
16	Pakistan	1.7
17	Philippines	1.2
18	Sri Lanka	1.1
19	Belarus	0.6
20	Brazil	0.6

[a] annual US$ value of net toy exports per person

129 Exports of Toys

The size of each territory indicates the dollar value of its net exports of toys. The data include sports equipment.

Toys, which here include sports equipment, make up 1% of worldwide exports measured by US dollar value. A piece of sports equipment is generally more expensive than the average toy. More toys are exported from East Asia, in terms of dollar value, than from any other region, and almost all of these are from China (including Hong Kong). Net earnings per person from toy exports vary considerably between territories in East Asia: in Hong Kong they are more than 10 times as large as in Taiwan, and almost 100 times as large as in China. (Hong Kong, being a part of China, of course has smaller net exports than its parent nation, but it can still have larger exports per person, since its population is much smaller than that of China as a whole.)

'At City Toys Ltd.,…Shenzhen, youngsters worked 16-hour days, seven days a week.' Agence France-Presse, 2000

NET EXPORTS OF TOYS
annual earnings, US$ billion

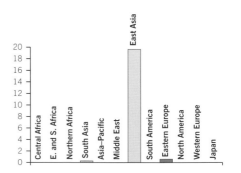

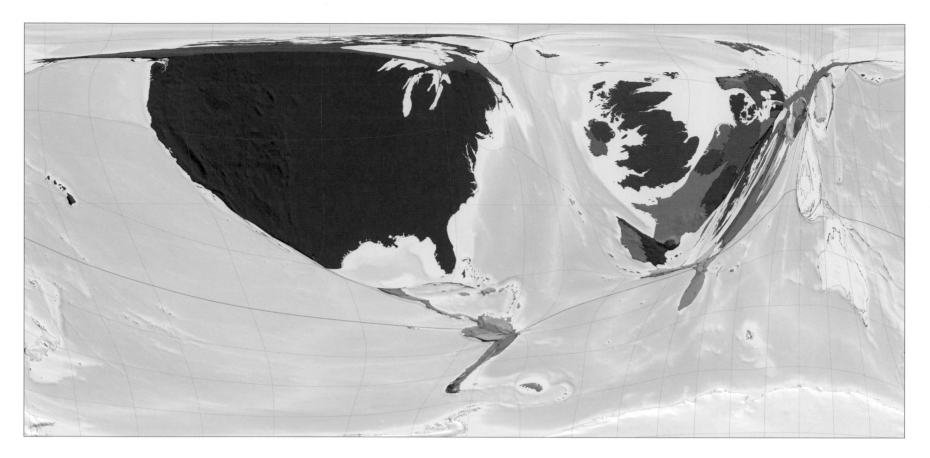

HIGHEST NET TOY IMPORTS

Rank	Territory	US$ᵃ
1	Andorra	340
2	Greenland	57
3	United States	51
4	Iceland	48
5	Norway	48
6	Niue	39
7	Luxembourg	36
8	New Zealand	36
9	Cyprus	35
10	Canada	33
11	Switzerland	32
12	United Kingdom	31
13	Bahamas	30
14	Australia	30
15	Singapore	29
16	Denmark	23
17	Brunei	23
18	Ireland	22
19	Qatar	21
20	Barbados	20

ᵃ annual US$ value of net toy imports per person

130 Imports of Toys

The size of each territory indicates the dollar value of its net imports of toys and sporting equipment. The map is, unsurprisingly, dominated by the richer territories.

Toys and sporting equipment are fun but they are not basic necessities, and import figures thus give an indication of disposable income. The United States has the highest net imports of toys in terms of dollar value, followed by the United Kingdom. The highest imports per person are to territories in Western Europe, North America, the Asia–Pacific and Australasia region, and Eastern Europe. The poorest territories in the world are located in Central Africa, East and Southern Africa, South Asia and Northern Africa, regions that are barely discernible on this map because so few territories there are net importers of toys.

NET IMPORTS OF TOYS
annual spending, US$ billion

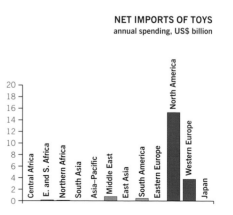

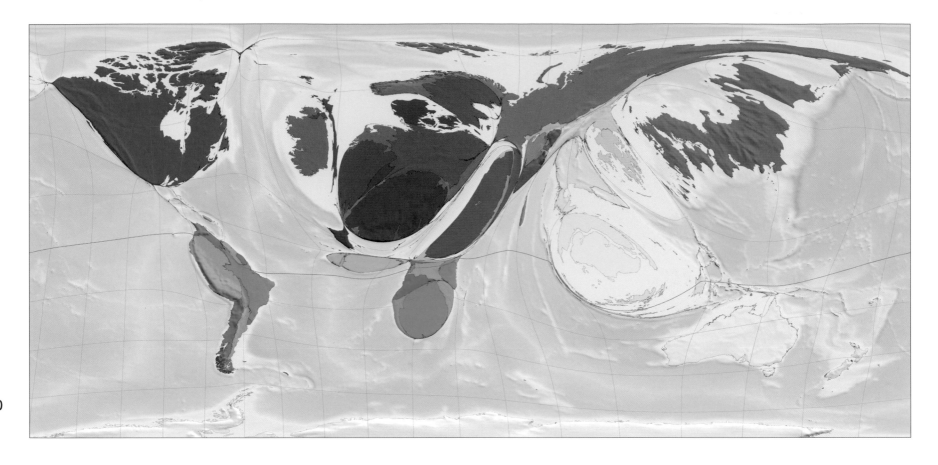

HIGHEST NET EXPORTS OF VALUABLES

Rank	Territory	US$[a]
1	Botswana	1,051
2	Singapore	926
3	Luxembourg	754
4	Switzerland	653
5	Ireland	506
6	Israel	498
7	Canada	373
8	Norway	370
9	Australia	245
10	Sweden	232
11	Hong Kong (China)	223
12	Belgium	216
13	Denmark	187
14	Guyana	148
15	Suriname	127
16	Cook Islands	112
17	Namibia	107
18	New Zealand	81
19	Japan	64
20	Russia	63

[a] **annual US$ value of net valuables exports per person**

131 Exports of Valuables

The size of each territory indicates the dollar value of its net exports of luxury items such as art, watches, jewelry and musical instruments, as well as precious stones and metals.

Earnings from the export of valuables constitute 3.4% of all export earnings. The highest net exports (in US dollars) per person are from Botswana, which is a major exporter of diamonds. If the net revenue from Botswana's valuables exports were divided equally among the members of its population, each person would earn US$1,051 a year. Currently more than a quarter of Botswana's 1.6 million inhabitants live on less than US$1 a day.

NET VALUABLES EXPORTS
annual earnings, US$ billion

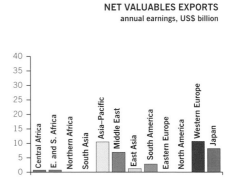

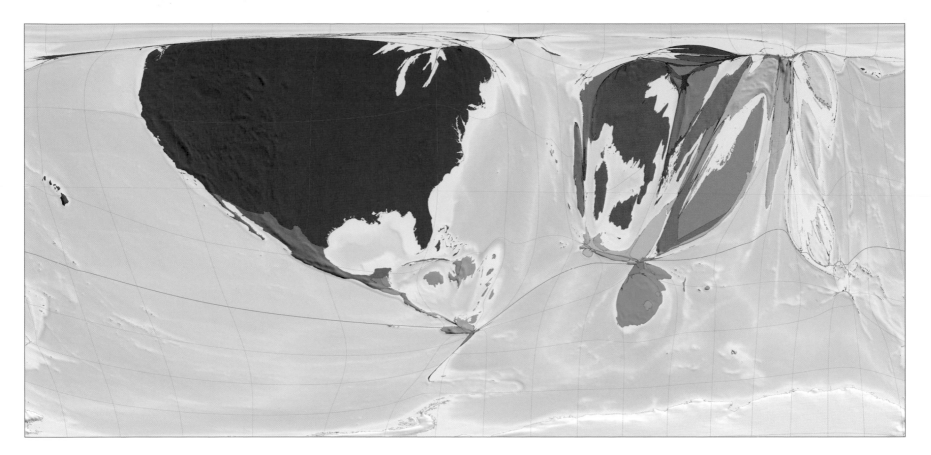

HIGHEST NET IMPORTS OF VALUABLES

Rank	Territory	US$[a]
1	United Arab Emirates	1,270
2	Andorra	1,186
3	Antigua and Barbuda	648
4	Greenland	644
5	Brunei	304
6	Niue	273
7	Bahamas	206
8	Macedonia, FYR of	200
9	Belize	199
10	Tuvalu	188
11	St Lucia	160
12	United States	156
13	Qatar	138
14	Bulgaria	131
15	Cyprus	128
16	Barbados	120
17	Italy	108
18	Kuwait	104
19	Oman	96
20	St Kitts–Nevis	95

[a] annual US$ value of net valuables
imports per person

132 Imports of Valuables

The size of each territory indicates the dollar value of its net imports of luxury items, as well as precious stones and metals. Worldwide, more money is spent importing valuables than on importing medicines.

The United States alone accounts for 59% of net imports of valuables. The second largest net importer, Italy, imports only a tenth as much as the US in terms of dollar value, while third place goes to the United Arab Emirates, which imports about two-thirds as much as Italy – but spends the most per person, at US$1,270 per year. One reason for this high figure is the wealth generated by petroleum exports.

South Africa is perhaps a surprising entry on this map. It has substantial exports of both diamonds and platinum, but nonetheless is not a net exporter of valuables. In fact it has relatively high net imports.

NET VALUABLES IMPORTS
annual spending, US$ billion

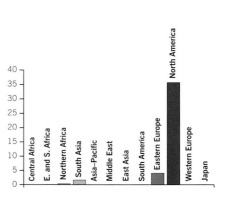

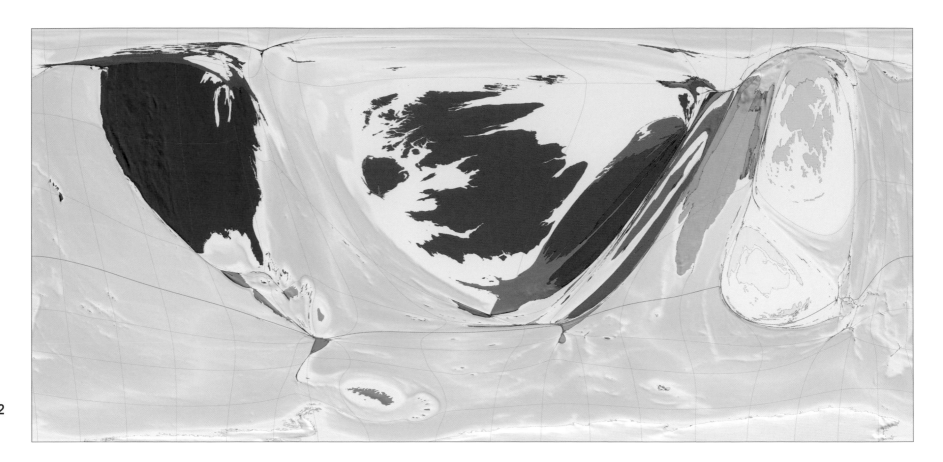

HIGHEST NET EXPORTS OF MERCANTILE AND BUSINESS SERVICES

Rank	Territory	US$[a]
1	Luxembourg	1,893
2	Cyprus	1,450
3	Hong Kong (China)	905
4	Singapore	901
5	Switzerland	461
6	Belgium	451
7	United Kingdom	437
8	Barbados	390
9	Israel	368
10	St Vincent and the Grenadines	109
11	Norway	94
12	Netherlands	90
13	United States	88
14	Canada	75
15	Malta	67
16	Iceland	64
17	Costa Rica	49
18	France	48
19	Swaziland	44
20	Denmark	42

[a] annual US$ value of net other services (those not shown in other maps) exports per person

133 Mercantile and Business Exports

The size of each territory indicates the dollar value of its net exports of mercantile and business services. Just three territories account for over half of all this trade.

Mercantile and business services include consultancy, accounting and communications, as well as other activities such as opinion polling and government service in consulates and military units.

Export earnings from mercantile and business services account for 8.9% of international trade. The United Kingdom, the United States and India are the biggest net exporters in this field, accounting for 60% of all net exports in US dollar terms. Most territories (136 out of 200) do not have net exports of mercantile and business services, meaning they buy more than they sell of these services. No territories in Central Africa, and only three territories in East and Southern Africa and Northern Africa, are net exporters.

'...management consultants are part of the day to day fabric of leading-edge world class companies. The more they are used, the more invaluable and acceptable their work becomes.'

Brian O'Rorke, executive director, Management Consultancies Association, 1996

NET MERCANTILE AND BUSINESS EXPORTS
annual earnings, US$ billion

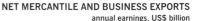
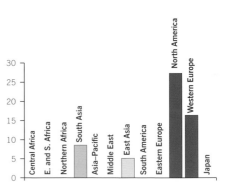

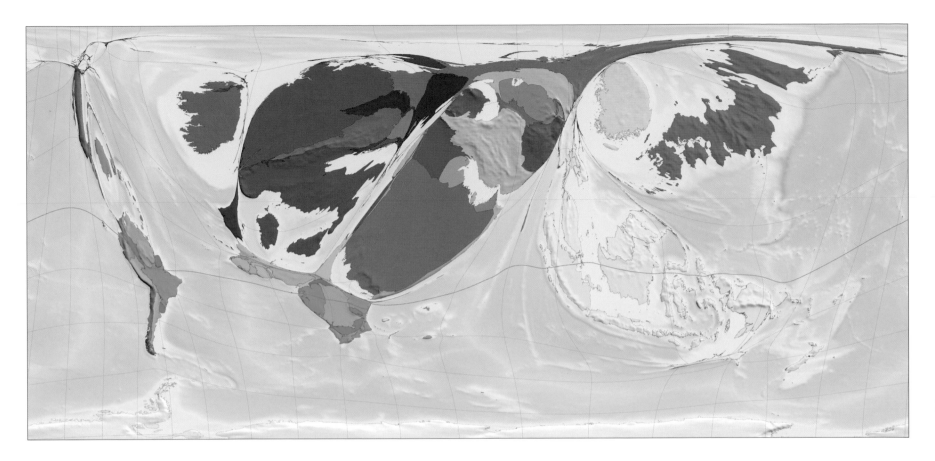

134 Mercantile and Business Imports

Most territories import more business services than they export. The size of each territory in this map indicates the dollar value of its net imports of mercantile and business services.

Saudi Arabia, Germany, Japan and Indonesia are the biggest net importers in US dollar terms. Of the 12 regions, the Middle East and Central Asia imports the most: US$76 (net) of mercantile and business services per inhabitant per year. Of individual territories, the highest per-person spending is in Ireland, the Seychelles, Saudi Arabia and Kuwait.

'In 1928, when the stock of Turkish Oil was redistributed, still at the centre of negotiations, Gulbenkian was again awarded his usual percentage and became known as "Mr. Five Percent".'

Fundação Calouste Gulbenkian (Calouste Gulbenkian Foundation), 2002

HIGHEST NET IMPORTS OF MERCANTILE AND BUSINESS SERVICES

Rank	Territory	US$ᵃ
1	Ireland	897
2	Seychelles	639
3	Saudi Arabia	517
4	Kuwait	293
5	St Kitts–Nevis	285
6	Austria	274
7	Antigua and Barbuda	251
8	Suriname	175
9	Congo	168
10	St Lucia	135
11	Italy	131
12	Jamaica	119
13	Kazakhstan	118
14	Grenada	116
15	Germany	114
16	Azerbaijan	109
17	Czech Republic	88
18	Malaysia	87
19	South Korea	74
20	Finland	71

ᵃ annual US$ value of net other services (those services not shown in other maps) imports per person

NET MERCANTILE AND BUSINESS EXPORTS
annual spending, US$ billion

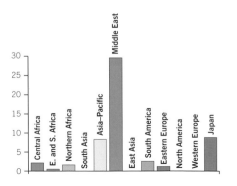

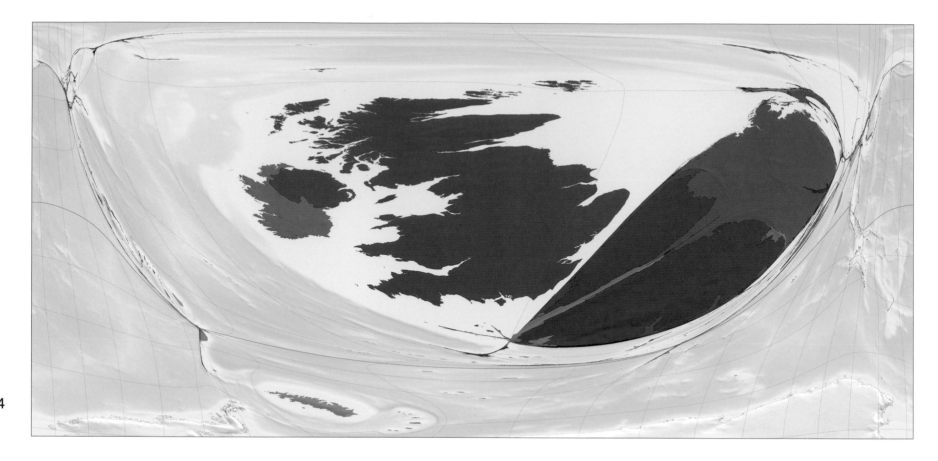

HIGHEST NET EXPORTS OF FINANCE AND INSURANCE SERVICES

Rank	Territory	US$ᵃ
1	Luxembourg	12,834.0
2	Switzerland	1,226.7
3	Ireland	431.3
4	United Kingdom	411.8
5	Cyprus	113.1
6	Barbados	96.0
7	St Vincent and the Grenadines	93.6
8	Germany	89.6
9	Norway	38.3
10	Sweden	17.9
11	Kuwait	15.1
12	Panama	10.6
13	Uruguay	10.1
14	Australia	5.4
15	Estonia	3.4
16	Slovenia	3.2
17	Bosnia Herzegovina	3.1
18	Latvia	3.1
19	Honduras	2.7
20	South Korea	1.9

ᵃ annual US$ value of net finance and insurance services exports per person

135 Finance and Insurance Exports

The size of each territory indicates the dollar value of its net exports of financial and insurance services, a market dominated by Western Europe.

Over 99% of net income from the export of financial and insurance services flows into territories in Western Europe. Despite this, almost half of the 24 territories in the region have no net financial and insurance services exports. The main exporting territories are the United Kingdom, Switzerland, Germany and Luxembourg. The tiny island of South Georgia is also large on this map because, as a part of UK territory, it is sized commensurately with the British mainland.

'Edward Lloyd opened a coffee shop on Tower Street in London in 1688, and that one turned into Lloyd's of London, insuring the world.' Jack Schofield, *Guardian Unlimited*, 2006

NET FINANCE AND INSURANCE EXPORTS
annual earnings, US$ billion

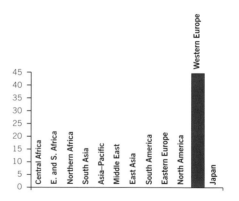

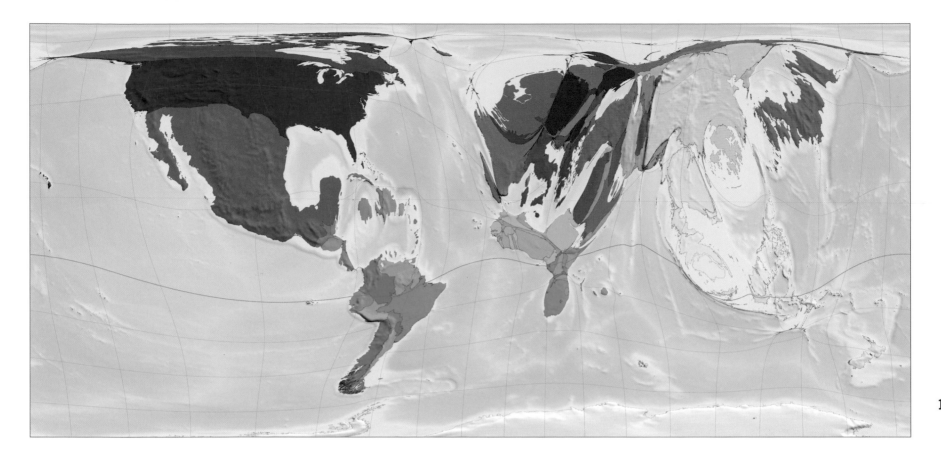

136 Finance and Insurance Imports

The size of each territory indicates the dollar value of its net imports of finance and insurance services.

Of the 200 territories in the world, 83.5% are net importers of insurance and financial services, meaning they buy more than they sell. Mexico, the United States and China import the most in US dollar terms, followed by Canada. That 3 of the 4 biggest importers are North American territories helps explain why the North American region as a whole is also a net importer.

'Egyptian culture still wrestles with the idea of investing money in something with no tangible returns. Insurers need to not only promote their policies but the idea of insurance itself.'
Sherine Abdel-Razek, *Al-Ahram*, 2005

NET FINANCE AND INSURANCE IMPORTS
annual spending, US$ billion

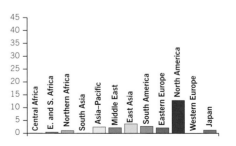

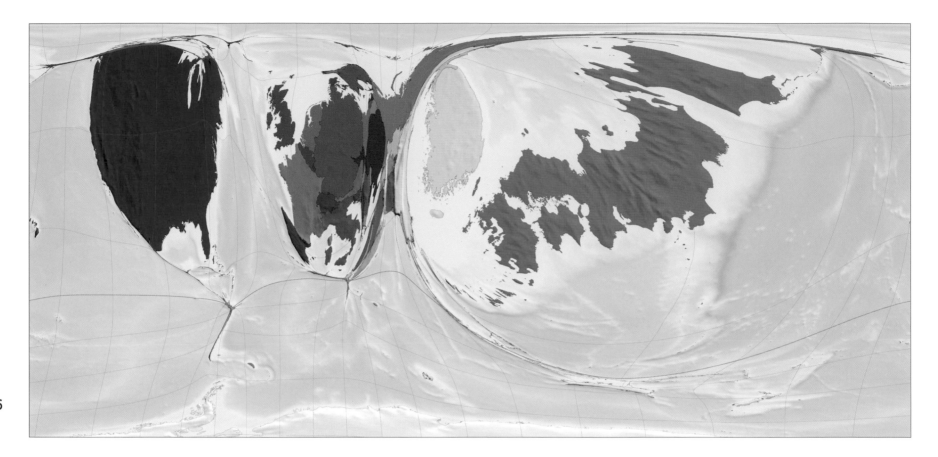

156

MOST AND FEWEST PATENTS GRANTED

Rank	Territory	Patents[a]
1	Japan	884
2	South Korea	490
3	United States	298
4	Sweden	235
5	Bahamas	208
6	Greenland	207
7	Germany	205
8	Switzerland	188
9	Netherlands	177
10	France	174
57	Bulgaria	18
57	Hungary	18
59	Macedonia, FYR of	17
60	Uzbekistan	16
61	Slovakia	15
62	Kyrgyzstan	13
63	Ireland	9
64	Iceland	7
65	Hong Kong (China)	6
66–200	135 countries	<5

[a] number of patents granted per million people, 2002

137 Patents Granted

Patents, the means of protecting and rewarding innovation, can be a double-edged sword. The size of each territory reflects the number of patents granted in a year.

Patents protect people's ideas and inventions by giving them the legal right to prohibit the use of those innovations by others, or to charge for their use. The aim is to reward creators for their work or intelligence. However, patents can also prevent people, particularly those in poorer territories, from benefiting from new inventions by putting unaffordable prices on access to new products or processes.

In 2002, which is the year shown on this map, 312,000 patents were granted worldwide. More than a third of those were granted in Japan and almost as many in the United States.

PATENT DISTRIBUTION
patents granted in 2002, thousands

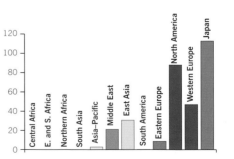

'...the world's poorest countries are still waiting for the promised benefits of stronger patent protection at home.' The Economist, 2001

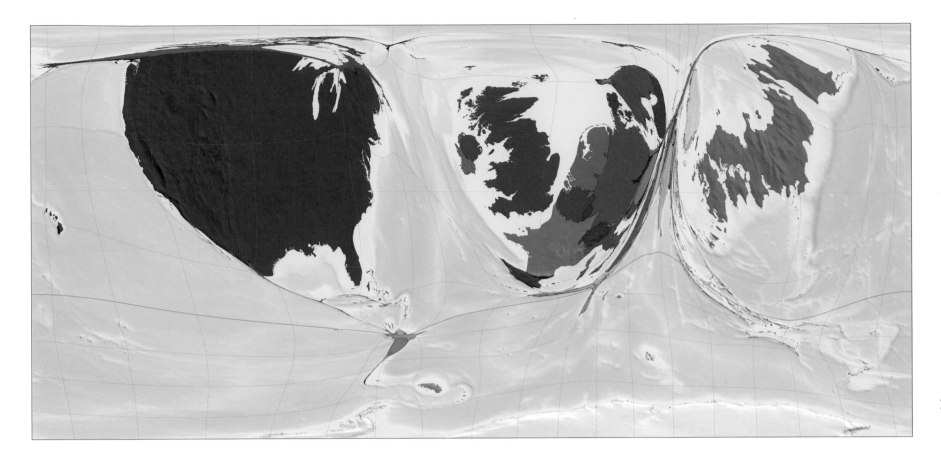

157

MANUFACTURED GOODS AND SERVICES

MOST AND FEWEST ROYALTY FEES RECEIVED

Rank	Territory	US$[a]
1	Luxembourg	275
2	Sweden	170
3	United States	152
4	United Kingdom	130
5	Netherlands	122
6	Greenland	108
7	Bahamas	108
8	Finland	108
9	Belgium	86
10	Japan	82
28	New Zealand	23
29	Croatia	19
30	South Korea	17
31	Australia	16
32	Austria	14
33	Maldives	12
34	Italy	9
35	Spain	9
36	Lesotho	6
37–200	164 countries	<5

[a] value of royalty fees received, per person, 2002

138 Royalty Fees

The size of each territory reflects its earnings from royalties and licence fees, adjusted for local purchasing power. Intellectual property can generate income over a long period.

Royalties and licence fees are payments made by someone for the use of an idea, invention or artistic creation that legally belongs to someone else. To receive these fees an inventor or artist must hold a copyright or patent on their creation, which typically remains active for some years after the initial invention. Thus most of the fees

for 2002 represented on this map result from work that was completed prior to that year.

Over half (53%) of all royalty and licence fees paid in 2002 were received by one territory, the United States. Large amounts were also received by Japan and the United Kingdom.

'The fight over royalties is a time-honored one in Hollywood, with actors, directors, and writers often at odds with the studios...'

Bryan Chaffin, *Mac Observer*, 2005

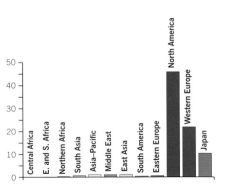

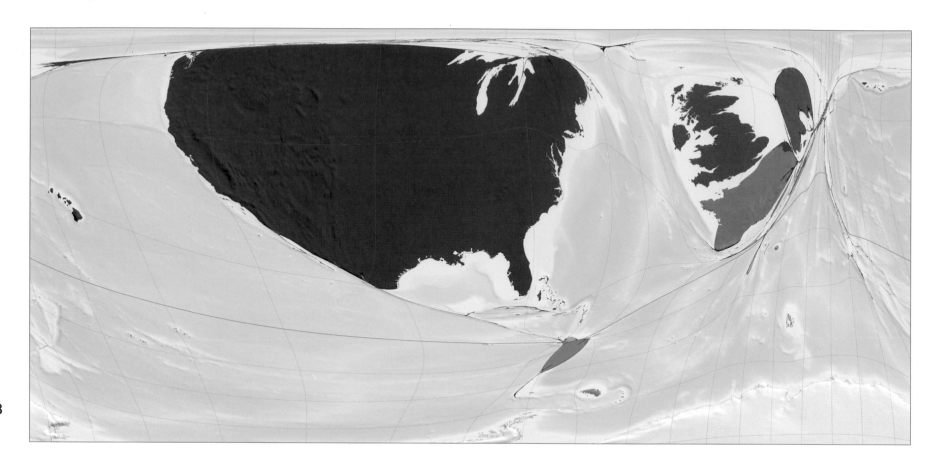

HIGHEST NET EXPORTS OF ROYALTIES AND LICENCE FEES

Rank	Territory	US$[a]
1	United States	85.86
2	Sweden	70.18
3	Greenland	52.93
4	Bahamas	52.93
5	Luxembourg	49.03
6	Cyprus	34.38
7	Paraguay	32.47
8	United Kingdom	29.62
9	France	24.30
10	Guyana	16.00
11	Maldives	12.77
12	Lesotho	5.83
13	Namibia	1.07
14	Tunisia	1.02
15	Cape Verde	0.73
16	Cuba	0.68
17	Kyrgyzstan	0.03
18	Moldova	0.02

[a] annual US$ value of net royalty and licence fee exports per person (the data source records only 18 net exporters)

139 Royalty and Licence Fee Exports

Economies profit from ideas as well as tangible trade. The size of each territory in this map indicates the dollar value of its net exports of royalties and licence fees.

The IMF defines royalties and licence fees as including 'payments and receipts for the authorized use of intangible, non-produced, non-financial assets and proprietary rights…with the use, through licensing agreements, of produced originals or prototypes…'. Exports of royalties and licence fees are thus earnings from international payments for past ideas, such as profits being made from owning copyrights and patents.

Just 18 territories make more than they spend on licence fees and royalties. Together they have earnings of more than US$30 billion per year, while the remaining 182 territories are net importers.

'Ideas shape our world. They are the raw materials on which our future prosperity and heritage depend.'
Kamil Idris, director general of WIPO, 2006

NET ROYALTIES AND LICENCE FEE EXPORTS
annual earnings, US$ billion

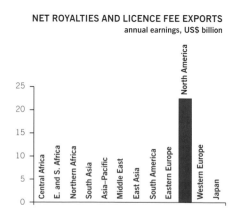

HIGHEST NET IMPORTS OF ROYALTIES AND LICENCE FEES

Rank	Territory	US$ᵃ
1	Ireland	2,748.3
2	Hong Kong (China)	181.2
3	Austria	116.4
4	Barbados	80.3
5	New Zealand	68.1
6	Canada	59.2
7	South Korea	45.7
8	Swaziland	41.3
9	Netherlands	40.8
10	Australia	36.4
11	Slovenia	35.5
12	Spain	35.2
13	Norway	35.0
14	Malta	28.2
15	Portugal	26.7
16	Malaysia	25.7
17	Greece	25.0
18	St Lucia	21.6
19	St Kitts–Nevis	19.4
20	Thailand	17.6

ᵃ annual US$ value of net royalty and licence fee imports per person

140 Royalty and Licence Fee Imports

The size of each territory indicates the dollar value of its net imports of royalties and licence fees.

Ireland spends more on royalties and licence fees than any other country in the world – more than three times as much as the next biggest importer, China. Ireland's spending is also the highest per person in the world; the runner-up, Hong Kong (which is part of China), has net per-person imports that are only a fifteenth of Ireland's.

'[The Singapore Treaty on Trademarks] will boost international trade and deliver an enhanced and harmonized trademark procedure that will benefit nations, brands and businesses.'
Burhan Gafoor, Singapore's permanent representative to the WTO and the UN, Geneva, 2006

NET ROYALTIES AND LICENCE FEE IMPORTS
annual spending, US$ billion

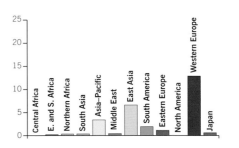

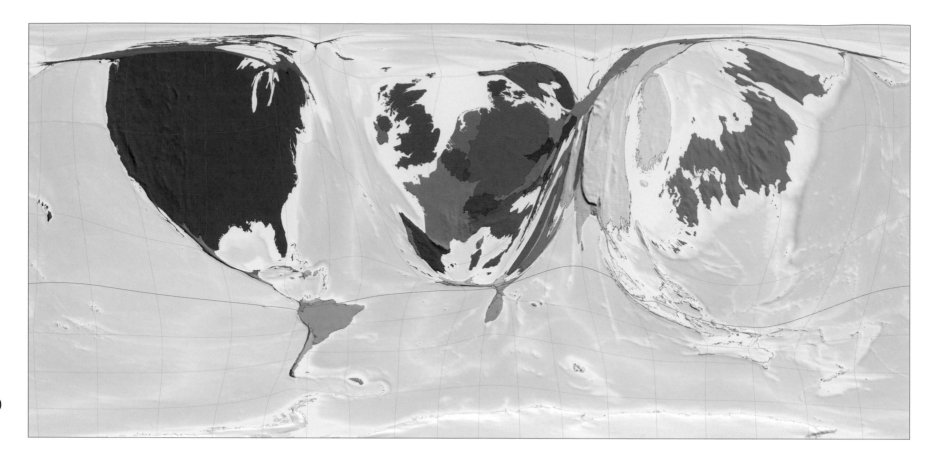

141

141 Research and Development Expenditure

The size of each territory reflects its spending on research and development – an area of both high costs and high potential rewards.

Research and development is undertaken by governments, manufacturers, and scientific, technological and medical companies to find new techniques and products. This can be an expensive pursuit, given the costs of materials, machines and skilled specialists, but it can also bring substantial financial rewards. In 2002, US$289 billion was spent on research

and development in the United States; in the same year there was practically no research and development spending in Angola. It is therefore not surprising that the number of patents granted, and the value of royalty and licence fees received (see Maps 137 and 139), are also vastly different in these countries.

'If we don't alleviate poverty and grow our economies, there will be no one left to do basic research. Once African economies grow, there will be enough time and money to go off and think deep thoughts.'

Asifa Nanyaro, director general of the Tanzania Industrial Research and Development Organization, 2004

HIGHEST AND LOWEST EXPENDITURE ON RESEARCH AND DEVELOPMENT

Rank	Territory	PPP US$ᵃ
1	Luxembourg	1,310
2	Sweden	1,202
3	Israel	1,017
4	United States	992
5	Finland	890
6	Iceland	840
7	Japan	833
8	Switzerland	790
9	Germany	678
10	Denmark	647
181	Nepal	2.7
182	Nigeria	2.5
183	Nicaragua	2.5
184	Mali	2.3
185	Niger	2.1
186	Burkina Faso	2.1
187	Guinea-Bissau	1.9
188	Sierra Leone	1.5
189	Madagascar	0.7
190–200	11 countries	<0.1

ᵃ **expenditure per person, 2002**

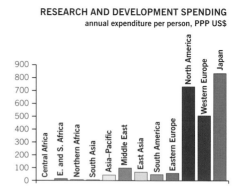

RESEARCH AND DEVELOPMENT SPENDING
annual expenditure per person, PPP US$

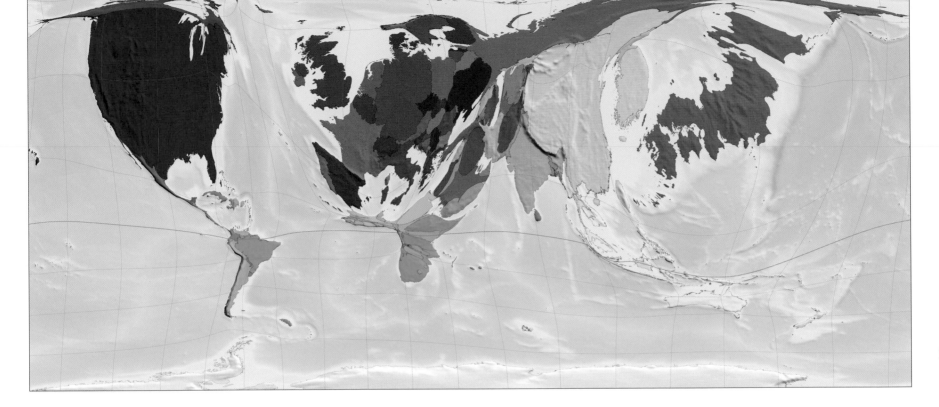

MOST AND FEWEST RESEARCH AND DEVELOPMENT EMPLOYEES

Rank	Territory	Employees[a]
1	Finland	7,057
2	Iceland	6,443
3	Japan	5,268
4	Sweden	5,155
5	Norway	4,339
6	United States	4,015
7	Singapore	3,875
8	Switzerland	3,557
9	Russia	3,476
10	Denmark	3,461
191	Syria	27
192	Uganda	23
193	Burundi	20
194	Burkina Faso	15
195	Madagascar	14
196	Nigeria	14
197	Jamaica	8
198	Oman	4
199	Cameroon	3
200	Senegal	2

[a] number of employees per million people, 2002

142 Research and Development Employees

The size of each territory indicates the number of people employed there in research and development. Population size, as well as wealth, has a significant effect on these numbers.

In most territories in the world there are at least some people working in research and development. However, in 80 territories there are fewer than 1,000 people in total working in the sector, and in 12 there are fewer than 20 – although the latter tend to be smaller territories to begin with: all of them except for Senegal have populations under 3 million. In the United States, by contrast, there are 1.2 million people working in research and development.

The territory with the highest proportion of research and development employees is Finland, where 7 people work in research and development for every 1,000 members of the population (including children, the retired and the unemployed).

RESEARCH AND DEVELOPMENT EMPLOYEES
number of research and development employees per 1,000 people

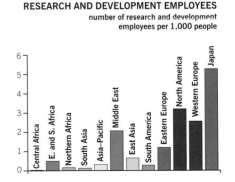

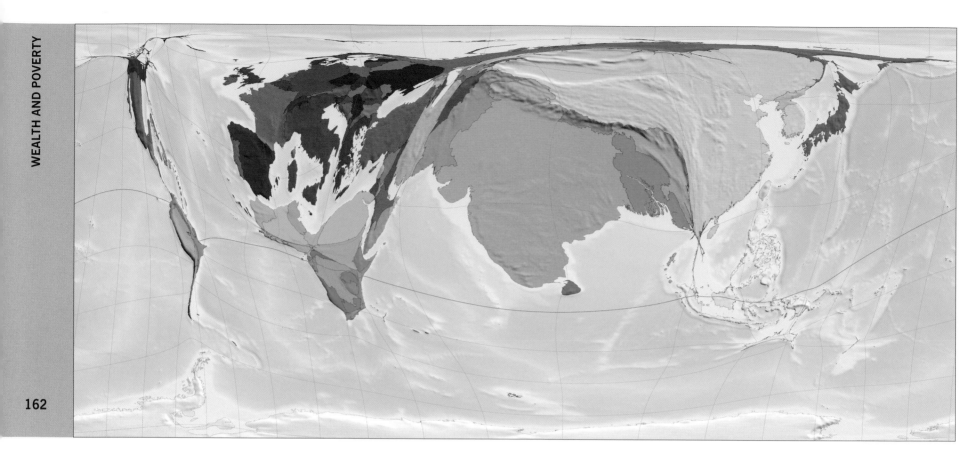

WEALTHIEST AND POOREST TERRITORIES IN 1 AD

Rank	Territory	PPP US$[a]
1	Bangladesh	454
2	India	453
3	Italy	453
4	Greece	452
5	France	452
6	South Korea	452
7	Pakistan	452
8	Spain	452
9	Germany	452
10	China	452
191	Paraguay	400
192	Brazil	400
193	Uruguay	400
194	Argentina	400
195	Canada	400
196	Panama	400
197	Greenland	400
198	Madagascar	0
198	Singapore	0
198	New Zealand	0

[a] GDP per person, in current PPP US$ (estimates and best guesses)

143 Wealth in the Year 1

The size of each territory shows the GDP, adjusted for local purchasing power, of the equivalent territory in AD 1.

This map shows how wealth was distributed across the world 2,000 years ago, in terms of modern boundaries. Wealth is measured in terms of gross domestic product (GDP), which is the total market value of all goods and services produced within a territory in a given year. The figures for GDP are also adjusted for purchasing power parity.

The best current estimates indicate that average GDP per person living in the year AD 1, expressed in current US dollars, was US$445. By 1990 the equivalent figure was US$5,248.

2,000 years ago there was probably much less variation between regions.

Indeed, since variations in GDP per person were low this map looks very similar to the population map for the year 1 (Map 003).

The Americas appear small on this map, partly because fewer people lived there in year 1, but also because South America had the lowest regional GDP per person, estimated as PPP US$400.

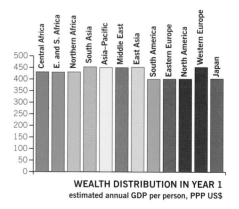

WEALTH DISTRIBUTION IN YEAR 1
estimated annual GDP per person, PPP US$

'Travellers and scholars who were attracted by the charms and fame of Bangladesh since time immemorial...showered effusive epithets on its bounties and wealth, affluence and prosperity...' UN mission to Bangladesh, 2006

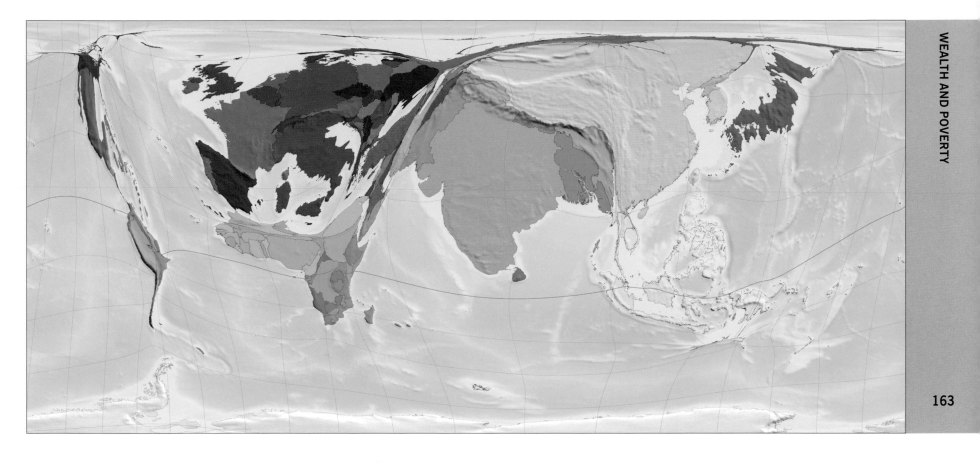

WEALTHIEST AND OTHER TERRITORIES IN 1500 AD

Rank	Territory	PPP US$[a]
1	Italy	1,100
2	Belgium	875
3	Netherlands	761
4	Denmark	738
5	France	727
6	United Kingdom	714
7	Austria	707
8	Sweden	695
9	Germany	688
10	Spain	661
11	Norway	640
12	Switzerland	632
21	Portugal	606
23	China	600
53	India	550
60	Ireland	526
61	Japan	500
65	Iraq	499
86	Turkey	496
106	Egypt	475

[a] **GDP per person, estimated in current PPP US$**

144 Wealth in the Year 1500

The size of each territory reflects its GDP in 1500, adjusted for local purchasing power. Europe was already among the wealthiest areas of the world.

In the year 1500 European territories were some of the richest on Earth in terms of GDP per person. The regions with the largest total GDP were East Asia and South Asia, although this is hardly surprising since these were also the most populous regions at the time.

The regions with the lowest GDP in 1500 were Central Africa and East and Southern Africa. These regions also had the lowest GDP per person. But in 2002 these regions had an even smaller proportion of the world's total GDP than they did in 1500.

'Slaves captured in raids and war grew in importance as a commodity...Kola nuts...were also important, as were the dyestuffs of northern Nigeria. All these goods were highly prized in and around the Mediterranean basin.' Richard Effland, 'The rest of the story about Africa', 2003

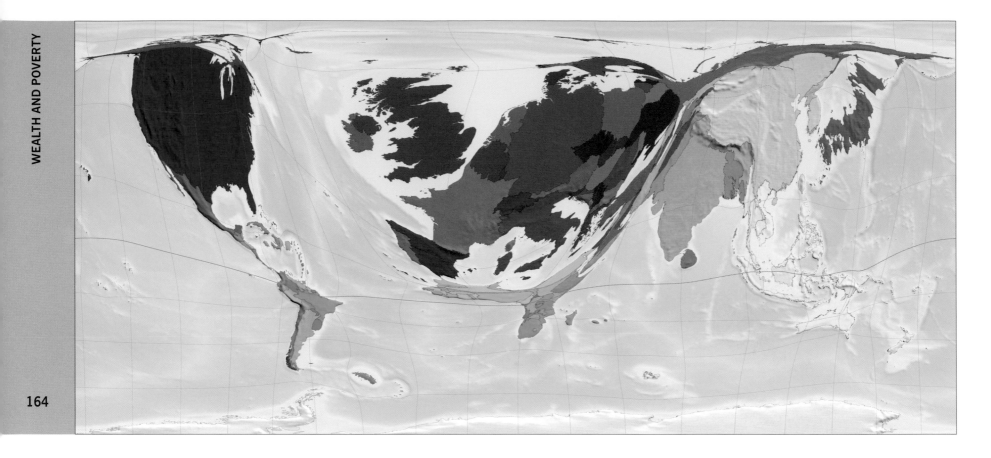

WEALTHIEST AND POORER TERRITORIES IN 1900

Rank	Territory	PPP US$[a]
1	United Kingdom	4,492
2	New Zealand	4,298
3	United States	4,091
4	Australia	4,013
5	Switzerland	3,833
6	Belgium	3,731
7	Netherlands	3,424
8	Denmark	3,017
9	Germany	2,985
10	Canada	2,911
141	Myanmar	685
142	Albania	685
143	Brazil	678
165	Bangladesh	607
169	India	599
170	Somalia	557
188	Mongolia	553
189	China	545
190	Nepal	539
193	DR Congo	483

[a] GDP per person, estimated in current PPP US$

145 Wealth in the Year 1900

The size of each territory reflects its GDP in 1900, adjusted for local purchasing power. By this time the effects of the industrial revolution were clearly apparent.

Between 1500 and 1900 the world average GDP per person doubled, primarily as a result of the increases in efficiency that came with the industrial revolution. A significant proportion of production moved to mechanized factories where jobs were more specialized, by contrast with earlier eras in which production took place in the home or in small workshops and a single worker might perform all stages of the production of a single item.

The world population also increased over this period, so that the rise in total GDP was even greater than the increase in GDP per person would suggest. By 1900 the world total GDP was US$2 trillion when expressed in 1990 dollars, adjusted for purchasing power parity.

'The industrial revolution...enormously increased the capacity of some groups, mostly Europeans at first, to produce goods and services. It greatly altered the distribution of wealth and poverty around the world...' San Diego State University / United States National Center for History, 2006

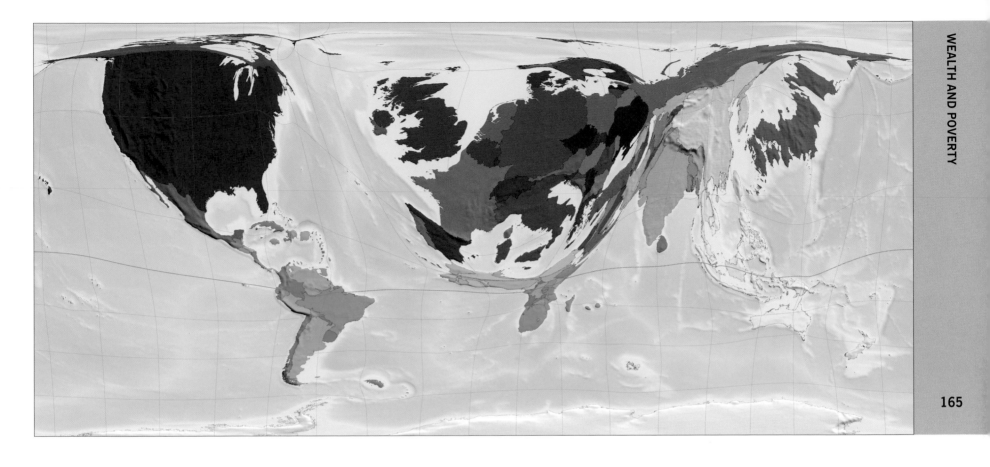

165

WEALTHIEST AND POOREST TERRITORIES IN 1960

Rank	Territory	PPP US$[a]
1	Qatar	33,104
2	Kuwait	28,813
3	United Arab Emirates	22,433
4	Switzerland	12,457
5	United States	11,328
6	France	9,785
7	Venezuela	9,646
8	New Zealand	9,465
9	Denmark	8,812
10	Australia	8,791
191	Mali	535
192	Cape Verde	508
193	Guinea-Bissau	501
194	Tanzania	459
195	Lesotho	458
196	Burundi	444
197	Ethiopia	439
198	Botswana	403
199	Malawi	394
200	Guinea	392

[a] GDP per person, estimated in current PPP US$

146 Wealth in the Year 1960

The size of each territory reflects its GDP in 1960, adjusted for local purchasing power.
By this time, most of the world's wealth was produced in North America and Western Europe.

Overall, the distribution of wealth among territories in 1960 was similar to that in 1900, except that the proportion in Asian territories had diminished somewhat while that in South American territories had increased. The highest levels of GDP per person in 1960 were in the smaller Middle Eastern territories of Qatar, Kuwait and the United Arab Emirates. The lowest were mainly in Northern Africa and East and Southern Africa.

TIMELINE OF WORLD WEALTH
annual GDP per person, PPP US$

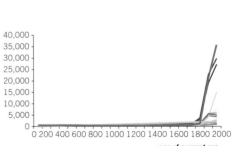

year of current era.
Each line represents one of the 12 regions

'The first Asian economic miracle was Japan's after World War II, rooted in the changes of the Meiji restoration...The Asian Tigers... began to emerge from 1960 onward...' Luis Alberto Moreno, *Latin Business Chronicle*, 2006

ᵃ **GDP per person, estimated in current PPP US$**

147 Wealth in the Year 1990

The size of each territory reflects its GDP in 1990, adjusted for local purchasing power. Although the existing pattern of wealth distribution persisted, some new features were evident.

Between 1960 and 1990 the world average GDP per person more than doubled, and total GDP, adjusted for purchasing power parity, rose from US$8 trillion to US$27 trillion. This increase in wealth was distributed in a pattern broadly similar to the established wealth of territories, although with some notable exceptions, such as the marked economic growth of Japan, China, South Korea and Taiwan.

The region with the lowest GDP in 1990 was Central Africa, which had just 0.8% of the GDP of the richest region, North America. Just 1% of North America's GDP, redistributed to Central Africa, would have more than doubled that region's wealth.

GDP IN 1990
GDP per person in PPP US$ (modelled) year 1990

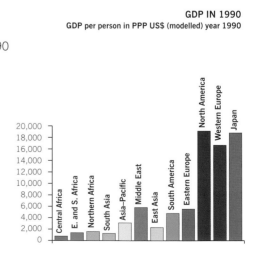

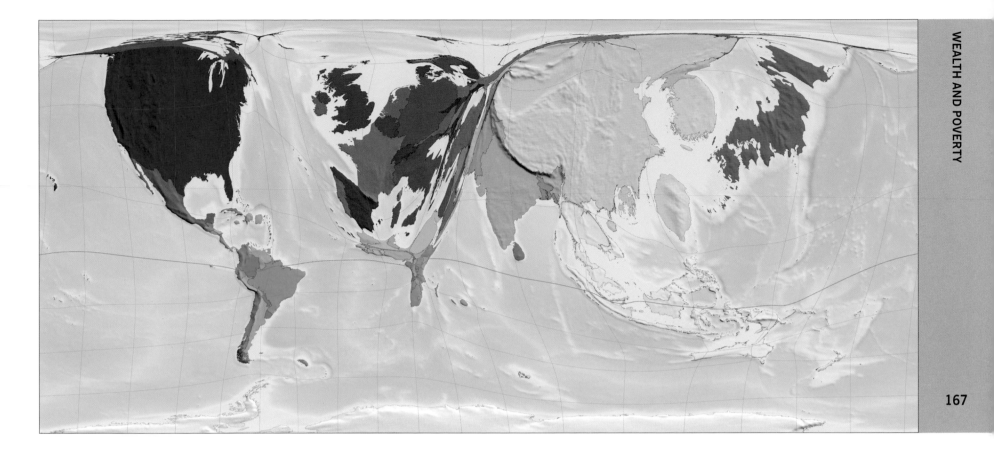

167

148 Projected Wealth in the Year 2015

The size of each territory reflects its projected GDP in the year 2015, adjusted for local purchasing power. Is China about to come full circle from 2,000 years ago?

If the projections shown on this map turn out to be correct, China will be producing 27% of the world's wealth by the year 2015. Interestingly, China produced almost exactly this fraction of the world's wealth – an estimated 26% – two millennia ago, in year AD 1. China's share of world production declined over the following centuries to just 5% by 1960, but now it is forecast to recover its former position.

The rest of the picture, however, will be quite different In 2015 from what it was in the year 1. 2,000 years ago the Americas had only a tiny fraction of the world's wealth, whereas today they have the lion's share – and are expected to retain it. African economies are predicted to remain small players on the international stage, and Eastern European territories are predicted to hold decreasing proportions of world wealth.

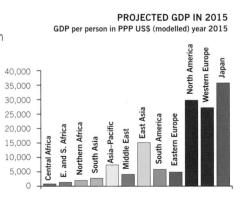

PROJECTED GDP IN 2015
GDP per person in PPP US$ (modelled) year 2015

'Asia's rise is the economic event of our age. Should it proceed as it has over the last few decades, it will bring the two centuries of global domination by Europe and, subsequently, its giant North American offshoot to an end.' Martin Wolf, *Financial Times*, 2003

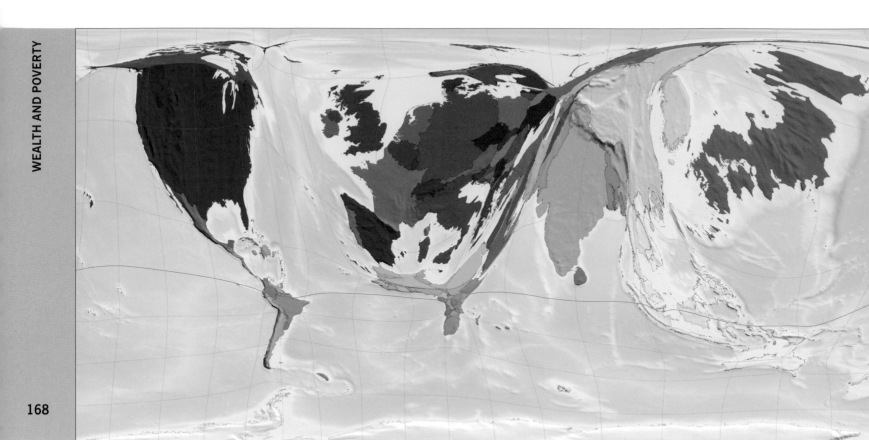

WHAT THE POOREST TENTH OF THE POPULATION EARNS

Rank	Territory	PPP US$ᵃ
1	Luxembourg	23,800
2	Norway	14,395
3	Japan	12,894
4	Finland	10,469
5	Ireland	10,231
6	Sweden	9,404
7	Austria	9,001
8	Germany	8,683
9	Netherlands	8,172
10	Belgium	8,021
191	Madagascar	137
192	Haiti	123
193	Eritrea	122
194	Lesotho	119
195	Burundi	116
196	Malawi	99
197	Zambia	88
198	Central African Republic	83
199	Niger	63
200	Sierra Leone	28

ᵃ **average annual individual earnings of poorest tenth of national population, 2002**

149 Earnings of the Poorest Tenth of the Population

The size of each territory indicates the total income of the poorest 10% of its population, measured in terms of local purchasing power.

Countries appear large on this map either if the poorest tenth of their population earns a high average wage or if there are simply a large number of people in the territory so that the wages of the poorest tenth add up to a large number. Thus Japan appears large because, although its population is not that big, even the poorest people there have relatively high incomes compared to other territories. India, by contrast, appears large because a huge number of people live there – about 8 times as many as in Japan – and their incomes add up to a large figure despite the fact that individually they earn relatively little.

Incomes are measured in terms of local purchasing power. Each person in the poorest tenth of the poorest country earns less than 1% of the income of a person in the poorest tenth in the richest country.

EARNINGS OF POOREST TENTH
average annual per person earnings, PPP US$

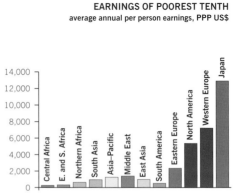

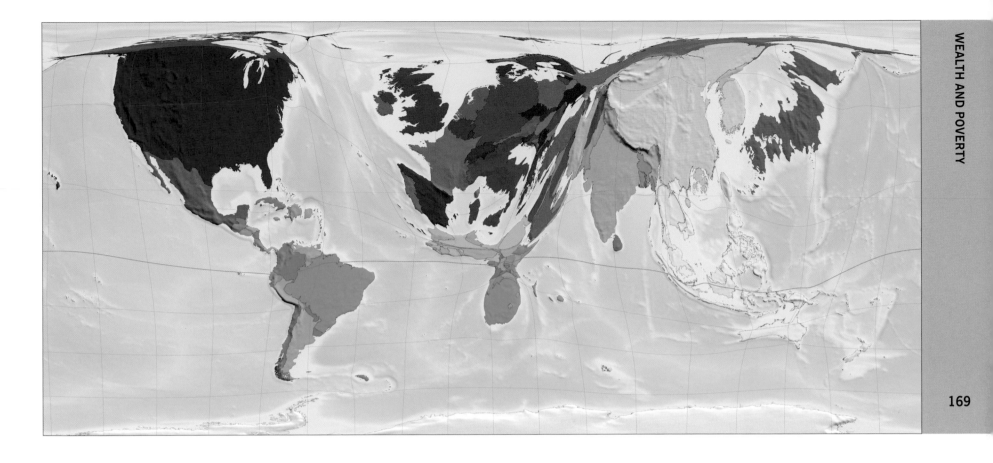

WHAT THE RICHEST TENTH OF THE POPULATION EARNS

Rank	Territory	PPP US$ᵃ
1	Luxembourg	161,840
2	United States	105,914
3	Ireland	100,846
4	Equatorial Guinea	96,653
5	Hong Kong (China)	91,039
6	Greenland	87,252
7	Norway	86,372
8	Seychelles	79,000
9	Singapore	78,173
10	Switzerland	76,580
191	Niger	2,801
192	Madagascar	2,642
193	Tajikistan	2,479
194	Sierra Leone	2,453
195	DR Congo	2,272
196	Burundi	2,236
197	Malawi	2,199
198	Yemen	2,174
199	Ethiopia	1,944
200	Tanzania	1,692

ᵃ average annual individual earnings of richest tenth of national population, 2002

150 Earnings of the Richest Tenth of the Population

The size of each territory indicates the total income of the richest 10% of its population, measured in terms of local purchasing power.

The country with the lowest figure for the combined incomes of the richest tenth of the population is Tanzania, resulting from a combination of its relatively small population (about 40 million), the relatively modest difference between the richest members of its population and the average, and its low wages. Per person, the richest tenth of the Tanzanian population earn little more than 1% of what their counterparts earn in Luxembourg.

The United States has the highest total earnings for the richest tenth in any territory, followed by China, then India. Together, the richest tenth in the United States earn 1.5 times as much as their counterparts in China, despite numbering less than a quarter as many persons, and over 4 times as much as their counterparts anywhere else. Among African nations, South Africa stands out: the richest tenth there earn nearly 3 times as much as those in any other African territory.

EARNINGS OF RICHEST TENTH
average annual per person earnings, PPP US$

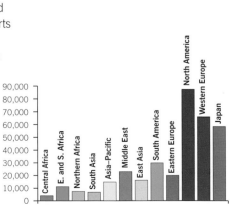

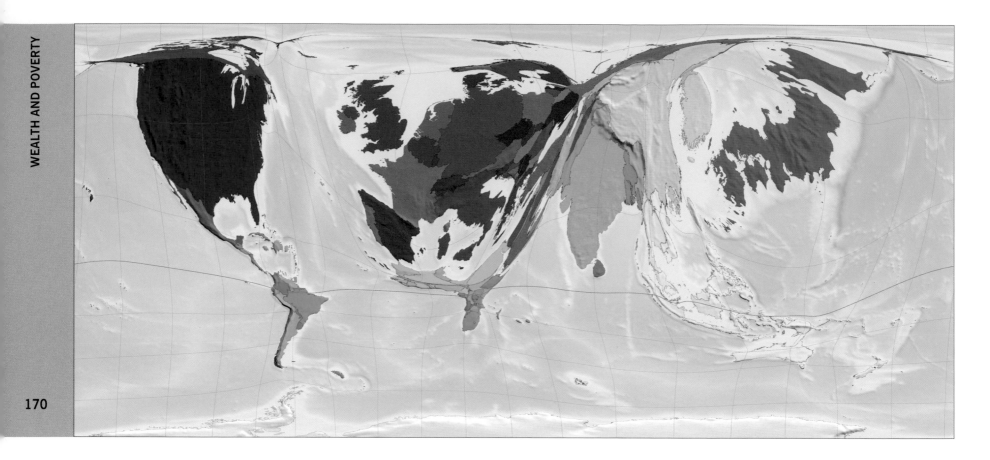

ª **average annual individual earnings of poorest fifth of national population, 2002**

151 Earnings of the Poorest Fifth of the Population

The size of each territory indicates the total income of the poorest 20% of its population, measured in terms of local purchasing power.

As a region, Japan has the richest poor people in the world. Central Africa, East and Southern Africa, and Northern Africa have the poorest. The poorest fifth of the population of South America have especially low incomes in relation to average incomes there.

EARNINGS OF POOREST FIFTH
average annual per person earnings, PPP US$

'Normally speaking, it may be said that the forces of a capitalist society, if left unchecked, tend to make the rich richer and the poor poorer and thus increase the gap between them.' Jawaharlal Nehru, statesman and prime minister of India 1947–64, 1960

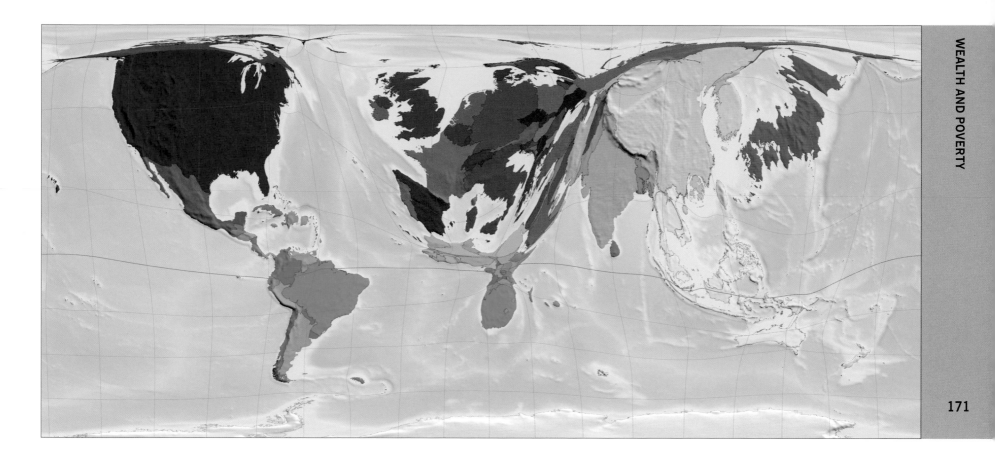

152 Earnings of the Richest Fifth of the Population

The size of each territory indicates the total income of the richest 20% of its population, measured in terms of local purchasing power.

As a region, Central Africa has the lowest average earnings per person among its rich, with the notable exception of Equatorial Guinea: this country has the world's fourth highest per-person earnings among the richest fifth of the population, principally because of income derived from oil. Equatorial Guinea is nonetheless quite small on the map because its population is tiny, just 500,000. In nearby Burundi and the Democratic Republic of Congo, by contrast, the richest fifth of the population are poorer than their counterparts in most of the rest of the world. The regions where the richest fifth make the most money are North America, Western Europe and Japan.

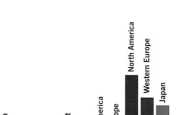

EARNINGS OF RICHEST FIFTH
average annual per person earnings, PPP US$

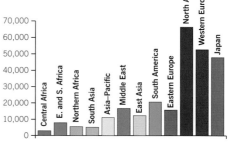

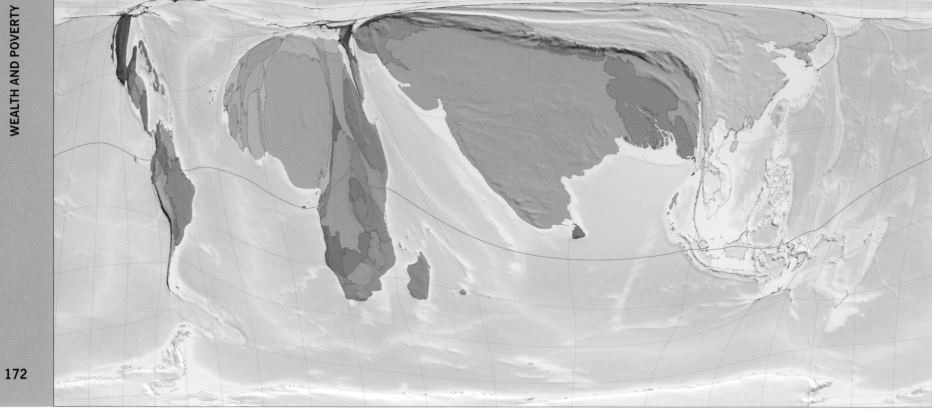

Rank	Territory	%ª
1	Mali	73
2	Nigeria	70
3	Central African Republic	67
4	Zambia	64
5	Niger	61
6	Gambia	59
7	Burundi	58
8	Sierra Leone	57
9	Madagascar	49
10	Nicaragua	45
11	Burkina Faso	45
12	Ghana	45
13	Malawi	42
14	Mozambique	38
15	Nepal	38
16	Lesotho	36
17	Bangladesh	36
17	Zimbabwe	36
19	Rwanda	36
20	Namibia	35

ª % of population living on less than US$1
a day, 2002 (no data for several Central
African countries)

153 Absolute Poverty (People Living on US$1 a Day)

The size of each territory shows the number of people living on US$1 a day or less, adjusted for local purchasing power. This is not enough for anyone to live on; yet many have to try.

The people represented on this map are those who live in absolute poverty.

In 2000 the United Nations laid out its Millennium Development Goals, a set of eight goals agreed upon by member states with the aim of improving the human condition in the new century. The first of them calls on the countries of the world to halve the proportion of people in the world with incomes of US$1 a day or less, or the local equivalent, by 2015.

In 2002 an estimated 17% of the world population had incomes of US$1 a day or less. In over 20 territories more than a third of the population earned this little. All but two of those territories are in Africa. Regionally, however, the largest numbers living on US$1 a day are in South Asia, most of them in India.

'The mass of the people struggle against the same poverty, flounder about making the same gestures...It is an underdeveloped world, a world inhuman in its poverty...' Frantz Fanon, *The Wretched of the Earth*, 1961

LIVING ON $1 A DAY OR LESS
millions of people

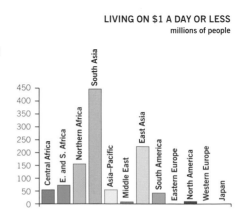

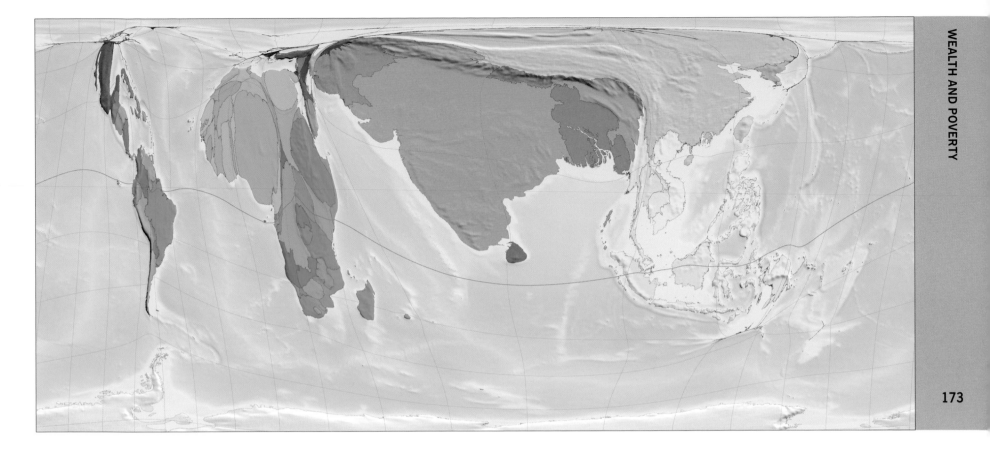

154 Abject Poverty (People Living on US$2 a Day)

The size of each territory shows the number of people living on US$2 a day or less, adjusted for local purchasing power: barely enough to survive, let alone thrive.

In 2002, 43% of the world population lived on US$2 a day or less. This figure includes all those shown in Map 153 who live on US$1 or less, plus those who earn between US$1 and US$2. This money has to cover the basics of food, shelter and water. With this amount to spend, medicines, new clothing and school books cannot be on a household's shopping list.

Where almost the entire population of a territory lives on incomes at this level, it is unsurprising if undernourishment is high, the level of education low and life expectancy short. In both Nigeria and Mali, 9 out of every 10 people subsist on less than US$2 a day.

LIVING ON US$2 A DAY OR LESS
millions of people

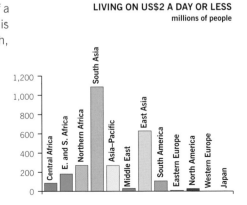

'Trickle-down theory – the less than elegant metaphor that if one feeds the horse enough oats, some will pass through to the road for the sparrows.' John Kenneth Galbraith, economist, undated

MOST AND FEWEST PEOPLE LIVING ON UP TO PPP US$10 A DAY

Rank	Territory	%ᵃ
1	Ethiopia	100
2	Tanzania	100
3	Burundi	100
4	Yemen	99
5	Malawi	99
6	DR Congo	99
7	Rwanda	99
8	Tajikistan	99
9	Guinea-Bissau	98
10	Madagascar	98
191	Czech Republic	0.072
192	Austria	0.039
193	Germany	0.019
194	Finland	0.004
195	Belgium	0.003
196	Sweden	0.001
197	Denmark	0.001
198	Japan	<0.001
199	Norway	<0.001
200	Luxembourg	<0.001

ᵃ % of population earning up to PPP US$10, 2002

155 Number of People Living on US$10 a Day or Less

The size of each territory indicates the number of people living on US$10 a day or less, adjusted for local purchasing power.

US$10 buys more in, say, Indonesia than it does in the United States, so earnings in US dollars alone do not allow us to make realistic comparisons between incomes in different places. This is why figures are adjusted for local purchasing power, using a principle called 'purchasing power parity' (PPP). When we say someone earns PPP US$10 in Indonesia we mean a sum that can buy the equivalent in that country of what US$10 would buy in the United States.

In 7 out of the 12 world regions more than half of the population live in households with income of less than PPP US$10 a day. In Central Africa 95% of households earn this little. In Western Europe and Japan less than 1% do.

LIVING ON UP TO US$10 A DAY
% of population

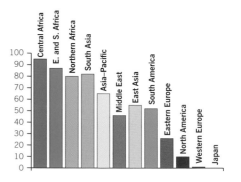

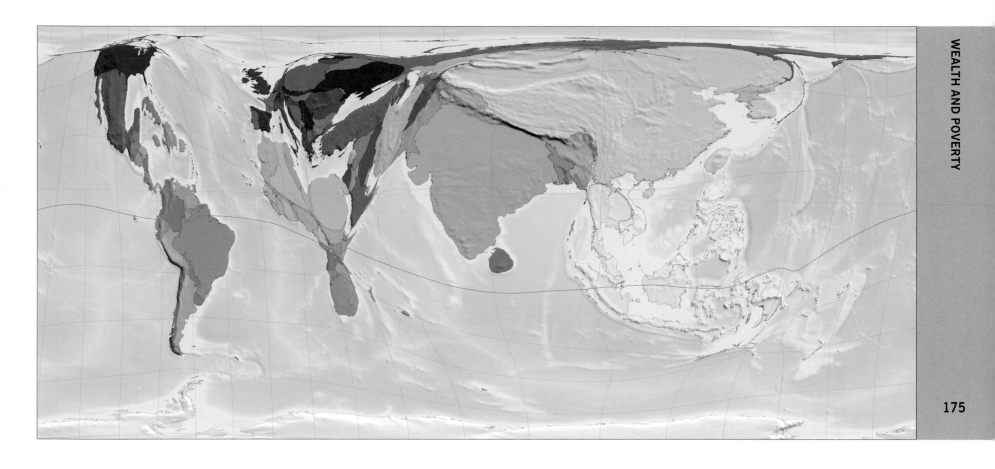

MOST AND FEWEST PEOPLE LIVING ON PPP US$10–20 A DAY

Rank	Territory	%ᵃ
1	Bosnia Herzegovina	55
2	Albania	48
3	Macedonia, FYR of	48
4	Ukraine	46
5	Belarus	45
6	Romania	45
7	Kazakhstan	44
8	Bulgaria	40
9	Maldives	40
10	Algeria	38
191	Yemen	0.51
192	Belgium	0.47
193	Burundi	0.47
194	Sweden	0.39
195	Denmark	0.24
196	Tanzania	0.22
197	Japan	0.12
198	Ethiopia	0.10
199	Norway	0.05
200	Luxembourg	0.01

ᵃ **% of population earning PPP US$10–20, 2002**

156 Number of People Living on US$10–20 a Day

The size of each territory indicates the number of people living on US$10–20 a day, adjusted for local purchasing power. These people are the comparatively rich in poor countries, and comparatively poor in rich ones.

China leads this category, with 332 million people, or 26% of its population, living on incomes of US$10–20 a day, adjusted for purchasing power parity (PPP). Among the regions, the highest proportions of people living on PPP US$10–20 a day are found in Eastern Europe, the Middle East and Central Asia, and East Asia: 34% in Eastern Europe and 25% in the Middle East and Central Asia and in East Asia.

In both the poorest and richest regions low proportions of the population live on PPP US$10–20 a day, but for opposite reasons. In poor regions most people earn less; in rich ones most people earn more.

LIVING ON US$10–20 A DAY
% of population

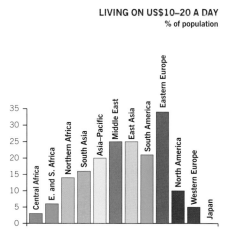

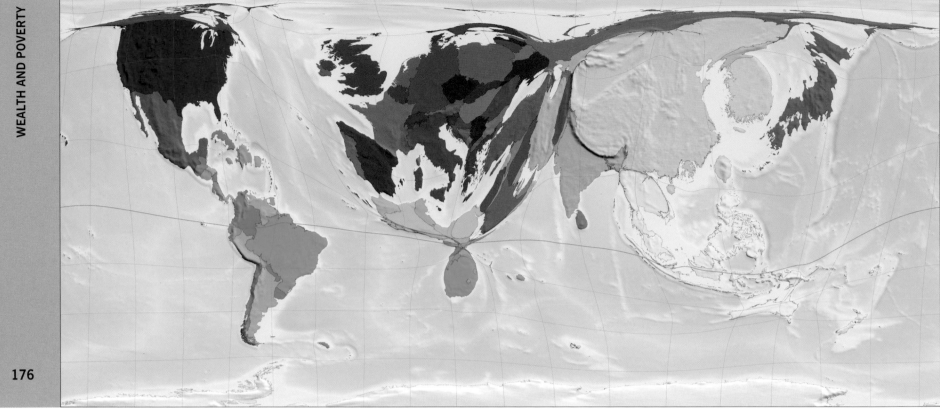

Rank	Territory	%ᵃ
1	Slovakia	67
2	Hungary	66
3	Czech Republic	65
4	Croatia	55
5	Slovenia	53
6	South Korea	52
7	Cyprus	51
8	Malta	50
9	Lithuania	50
10	Poland	49
191	Malawi	0.095
192	Kyrgyzstan	0.082
193	DR Congo	0.069
194	Uzbekistan	0.045
195	Tajikistan	0.042
196	Burundi	0.022
197	Yemen	0.014
198	Rwanda	0.014
199	Tanzania	0.009
200	Ethiopia	0.001

ᵃ **% of population earning PPP
US$20–50, 2002**

157 Number of People Living on US$20–50 a Day

The size of each territory indicates the number of people living on US$20–50
a day, adjusted for local purchasing power.

Once again China has the largest number
of people in this category: 192 million
living on incomes of US$20–50 a day,
adjusted for purchasing power parity.
Behind China comes the United States
(with 80 million) followed by Russia
(41 million).

As a percentage of population,
the highest numbers in this range are
found in Eastern and Western Europe.
However, an even higher percentage of
Eastern Europeans live on PPP US$10–20
a day, while a higher proportion of Western
Europeans have incomes between
PPP US$50 and US$100 a day.

LIVING ON US$20–50 A DAY
% of population

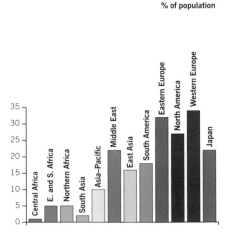

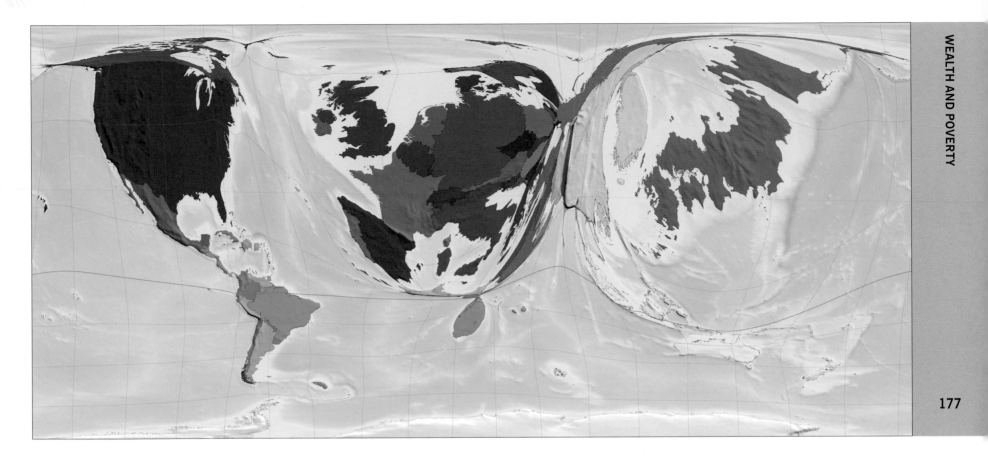

158 Number of People Living on US$50–100 a Day

The size of each territory indicates the number of people living on US$50–100 a day, adjusted for local purchasing power. These more prosperous people are concentrated in a few wealthier regions.

A majority (61%) of the Japanese population live in households with daily incomes between US$50 and US$100, adjusted for purchasing power parity. North America and Western Europe are also particularly large on this map, while South Asia and Central Africa have almost completely disappeared. Indonesia, Viet Nam, Cambodia, Laos, Myanmar, Timor-Leste and much of East and Southern Africa are entirely invisible because virtually no one in those territories earns an income in this range.

'It's as though the people of India have been rounded up and loaded onto two convoys of trucks…The tiny convoy is on its way to a glittering destination…The other convoy just melts into the darkness and disappears.' Arundhati Roy, author and campaigner, *The Nation*, 2002

LIVING ON US$50–100 A DAY
% of population

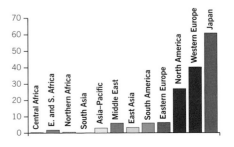

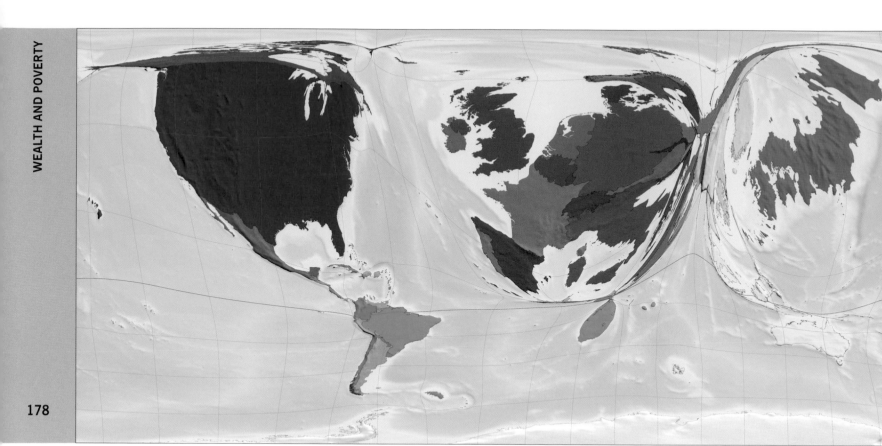

159 Number of People Living on US$100–200 a Day

The size of each territory indicates the number of people living on US$100–200 a day, adjusted for local purchasing power. Much of the world approaches invisibility on this scale.

In all regions except North America, Western Europe and Japan, the percentage of people living in households with incomes of US$100–200 a day, adjusted for purchasing power parity, is less than 2.1%. Within North America, Western Europe and Japan, on the other hand, 16–19% of people live on this amount.

In 75 of the 200 territories less than 1 in 1,000 people live on this much income, even after PPP adjustment (where a higher value is given to the currency in territories where it is cheaper to live).

LIVING ON US$100–200 A DAY
% of population

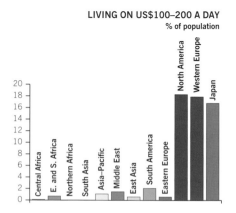

'Every man is rich or poor according to the degree in which he can afford to enjoy the necessaries, conveniences, and amusements of human life.' Adam Smith, *An Inquiry into…the Wealth of Nations*, 1776

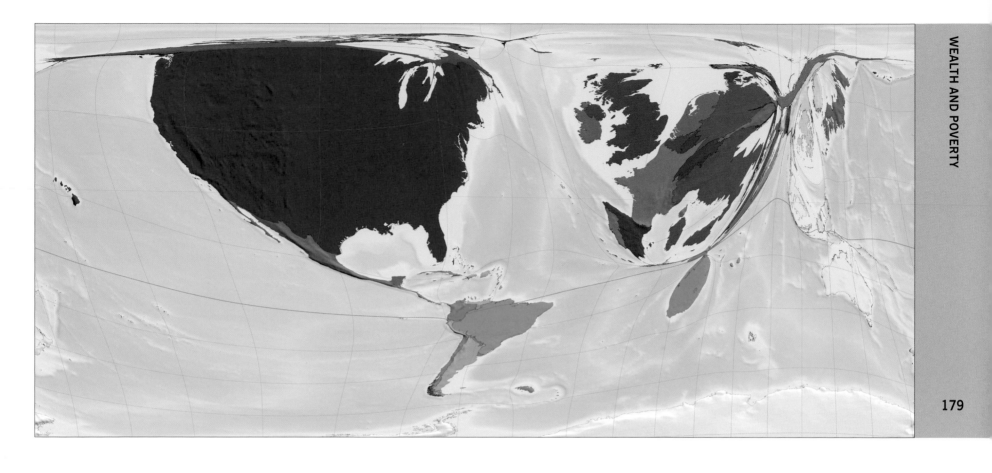

MOST PEOPLE LIVING ON OVER PPP US$200 A DAY

Rank	Territory	%ᵃ
1	Luxembourg	35
2	United States	11
3	Ireland	10
4	Greenland	8
5	Equatorial Guinea	7
6	Hong Kong (China)	6
7	Australia	5
8	Switzerland	5
9	Canada	5
10	Singapore	5
11	Norway	4
12	United Kingdom	4
13	Italy	4
14	Netherlands	4
15	United Arab Emirates	4
16	Iceland	4
17	France	3
18	St Kitts–Nevis	3
19	New Zealand	3
20	Qatar	3

ᵃ % of population earning over PPP US$200, 2002

160 Number of People Living on more than US$200 a Day

The size of each territory indicates the number of people living on more than US$200 a day, adjusted for local purchasing power. Over half of these high earners live in the United States.

In 2002, 53 million people in the world lived in households with incomes of more than US$200 a day, after adjusting for purchasing power parity. Of these 53 million, 58% lived in the United States. Western Europe and South America are also home to quite large populations of very high earners. Within Western Europe most of the high earners are in the United Kingdom, Italy and France. The highest earners of South America live primarily in Brazil and Argentina. Few very high earners live in South Asia, Northern Africa, Eastern Europe or Central Africa.

LIVING ON OVER US$200 A DAY
% of population

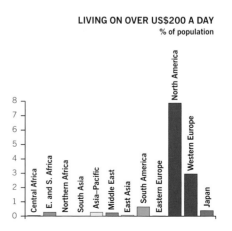

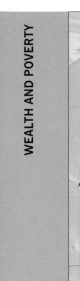

180

HIGHEST AND LOWEST EXCHANGE RATE BASED GDPs

Rank	Territory	US$ᵃ
1	Luxembourg	47,354
2	Norway	41,974
3	Switzerland	36,687
4	United States	36,006
5	Denmark	32,179
6	Japan	31,407
7	Ireland	30,982
8	Iceland	29,749
9	Qatar	28,634
10	Greenland	27,648
191	Tajikistan	193
192	Niger	190
193	Myanmar	185
194	Malawi	177
195	Sierra Leone	150
195	Eritrea	150
197	Guinea-Bissau	141
198	DR Congo	111
199	Burundi	102
200	Ethiopia	90

ᵃ **GDP per person, based on exchange rate conversions, 2002**

161 Unadjusted Wealth

The size of each territory indicates its GDP, not adjusted for local purchasing power. The differences in what money can buy have implications for international mobility.

This map shows raw figures for the GDP of each territory in 2002, in US dollars. The numbers have been converted from local currencies at prevailing exchange rates for the purposes of comparison, but no adjustments have been made to account for purchasing power – the variation in dollar cost of common commodities in different areas. For some, their money gains value when they go to another territory because the cost of goods is lower there; for others, it loses value because prices are higher. This difference in purchasing power facilitates the movement of some people, while severely limiting that of others.

Wealth, as reflected by GDP per person, is highest in Luxembourg, Norway and Switzerland. It is lowest in Ethiopia, Burundi and the Democratic Republic of Congo.

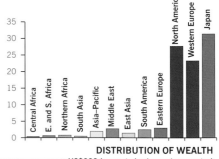

DISTRIBUTION OF WEALTH
GDP per person per year, US$000 (converted using exchange rates)

'...for the citizens of most countries today, the success of their economy in the harsh world of global competition is of paramount importance.' Deanne Julius, 'US economic power: waxing or waning?', 2005

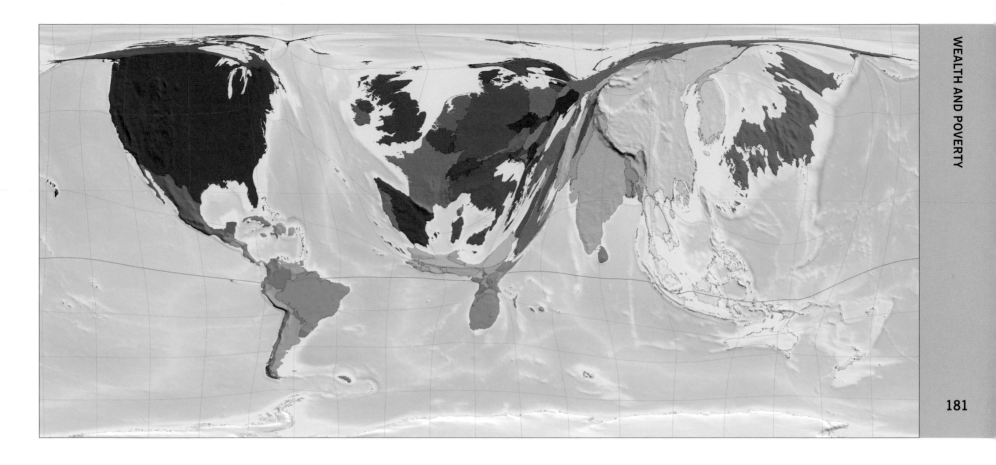

162 Adjusted Wealth

The size of each territory indicates its GDP, adjusted for local purchasing power. This adjustment does not even out the discrepancies in wealth across the world.

When differences in local purchasing power are taken into account, 46% of world wealth in 2002 was held in North America and Western Europe. The regions with the most purchasing power per person, by a wide margin, are North America, Japan and Western Europe. The proportion of world wealth found in many poorer areas is greater when measured in terms of local purchasing power than when measured in simple dollar terms, but even after this adjustment Central Africa has the lowest regional GDP per person, at only half that of the next lowest-scoring region.

DISTRIBUTION OF PURCHASING POWER
GDP per person per year, PPP US$000

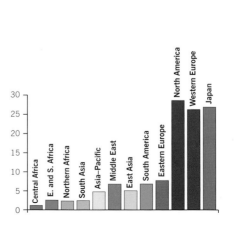

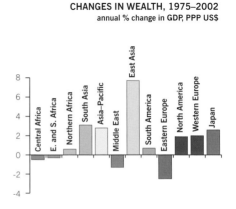

LARGEST INCREASES IN WEALTH

Rank	Territory	PPP US$[a]
1	Luxembourg	39,968
2	Equatorial Guinea	28,600
3	Ireland	24,991
4	Norway	19,235
5	Hong Kong (China)	18,496
6	Singapore	17,601
7	United States	14,805
8	South Korea	13,523
9	Japan	13,468
10	Cyprus	12,898
11	Austria	12,548
12	Malta	12,124
13	Netherlands	11,594
14	Australia	11,259
15	United Kingdom	11,230
16	Germany	11,223
17	Greenland	11,101
18	Belgium	10,984
19	Italy	10,946
20	Iceland	10,878

[a] rise in GDP per person 1975–2002

163 Growth in Wealth

The size of each territory shows the growth of GDP from 1975 to 2002, adjusted for local purchasing power. Two-thirds of the territories in the world experienced increases in their wealth over this period.

China saw the biggest increase in wealth between 1975 and 2002, followed by the United States, Japan, India and Germany. China and India do well mainly because of their very large populations: they experienced relatively small increases in GDP per person over the period in question, but those translate into large growth in total GDP when multiplied by the population size.

Unfortunately, most territories with small GDPs to begin with have experienced only small increases at best. Although distributions of wealth do change, most growth in wealth has occurred in places that are already relatively wealthy.

CHANGES IN WEALTH, 1975–2002
annual % change in GDP, PPP US$

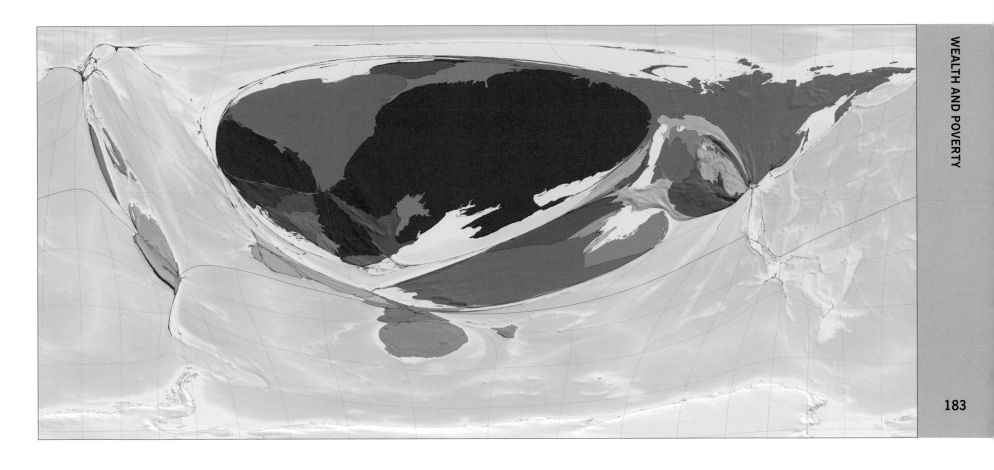

LARGEST DECLINES IN WEALTH

Rank	Territory	PPP US$ [a]
1	Ukraine	25,903
2	United Arab Emirates	25,847
3	Slovenia	17,826
4	Czech Republic	15,172
5	Saudi Arabia	12,409
6	Poland	10,153
7	Turkmenistan	10,073
8	Lithuania	9,922
9	Croatia	9,845
10	Tajikistan	8,608
11	Qatar	8,477
12	Georgia	7,296
13	Kuwait	6,258
14	Macedonia, FYR of	6,221
15	Bosnia Herzegovina	5,740
16	Belarus	5,307
17	Moldova	5,110
18	Djibouti	5,107
19	Russia	3,516
20	Gabon	3,321

[a] fall in GDP per person 1975–2002

164 Decline in Wealth

The size of each territory shows the decline in GDP from 1975 to 2002, adjusted for local purchasing power. Over half of the territories that became poorer over this period were in Eastern Europe.

Just under one-third of all territories experienced a decline in GDP, adjusted for purchasing power parity, between 1975 and 2002. More than half of the territories experiencing a decline were in Eastern Europe; almost all of the remainder were in the Middle East and Central Asia. Ukraine, Russia, Poland and Saudi Arabia experienced the largest declines, with Ukraine registering a decline more than twice that undergone by any other territory.

In some regions no territories have experienced a decline in GDP. These regions are South Asia, East Asia, Asia–Pacific and Australasia, North America, Western Europe and Japan.

CHANGES IN WORLD WEALTH 1975–2002
overall % change in GDP, PPP US$

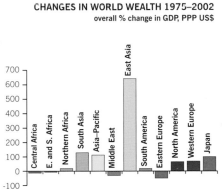

165 National Income

The size of each territory indicates its GNI in US dollars, showing the importance of foreign earnings in national wealth.

Gross national income (GNI) is the total income received in return for goods and services produced in a territory, plus income from abroad such as that received by domestic companies operating elsewhere or from foreign investments. It is the inclusion of income from abroad that distinguishes GNI from GDP. Large multinational companies can earn tremendous amounts of money, which are not properly accounted for if one looks only at GDP (Map 161). The GNI registers this income, attributing it to the home territory of the company in question.

GNI per person is highest in Western Europe, North America and Japan, and lowest in Burundi and Ethiopia. Of the total worldwide GNI, the United States receives 33%, Western Europe 28% and Japan 13%.

'[The gross national product] does not include the beauty of our poetry or the strength of our marriages; the intelligence of our public debate or the integrity of our public officials...it measures everything...except that which makes life worthwhile.' Robert Francis Kennedy, University of Kansas, 1968

GROSS NATIONAL INCOME
US$ billion, 2003

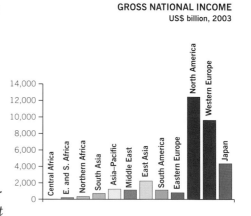

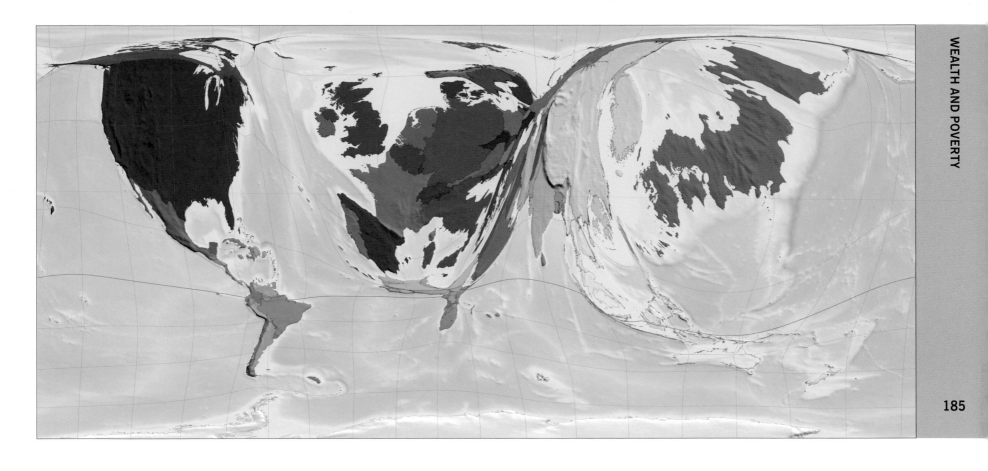

166 National Savings

The size of each territory indicates its gross national savings in US dollars, showing how much it has at its disposal to invest in its own future.

Gross national savings is the label given to the difference between gross national income and (public and private) expenditure. In effect it represents how much more money a country earns than it spends – money that is typically then reinvested to help sustain and expand the country's economy and infrastructure.

The territories with the largest national savings are the United States, Japan, China and Germany. The United States' savings were 10 times those of India, 100 times those of Taiwan, 1,000 times those of Azerbaijan, 10,000 times those of Benin and 100,000 times those of Sierra Leone.

'...poverty in the world is an artificial creation. It doesn't belong to human civilization, and we can change...the only thing we have to do is to redesign our institutions and policies...'

Muhammad Yunus, winner of the Nobel Peace Prize, 2006

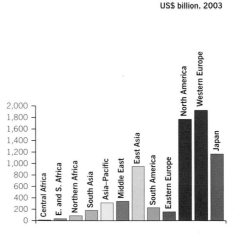

GROSS NATIONAL SAVINGS
US$ billion, 2003

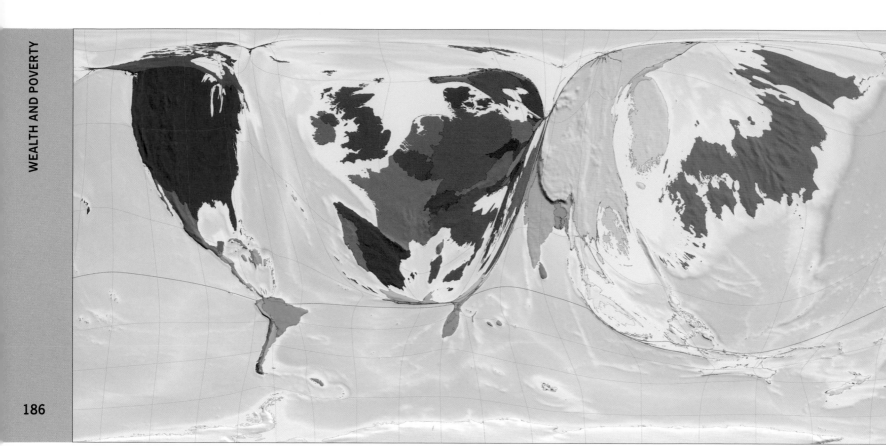

167 Adjusted Savings Rate

The size of each territory indicates the savings rate, adjusted to reflect the estimated value of non-monetary social and environmental investments.

Map 166 shows the annual difference between the amount of money a territory earns and the amount it spends: money that is normally invested to help sustain and expand the economy and infrastructure.

However, there are other types of investments than purely monetary ones – education, for example. There are also 'negative' investments, that is, actions now that will incur costs later, such as depletion of fossil fuels, minerals and forests, carbon and particulate emissions, and wear and tear on buildings and infrastructure. This map shows the World Bank's adjusted savings measure, which includes an estimate of the monetary equivalent of such social and environmental investments, both positive and negative. Territories with a negative savings rate have size zero on this map and so do not appear.

In 2003 there were high adjusted savings in Japan, the United States and China. These territories accounted for 48% of adjusted savings. Territories in the Middle East and Central Asia, and in Africa, by contrast, are barely visible on this map because they have low adjusted savings.

NET ADJUSTED SAVINGS
US$ thousand, per person, 2003

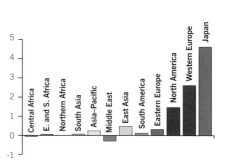

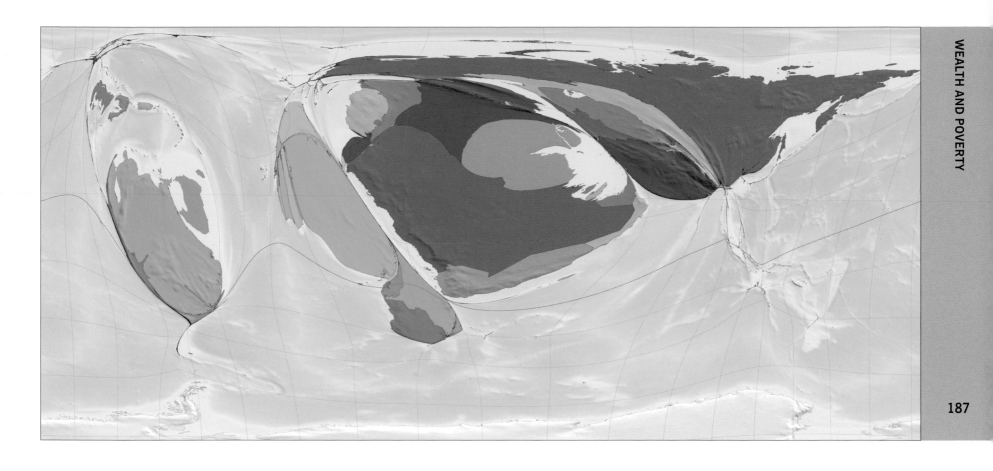

168 Negative Adjusted Savings

The size of each territory indicates negative savings, adjusted for environmental and social benefits and costs. Negative savings may take the form of financial debt or depletion of natural resources.

This map is based on the same data as Map 167, but shows those territories with negative adjusted savings rates. Territories with a positive savings rate have size zero on this map and so do not appear.

A negative adjusted savings rate suggests a territory that is on an unsustainable path. A negative savings rate may be reflected in growing debt, although it is also possible for a territory with negative savings to appear financially healthy because its losses are mostly in the form of, for example, depletion of natural resources. Thus, 75% of negative savings in 2003 were in territories in the Middle East and Central Asia, where fossil fuel reserves are rapidly becoming depleted. The largest negative savings rates were in Russia, Saudi Arabia and Venezuela.

'Many people no longer view debt as a repayable, finite concept but as a permanent condition akin to indentured servitude.'

Lezak Shallat, *New Internationalist*, 2006

NET POSITIVE AND NEGATIVE ADJUSTED SAVINGS
US$ billion, 2003

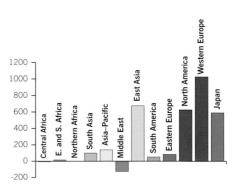

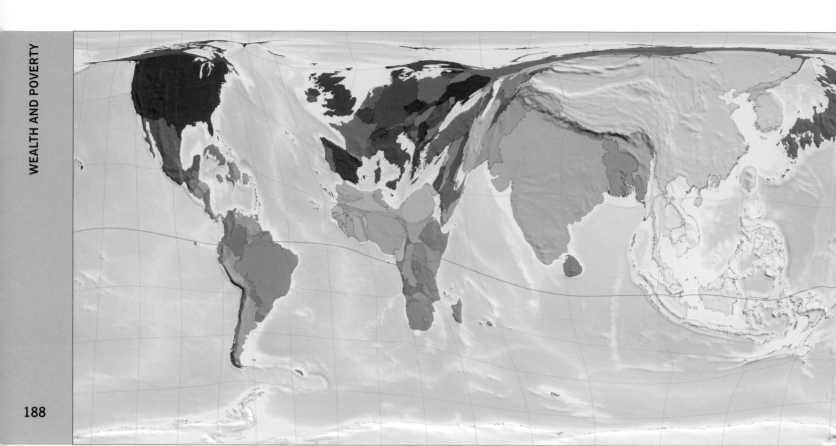

169 Human Development

The size of each territory shows the overall level of human development, quantified as the population multiplied by the UN's Human Development Index.

The Human Development Index is a measure used by the UNDP to assess quality of life in different areas. It combines measures of health, wealth and education to give an overall score to a territory representing the quality of life there. The highest possible score is 1,000, which corresponds to an average life expectancy of 85 years or more, 100% adult literacy,

100% school enrolment and a GDP over US$40,000 per person per year. Norway has the highest Human Development Index score at 956; Sierra Leone has the lowest, at 273. The world average is 698.

Territories appear large on this map if a large number of people who live there have a high quality of life, as measured by the index.

HUMAN DEVELOPMENT
Human Development Index score, on a scale of 0–1000, 1000 is optimum

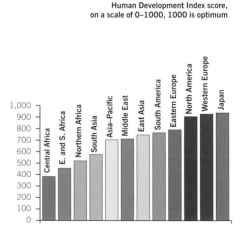

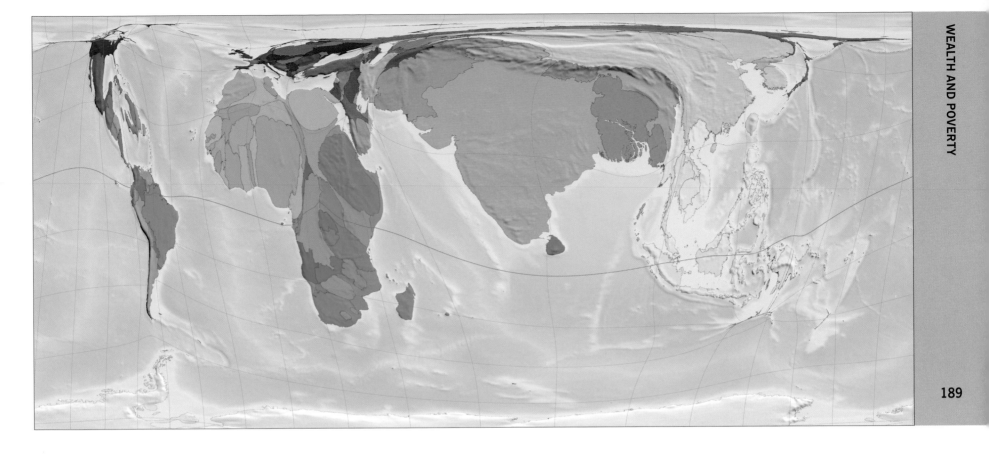

MOST AND FEWEST PEOPLE LIVING IN POVERTY

Rank	Territory	%[a]
1	Burkina Faso	66
2	Niger	61
3	Mali	59
4	Ethiopia	56
5	Zimbabwe	52
6	Zambia	50
7	Mozambique	50
8	Chad	50
9	Mauritania	48
10	Guinea-Bissau	48
191	Japan	1.1
192	Spain	1.1
193	France	1.1
194	Luxembourg	1.1
195	Germany	1.0
196	Denmark	0.9
197	Finland	0.8
198	Netherlands	0.8
199	Norway	0.7
200	Sweden	0.7

[a] % of population living in poverty, 2002, calculated using the Human Poverty Indices 1 and 2

170 Human Poverty

The size of each territory shows the overall level of poverty, quantified as the population of the territory multiplied by the Human Poverty Index.

The Human Poverty Index is used by the UNDP to measure the level of poverty in different territories. Poverty is difficult to assess using financial data alone: many factors affect standards of living, and people who have similar incomes but live in different areas may have very different experiences. The Human Poverty Index attempts to capture these differences by incorporating non-financial measures such as life expectancy and adult literacy as well as financial ones like household income and unemployment rates. There are two indices, one for poorer countries, one for richer. The result is a single overall score for the poverty level in a territory. The score is high for territories where there is great poverty and low where there is little. The highest Human Poverty Index scores are in Central Africa. The lowest are in Japan.

Territories appear large on this map where there are many people living in poverty, as measured by the index, and small where there are few.

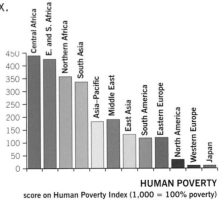

HUMAN POVERTY
score on Human Poverty Index (1,000 = 100% poverty)

'My field experience of the complexity and variety of country situations made me chary of stylised generalisations about "the third world".' Angus Maddison, specialist in historical demography and economy, 2002

LARGEST INCREASES IN HUMAN DEVELOPMENT

Rank	Territory	HDI[a]
1	Libya	417
2	Maldives	352
3	Cape Verde	340
4	Seychelles	324
5	Brunei	310
6	Equatorial Guinea	307
7	Oman	277
8	Slovenia	267
9	Bahrain	259
10	Mauritius	256
11	Cyprus	255
12	Gabon	252
13	Sao Tome and Principe	249
14	Qatar	249
15	Czech Republic	240
16	Yemen	239
17	Tonga	230
18	Tunisia	229
19	Indonesia	225
20	Estonia	225

[a] **improvement in Human Development Index (HDI) points 1975–2002, on a range 0–1,000**

171 Increase in Human Development

The size of each territory shows how far human development increased overall between 1975 and 2002, quantified using the Human Development Index.

This map shows increases in the level of human development, quantified using the same index employed on Map 169. Territories in which the level of human development decreased have size zero on this map and so do not appear, but the vast majority of territories did show increases. Since the Human Development Index combines a variety of different measures,

however, it is possible for a country to show an overall increase by improving on some measures while worsening on others.

The biggest total improvement to people's lives has been in China and India, where even a small improvement in the average quality of life can affect a very large number of people – up to 2.3 billion in these two territories alone.

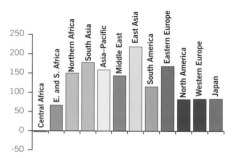

CHANGES IN HUMAN DEVELOPMENT SCORE
points change in Human Development Index 1975–2002, on a scale of 0–1000

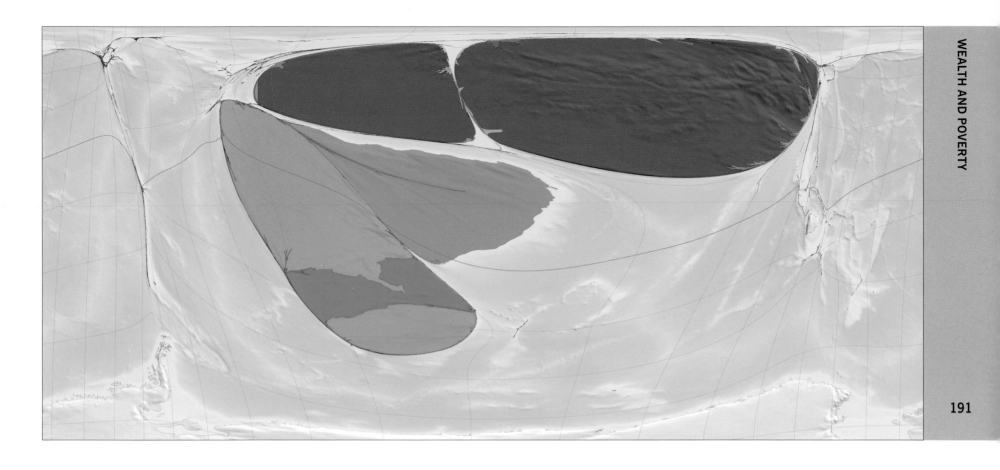

LARGEST DECREASES IN HUMAN DEVELOPMENT

Rank	Territory	HDI[a]
1	Afghanistan	0·238
2	Somalia	0·235
3	Iraq	0·086
4	Zambia	0·077
5	Zimbabwe	0·056
6	DR Congo	0·045
7	Timor-Leste	0·032

[a] reduction in Human Development Index (HDI) points 1975–2002, measured using a range of 0 being low, and 1 being high

172 Decrease in Human Development

The size of each territory shows how far human development decreased between 1975 and 2002. Overall, few territories experienced a decline in quality of life, but those few have suffered gravely.

Human Development Index scores fell in just seven territories between 1975 and 2002. These were Afghanistan, Somalia, Iraq, Zambia, Zimbabwe, the Democratic Republic of Congo and Timor-Leste (East Timor). Among them, these territories have in the years since 1975 experienced civil and international wars, coups, droughts, disease, economic failure and heavy international debt. The result has been decreases in life expectancy, adult literacy, school enrolment and GDP, the indicators of human development. Central Africa is the only region to have experienced an overall decline in human development, measured according to the index.

'For the next couple of years, the words economic development and reform danced on a lot of lips in Kabul. ...For a while, a sense of rejuvenation and purpose swept across the land.'

Khaled Hosseini, *The Kite Runner*, 2003

MOST UNDERNOURISHED POPULATIONS IN 1990

Rank	Territory	%ª
1	Mozambique	69
2	Haiti	65
3	Angola	61
4	Chad	58
5	Central African Republic	50
6	Burundi	49
6	Malawi	49
8	Sierra Leone	46
9	Zambia	45
10	Kenya	44
11	Rwanda	43
11	Zimbabwe	43
11	Cambodia	43
14	Niger	42
15	Peru	40
15	Guinea	40
15	Sao Tome and Principe	40
15	Equatorial Guinea	40
19	Congo	37
20	5 countries	35

173 Undernourishment in 1990

The size of each territory shows the number of undernourished people in 1990. In at least three-quarters of the world's territories there were some people who did not have enough to eat.

A person will be undernourished if they do not receive the minimum level of dietary energy consumption required to undertake basic daily tasks for a prolonged period of time – in simple terms, if they do not have enough to eat. In practice, undernourishment is defined in terms of a minimum healthy body weight, which varies by height, age and sex.

840 million people, which is 16% of the world population, were undernourished in 1990. The highest number of undernourished people lived in India and China. The highest *proportion* of undernourished people was in Mozambique, where 69% of the population did not have enough to eat in 1990. Almost a quarter of all territories recorded no undernourishment at all in 1990. The regions of Japan, Eastern Europe and Western Europe are barely visible on the map because collectively they recorded very little undernourishment.

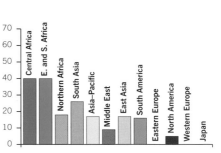

UNDERNOURISHMENT IN 1990
% of population

'Should we really let our people starve so we can pay our debts?'

Julius Nyerere, president of Tanganyika (1962–4) and of Tanzania (1964–85), 1985

MOST UNDERNOURISHED POPULATIONS IN 2000

Rank	Territory	%ᵃ
1	DR Congo	75
2	Tajikistan	71
3	Burundi	70
4	Eritrea	61
5	Mozambique	53
6	Armenia	51
7	Sierra Leone	50
7	Zambia	50
9	Angola	49
9	Haiti	49
11	Central African Republic	44
12	Tanzania	43
13	Ethiopia	42
14	Rwanda	41
15	Zimbabwe	39
16	Cambodia	38
16	Mongolia	38
18	Kenya	37
19	Madagascar	36
20	Chad	34

ᵃ % of population undernourished, 2000

174 Undernourishment in 2000

The size of each territory shows the number of undernourished people in 2000. The increase over the previous decade is masked by the greater increase in population overall.

Over the decade from 1990 to 2000, the number of undernourished people in the world increased from 840 million to 858 million. However, the total population of the world increased substantially faster over the same period so that the percentage of undernourished in the world actually decreased, from 16% to 14%.

Over 60% of all people currently living in Central Africa are undernourished. The Democratic Republic of Congo has the highest levels: 3 in every 4 people in the DRC are undernourished.

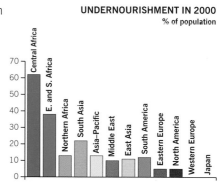

UNDERNOURISHMENT IN 2000
% of population

'The number of hungry people remains intolerably high, progress in reaching them unconscionably slow and the costs in ruined lives and wasted resources incalculably large.' Lynn Brown, FAO, 2004

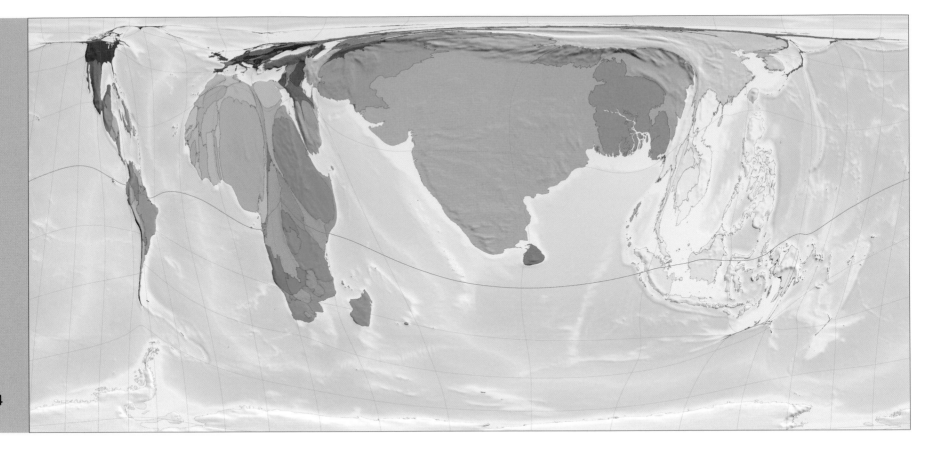

UNDERWEIGHT CHILDREN

Rank	Territory	%ᵃ
1	Bangladesh	48
2	Nepal	48
3	India	47
3	Ethiopia	47
5	Yemen	46
6	Burundi	45
6	Cambodia	45
8	Eritrea	44
9	Timor-Leste	43
10	Niger	40
11	Laos	40
12	Pakistan	38
13	Nigeria	36
14	Myanmar	35
14	Papua New Guinea	35
16	Burkina Faso	34
17	Mali	33
18	Madagascar	33
19	Viet Nam	33
20	Mauritania	32

ᵃ % of underweight children aged
0–14 years

175 Underweight Children

The size of each territory indicates the number of those aged 15 and under who
are significantly below the average weight for a child of a given age in a given area.

Typical healthy weights vary between
populations, so what counts as
underweight varies from one territory to
another. Technically, children are defined
as being underweight when they are more
than two standard deviations below the
median weight.

There are some underweight children
in all territories, although the percentage
can be as low as 1%, as it is in Chile and
Japan. At the other end of the scale,

in Bangladesh, Nepal and India almost
half of all children under 5 years of age
are underweight. East and Southern Africa,
Northern Africa, East Asia, and the
Asia–Pacific and Australasia region are
also home to relatively large numbers of
underweight children. Within these regions
the territories with the largest populations
of underweight children are Ethiopia,
Nigeria, China and Indonesia.

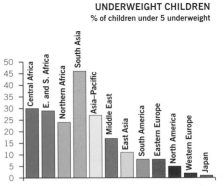

UNDERWEIGHT CHILDREN
% of children under 5 underweight

*'Poor nutrition is implicated in more than half of all child deaths
worldwide – a proportion unmatched by any infectious disease
since the Black Death.'* Jean-Louis Sarbib, World Bank, 2006

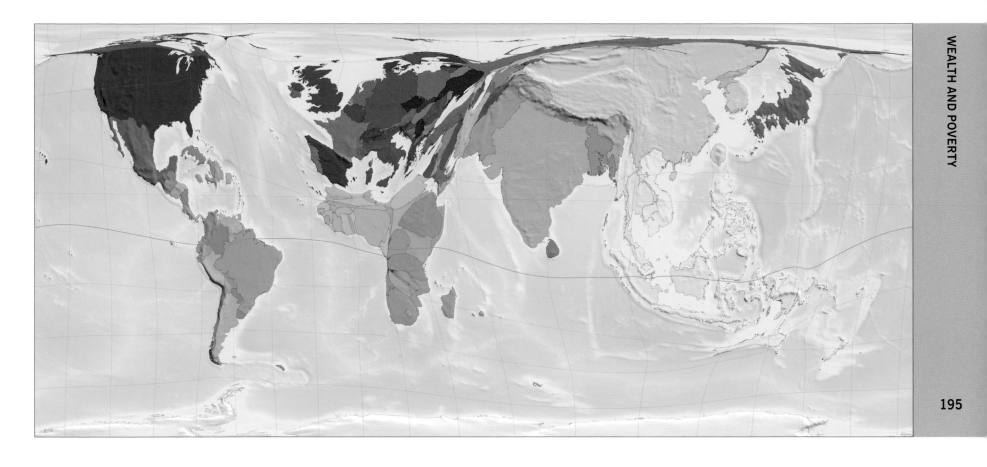

HIGHEST LEVELS OF FEMALE EMPOWERMENT

Rank	Territory	GEM[a]
1	Norway	0.91
2	Sweden	0.85
3	Denmark	0.85
4	Finland	0.82
5	Netherlands	0.82
6	Iceland	0.82
7	Belgium	0.81
8	Australia	0.81
9	Germany	0.80
10	Canada	0.79
11	New Zealand	0.77
12	Switzerland	0.77
13	Austria	0.77
14	United States	0.77
15	Greenland	0.72
16	Spain	0.72
17	Ireland	0.71
18	Bahamas	0.70
19	United Kingdom	0.70
20	Costa Rica	0.66

[a] Gender Empowerment Measure, 2002, on a scale from 0 to 1

176 Equality of the Sexes

The size of each territory shows gender equality, quantified as the population multiplied by the Gender Empowerment Measure. By this measure, no territory offers women opportunities as good as those it offers to men.

The Gender Empowerment Measure is an index used by the UNDP to measure the level of equality in a territory between opportunities for women and opportunities for men. It combines a number of individual indicators, including parliamentary representation by females and males; the respective proportions of women and men serving as legislators, senior officials, managers, and professional and technical employees; and the ratio of female to male earnings.

This map shows the population of each territory multiplied by the Gender Empowerment Measure, so that territories appear large where there are many women with good opportunities.

The territories where women have the most opportunities are in Western Europe. Those where they have fewest are the Middle Eastern territories of Yemen and Saudi Arabia. No data were available, however, for any territory in Central Africa.

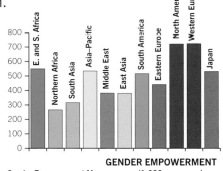

GENDER EMPOWERMENT
Gender Empowerment Measure score (1,000 = women have equal opportunities with men)

'In many countries, women own nothing, inherit nothing and earn nothing. Three out of four of the poorest billion people of the world are women.' Gro Harlem Brundtland, director general of the WHO, 2006

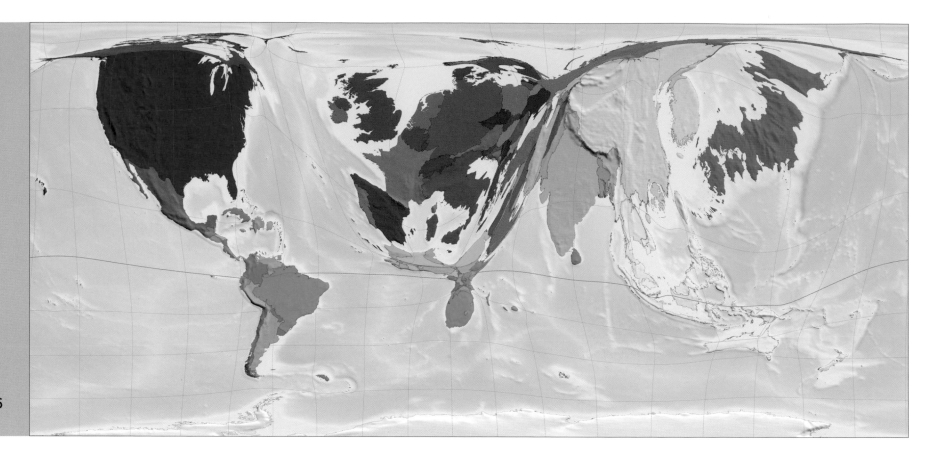

[a] average annual individual earnings of all men (including male children and some non-working adults), 2002

177 Males' Income

The size of each territory indicates the total income of male members of the population, measured in terms of local purchasing power. The total income of all males in the world is more than double that of all females.

Males in Central Africa have the lowest average earnings, with annual incomes only 3.7% of those of males in North America and Western Europe – and the inequality would be far greater still if income were measured in simple dollar terms rather than being adjusted for local purchasing power.

The greatest income inequality between males and females is in South Asia where, on average, males earn five times as much as females. The neighbouring region of East Asia has the smallest difference between males' and females' income, though males there still earn twice what females earn.

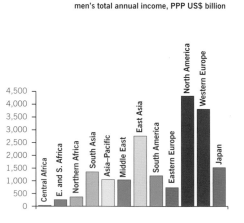

MEN'S INCOME DISTRIBUTION
men's total annual income, PPP US$ billion

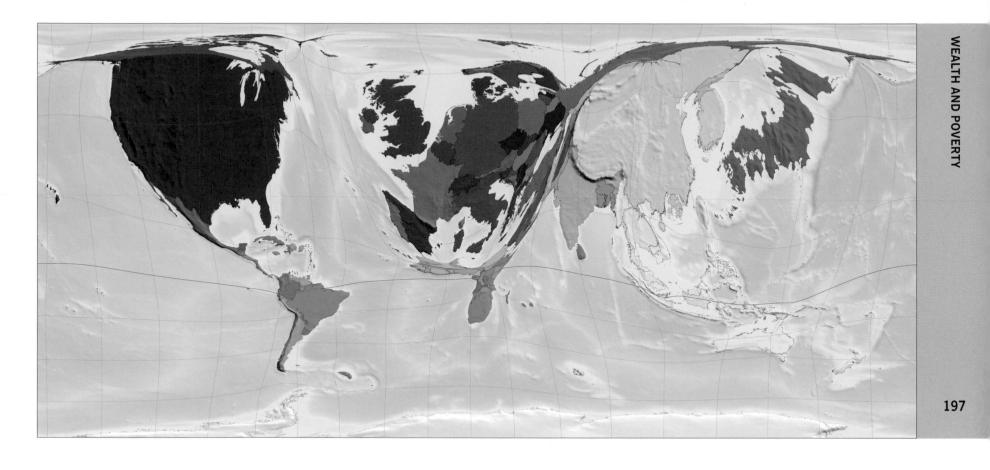

Females' Income

The size of each territory here indicates the total income of female members of the population, measured in terms of local purchasing power. In every region females earn less than males do.

Total income of females is highest in the United States and second highest in China. Females in the United States have high total income because on average they are some of the highest-paid in the world. Females in China earn below the world average; they have high total income

simply because China has such a large population.

The highest female income per person is found in Norway and Denmark, and the lowest in Yemen and Sierra Leone, where women earn 150 to 250 times less than in Norway.

WOMEN WITH THE LARGEST AND SMALLEST EARNINGS

Rank	Territory	PPP US$ [a]
1	Norway	15,082
2	Denmark	13,133
3	United States	12,710
4	Sweden	12,242
5	Iceland	11,667
6	Canada	11,337
7	Australia	10,654
8	Luxembourg	10,371
9	Finland	10,124
10	Greenland	9,870
191	Zambia	196
192	Pakistan	194
193	Sudan	185
194	Malawi	178
195	Ethiopia	160
196	DR Congo	150
197	Nigeria	149
198	Guinea-Bissau	140
199	Sierra Leone	85
200	Yemen	61

[a] average annual individual earnings of all women (including female children and some non-working adults), 2002

WOMEN'S INCOME DISTRIBUTION
women's total annual income, PPP US$ billion

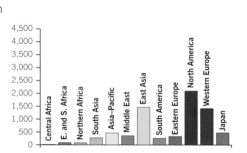

'In a growing number of marriages, it's the woman who is bringing home the big paycheck. Is she stressed? Yes. Resentful? A little. Would she trade places with her husband? Not on your life.'

Kimberly Goad, 'Big-earning wives (and the men who love them)', 2006

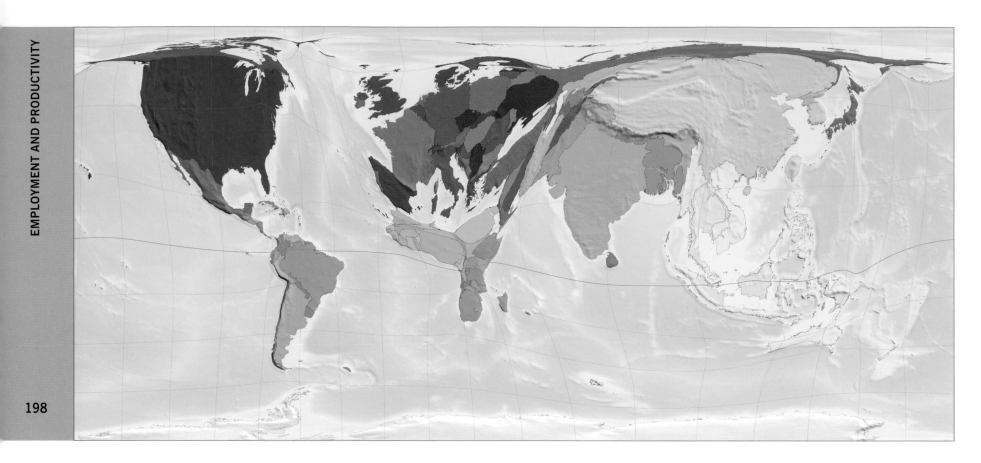

179 Cereals Production

Cereals are the essential staple foods that support all societies. The size of each territory in this map indicates the amount of cereal produced per year, in tonnes.

Worldwide, a total of 2 billion tonnes of cereal were produced in 2002 (the year shown on this map). This includes both cereals that are consumed by the producing territory and those that are exported. As a region, East Asia produces the most cereals, but North America produces the most per person. North American cereals production per person is 20 times that of Central Africa.

All territories produce some cereals. The territories with the lowest production of cereals tend to be islands. Of the 10 territories with the lowest per person cereal production, 8 are islands.

'For about a third of the world's population, rice equals life. The cereal provides more than half the daily calories that these people consume.'

John Travis, 'Hot cereal: rice reveals bumper crop of genes', 2002

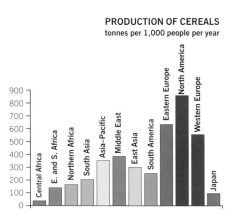

PRODUCTION OF CEREALS
tonnes per 1,000 people per year

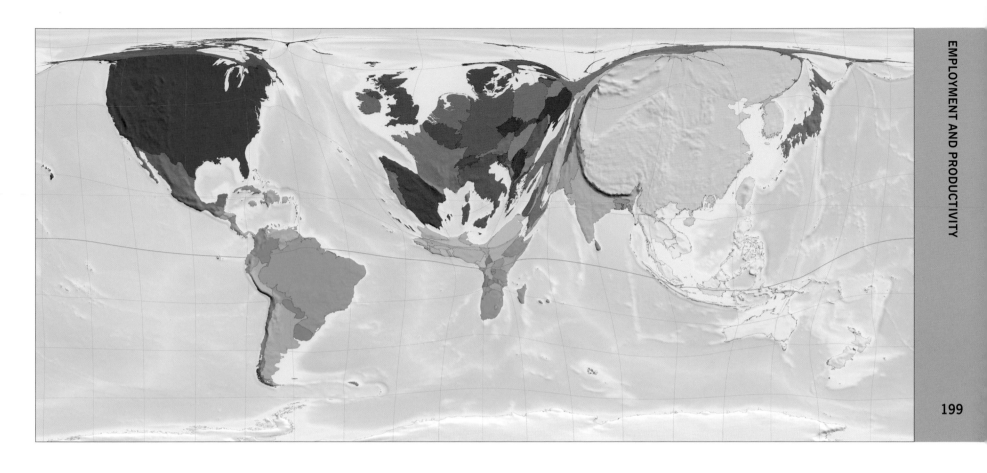

180 Meat Production

Just three territories produce half of all the meat in the world. The size of each
territory on this map indicates the amount of meat produced there per year, in tonnes.

China, the United States and Brazil produce the most meat. Even the smallest producer of the three, Brazil, produces twice as much as the next biggest producer.

The territories that produce the least meat are the Pacific islands of Niue, Nauru and Tuvalu, along with the minute Vatican City. Per person production of meat is lowest in the regions of South Asia and Central Africa.

In terms of total weight, the world produces ten times as much cereals as meat. However, meat has substantially higher calorific value than cereals, with as much as 5 to 10 times the calories weight for weight, so that total calorie consumption in the form of meat and vegetables is roughly the same.

'British beef producers would be extinct were it not for subsidies and European tariffs. Brazilian meat threatens them only because it is so cheap that it can out-compete theirs even after trade taxes have been paid.' George Monbiot, *Guardian*, 2005

MEAT PRODUCTION
tonnes per 1,000 people per year

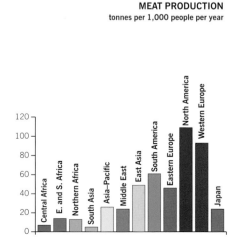

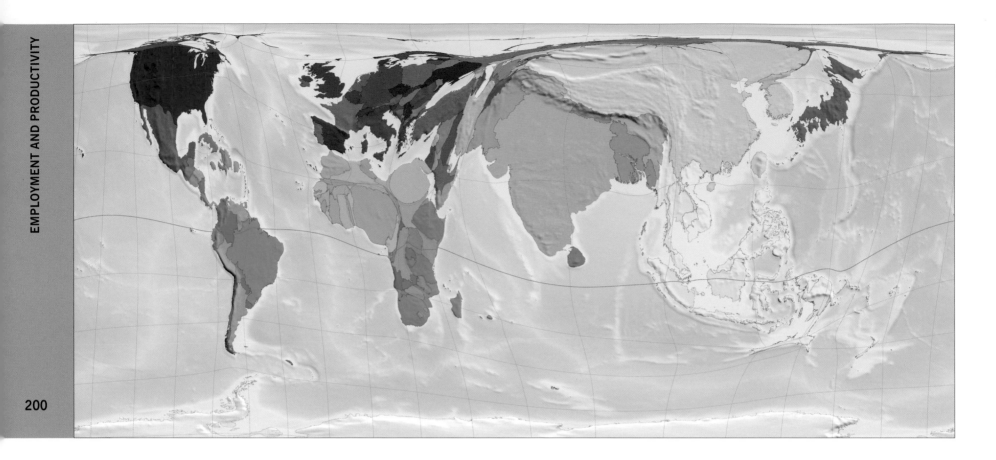

Rank	Territory	Calories[a]
1	Turkey	3,128
2	Egypt	3,066
3	Tunisia	3,003
4	Syria	2,953
5	Libya	2,948
6	Morocco	2,868
7	Greece	2,851
8	Lebanon	2,829
9	Indonesia	2,752
10	Jordan	2,748
191	Kenya	1,736
192	DR Congo	1,708
193	Bahamas	1,693
194	North Korea	1,688
195	Burundi	1,641
196	Afghanistan	1,606
197	Antigua and Barbuda	1,594
198	Eritrea	1,530
199	Mongolia	1,071
200	Somalia	828

[a] daily calories per person, 1997

181 Vegetable and Vegetable Product Consumption

The size of each territory indicates its total consumption of vegetables and vegetable products in calories. The average North American could give up meat altogether without going hungry.

Vegetable products are defined broadly to include all edible plant matter. Thus this map includes, for example, calories derived from fruit, bread and sugar.

People in China and India collectively consume the most vegetables, although this is not particularly surprising since these territories also have by far the largest populations. Vegetable consumption per person varies from country to country far less than many things mapped in this atlas. The territories with highest and lowest consumption per person, Turkey and

Somalia, differ by a factor of less than four. The reason is that people need to attain a certain calorific intake to survive, which sets a lower limit on their intake. Vegetable intake is lowest where people go hungry. It is highest where food is more plentiful but less meat is eaten.

The region with the highest per person consumption of vegetables and vegetable products is North America, at 2,642 calories per person per day – sufficient for an averagely active person to live on.

DAILY VEGETABLE INTAKE PER PERSON
calories consumed per person per day

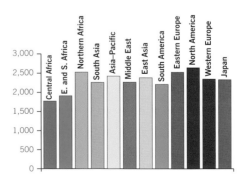

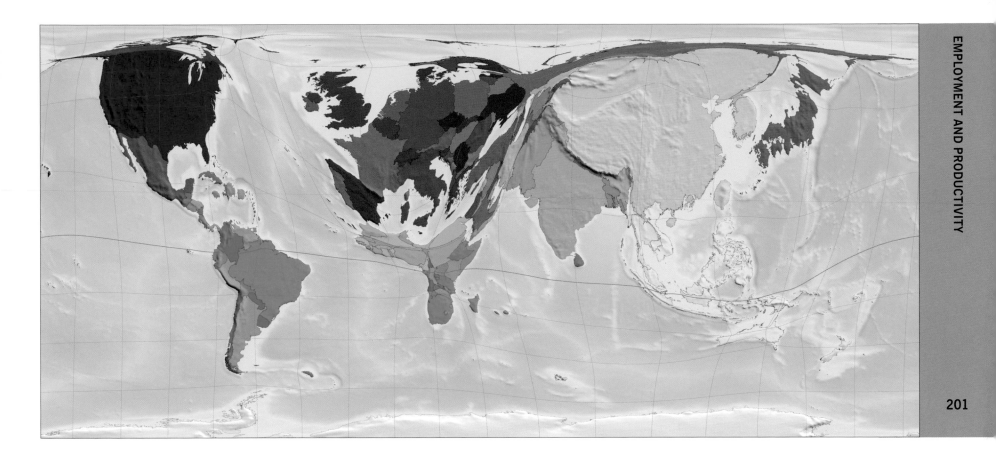

182 Meat and Meat Product Consumption

The size of each territory indicates its total consumption, in calories, of meat,
including all animal foods except for dairy products like milk and yoghurt.

Meat consumption per person is highest in Western European territories, which account for 9 of the top 10 meat-consuming territories. The tenth is New Zealand, which is famous for its high ratio of sheep to people and for the production of lamb. The largest total amount of meat is consumed in China, where a quarter of all the meat eaten in the world is consumed. A fifth of the world's population lives in China, eating on average 510 calories of meat per person a day, which is just above the world daily average of 432 calories per person. By contrast, over twice as much (1,058 calories) is consumed daily by people in Western Europe.

MEAT CONSUMED PER PERSON PER DAY
calories consumed per person per day

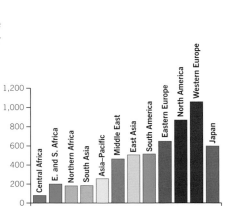

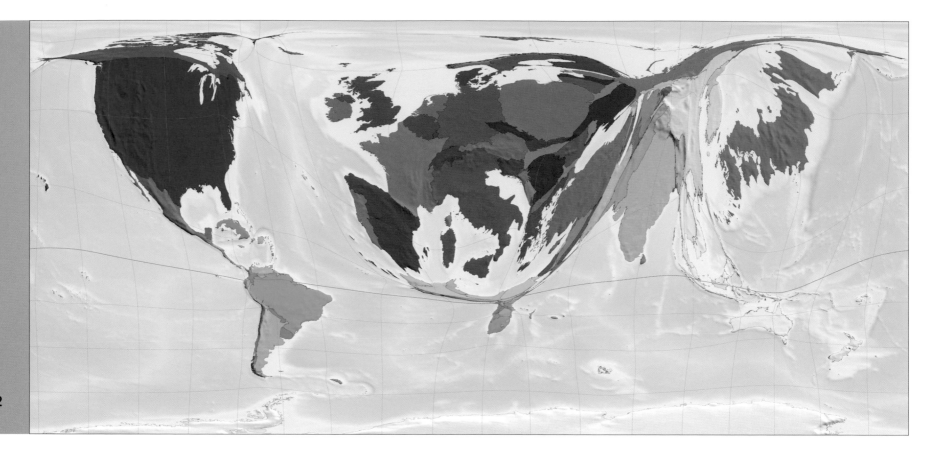

183 Tractors Working

The size of each territory indicates the number of working tractors used for farming. Tractors used, for example, to mow lawns are excluded.

Tractors have a lopsided distribution that splits the world into two groups. A majority of the world's population lives in countries with fewer than 4 tractors per 1,000 people. The rest live in countries with more than 13 tractors per 1,000 people. The cost of purchasing, maintaining and fuelling tractors still makes mechanization prohibitively expensive across much of the globe, and as a result the bulk of the world's agricultural labour is still performed manually, sometimes assisted by ox, yak, horse or buffalo.

The 10 territories with the highest rate of tractor ownership are all European; the 10 territories with the fewest tractors per resident are mainly African. This map shows only working tractors. Broken tractors may still be useful for parts, but working ones are a better indicator of agricultural practices.

'The most successful farmers...live in large, roomy houses made of cement and bricks. All of them own at least one tractor.'

Bharat Dogra, 'India: "green revolution" – bad news for poor labourers', 2000

WORKING TRACTORS
tractors per 1,000 people

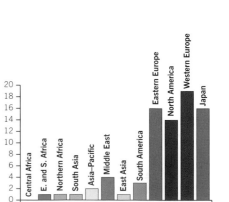

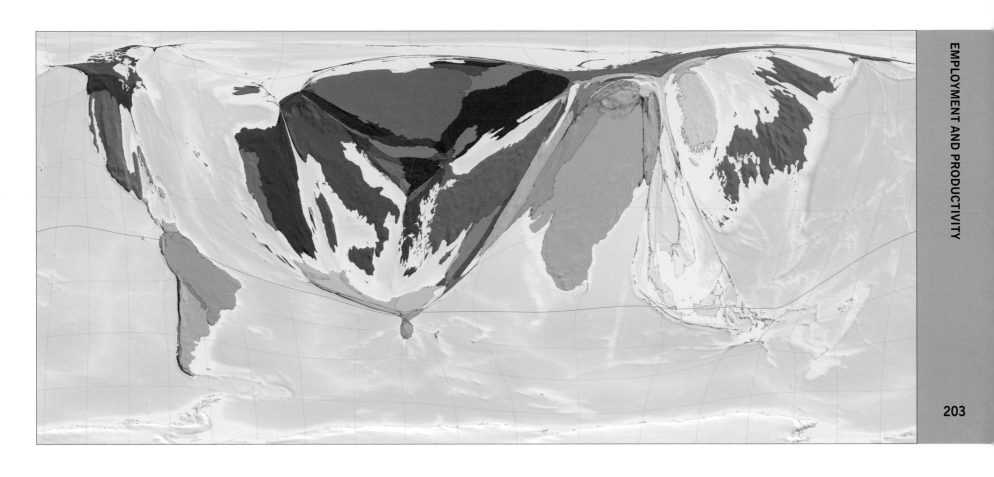

INCREASE IN TRACTORS

Rank	Territory	Extra tractors[a]
1	Poland	17.9
2	Italy	10.0
3	Greece	10.0
4	Spain	9.8
5	Portugal	8.4
6	Cyprus	7.9
7	Turkey	7.3
8	Hungary	5.9
9	Japan	4.4
10	South Korea	4.2
11	Syria	4.2
12	Ireland	3.7
13	Argentina	3.5
14	Fiji	3.4
15	Thailand	3.2
16	Libya	3.1
17	Switzerland	2.4
18	Canada	2.4
19	Iran	2.3
20	Botswana	2.1

[a] **increase in tractors per 1,000 people, 1980–2001**

184 Increase in Tractors

The size of each territory indicates the increase in the number of working tractors between 1980 and 2001.

Between 1980 and 2001 the number of working tractors worldwide increased by 6.5 million, from 20.4 million to 26.9 million.

Tractor ownership rose in most places over these 20 years. Only 36 territories did not see an increase, though East and Southern Africa as a region experienced an overall decrease. The largest increase in tractor numbers was in Eastern Europe, which added an extra 8 tractors per 1,000 people between 1980 and 2001.

CHANGE IN NUMBER OF TRACTORS OWNED
increase or decrease per 1,000 people, 1980–2001

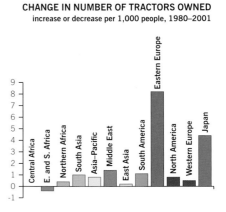

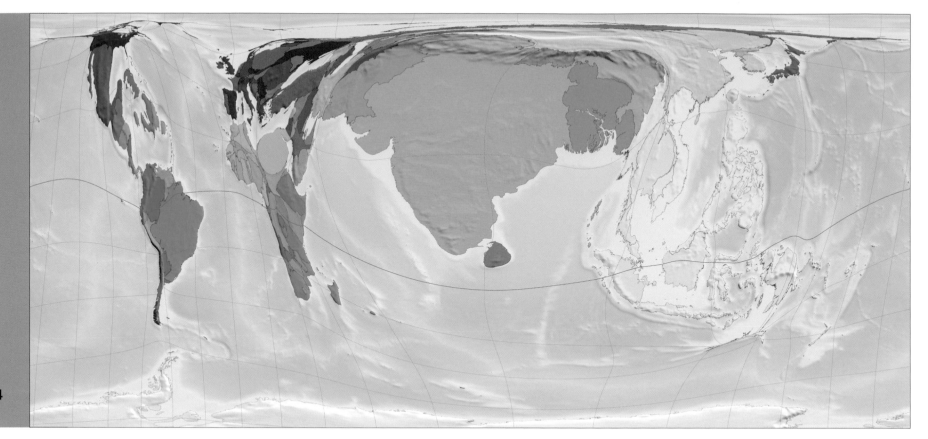

MOST AND FEWEST MEN WORKING IN AGRICULTURE

Rank	Territory	%ᵃ
1	Thailand	16.1
2	United Arab Emirates	16.0
3	Georgia	15.3
4	Haiti	14.9
5	Moldova	14.6
6	Bangladesh	14.4
7	India	13.7
8	Kyrgyzstan	12.5
9	Bhutan	12.4
10	Guatemala	12.3
191	Nigeria	0.93
192	Sweden	0.90
193	Germany	0.83
194	United Kingdom	0.82
195	France	0.72
196	United States	0.67
197	Hong Kong (China)	0.50
198	Belgium	0.48
199	Argentina	0.42
200	Singapore	0.31

ᵃ men working in agriculture as % of population, 2002

185 Male Agricultural Workers

The size of each territory indicates the number of male agricultural workers there.
In total, there are nearly one and a half times as many male as female agricultural workers.

This map shows a distribution broadly similar to that for female agricultural workers opposite. However, a larger proportion of male workers is found in Brazil, Colombia and Mexico.

The territory with the second highest proportion of male agricultural workers is the United Arab Emirates, which also has the lowest proportion of female agricultural workers. This contrasts with other low-scoring territories such as Singapore and Argentina, which have low proportions of both male and female workers.

Worldwide there are 1.4 male agricultural workers for every female agricultural worker. Many of the people represented on this map engage in peasant and subsistence farming. Mechanized farming does not require a large labour force, so the areas where machines are most used in agriculture are small on this map.

DISTRIBUTION OF MALE AGRICULTURAL WORKERS

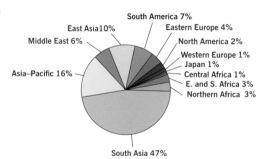

South America 7%
East Asia 10%
Eastern Europe 4%
North America 2%
Middle East 6%
Western Europe 1%
Japan 1%
Central Africa 1%
Asia–Pacific 16%
E. and S. Africa 3%
Northern Africa 3%
South Asia 47%

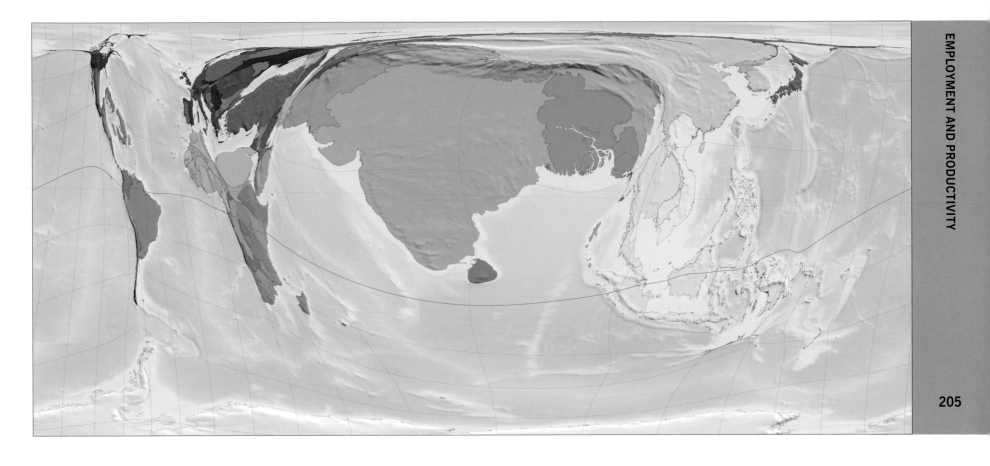

186 Female Agricultural Workers

The size of each territory indicates the number of female agricultural workers
there. Most of these women work in the poorer regions and territories.

This map of female agricultural workers
(where Asian territories are large) is
almost the opposite of Map 183, 'Tractors
Working', which shows that most tractors
are found in richer regions.

Over half of the world's female
agricultural workers live and work in South
Asia, with 39% in India. Arable agricultural
produce in India includes jute, rice,
oilseed, cotton, sugar cane, wheat, tea
and potatoes. Indian pastoral agriculture
includes the farming of cattle, water
buffalo, goats, poultry, sheep and fish.

As a percentage of total population, the
lowest proportion of female agricultural
workers is found in the United Arab
Emirates.

**DISTRIBUTION OF FEMALE
AGRICULTURAL WORKERS**

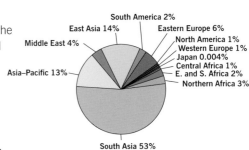

South America 2%
East Asia 14%
Eastern Europe 6%
Middle East 4%
North America 1%
Western Europe 1%
Japan 0.004%
Central Africa 1%
Asia–Pacific 13%
E. and S. Africa 2%
Northern Africa 3%
South Asia 53%

*'...whoever could make two ears of corn or two blades of grass grow
on a patch of land where only one grew before, does a greater service
to mankind...than the whole gang of politicians put together.'*

King of Brobdingnag, in *Gulliver's Travels* by Jonathan Swift, 1754

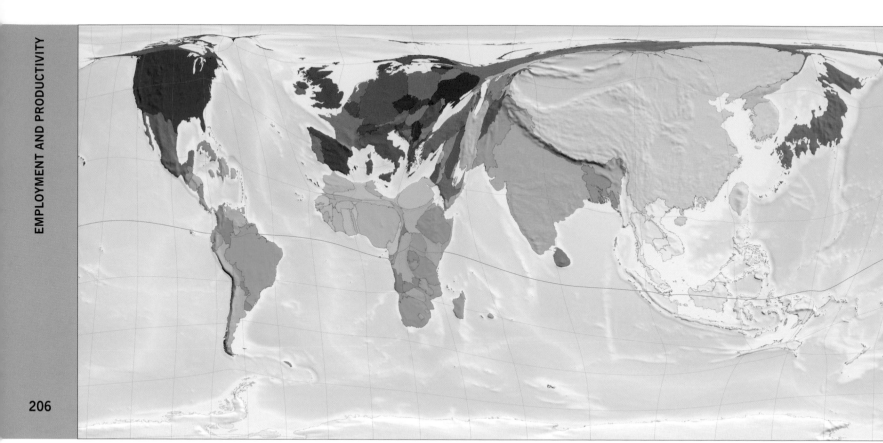

MOST AND FEWEST MEN WORKING IN INDUSTRY

Rank	Territory	%ᵃ
1	Czech Republic	15.3
2	Slovakia	14.9
3	Portugal	13.5
4	Slovenia	13.4
5	Germany	13.1
6	Estonia	12.8
7	Suriname	12.8
8	China	12.5
9	Japan	12.3
10	Spain	12.3
191	Namibia	4.1
192	Bhutan	4.0
193	Nepal	4.0
194	Maldives	3.7
195	Georgia	3.7
196	Haiti	3.5
197	Azerbaijan	3.5
198	Kyrgyzstan	3.2
199	Yemen	3.1
200	Bangladesh	3.0

ᵃ men working in industry as % of population, 2002

187 Males in Industry

The size of each territory indicates the number of male industrial workers there. Industrialization may be a matter of history in some places, but in others it is a present reality.

Industry here means manufacturing – the production of tangible goods. In total, 519 million males work in industry around the world. A third of them live in East Asia, and the majority of those in China. The territories with the highest percentages of the male population working in industry are found in the Eastern European territories of the Czech Republic and Slovakia; the territory with the lowest percentage is Bangladesh.

Although industrialization began as long ago as the eighteenth century in some countries, it is a relatively recent phenomenon elsewhere and is still just beginning for many people. As a consequence it has been claimed that the fastest-growing social class worldwide is the 'working' class.

DISTRIBUTION OF MALE INDUSTRIAL WORKERS

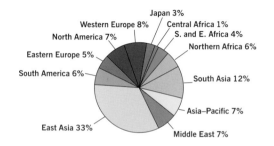

Japan 3%
Western Europe 8%
North America 7%
Eastern Europe 5%
South America 6%
East Asia 33%
Central Africa 1%
S. and E. Africa 4%
Northern Africa 6%
South Asia 12%
Asia–Pacific 7%
Middle East 7%

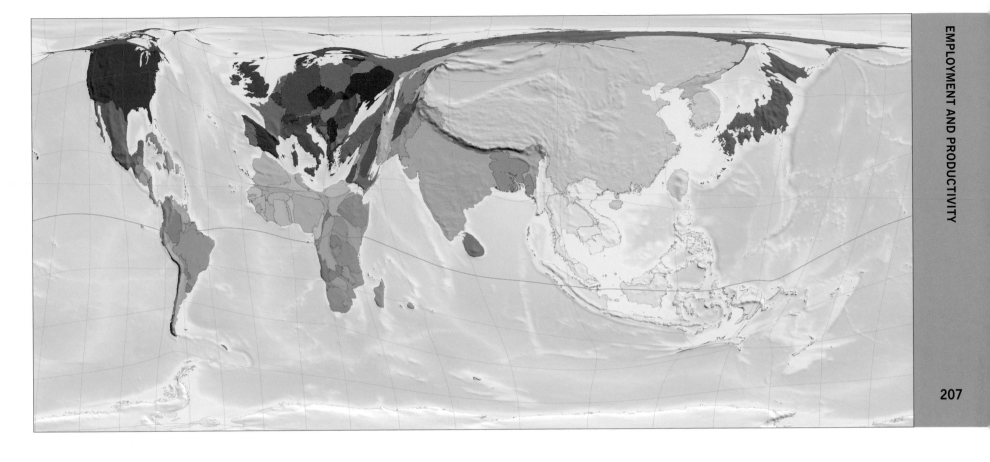

MOST AND FEWEST WOMEN WORKING IN INDUSTRY

Rank	Territory	%[a]
1	Czech Republic	7.2
2	Slovenia	6.7
3	Slovakia	6.7
4	Mauritius	6.2
5	Estonia	5.8
6	Morocco	5.7
7	Russia	5.7
8	Hungary	5.3
9	Ukraine	5.1
10	Armenia	5.1
191	Peru	1.2
192	Paraguay	1.1
193	Namibia	1.1
194	Belize	1.0
195	Haiti	1.0
196	Pakistan	1.0
197	Egypt	0.8
198	Gaza and West Bank	0.3
199	Yemen	0.2
200	Suriname	0.1

[a] women working in industry as % of population, 2002

188 Females in Industry

The size of each territory indicates the number of female industrial workers there. More than a third of all females working in industry across the world are in China.

Numerically, China tops the world in female industrial workers. China's main industries include the production of iron, steel, machines, armaments, textiles, chemical fertilizers, processed food, toys, automobiles and electronics.

The territories with the highest percentages of the female population working in industry are in Eastern Europe. Some territories in South Asia and the Asia–Pacific and Australasia also have relatively high figures. Industrial labour has never been the sole preserve of males, despite its reputation, and female industrial workers are found in both the richest and the poorest regions. Females were key elements of the labour force in the earliest industrial mills, and still form a large part of the workforce worldwide.

'In terms of the distribution of women within the industrial sector, women are highly concentrated in garments and textiles industries...' UNDP China, 2003

DISTRIBUTION OF FEMALE INDUSTRIAL WORKERS

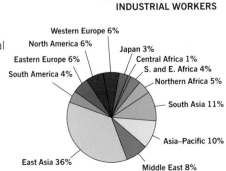

Western Europe 6%
North America 6%
Eastern Europe 6%
South America 4%
Japan 3%
Central Africa 1%
S. and E. Africa 4%
Northern Africa 5%
South Asia 11%
Asia–Pacific 10%
Middle East 8%
East Asia 36%

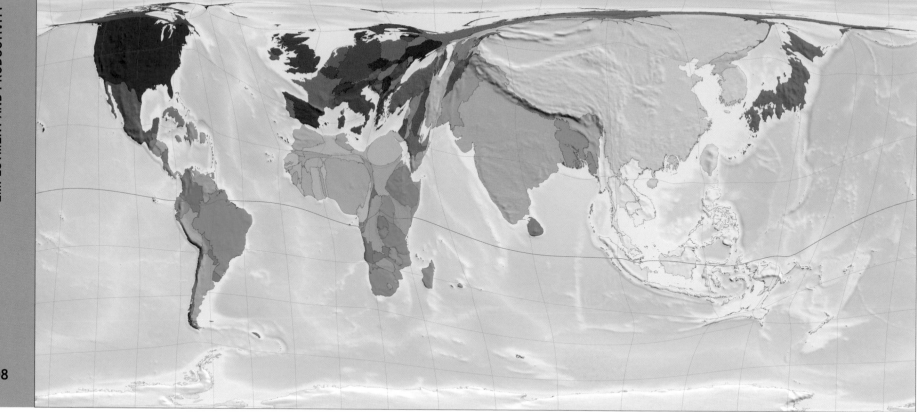

MOST AND FEWEST MEN WORKING IN SERVICES

Rank	Territory	%[a]
1	Hong Kong (China)	24.2
2	Singapore	21.0
3	Bahamas	19.7
4	Switzerland	19.3
5	China	19.1
6	Argentina	19.0
7	Canada	18.9
8	Australia	18.6
9	United States	18.4
10	Japan	18.4
191	Yemen	9.2
192	Moldova	8.7
193	Pakistan	8.7
194	Bhutan	8.5
195	Nepal	8.4
196	Kyrgyzstan	8.2
197	Romania	8.2
198	Bangladesh	8.1
199	Guatemala	6.8
200	Haiti	5.5

[a] men working in services as % of population, 2002

189 Service Sector Males

The size of each territory indicates the number of male service sector workers there. Overall, more males than females work in services.

Services means work that does not produce a material object, such as hospitality, call-centre work, armed forces labour and transportation.

Males working in the service sector account for 14% of the world population. China has the largest number of males working in services, followed by India and the United States.

Many services, such as cleaning, teaching and security services, need to be or are better performed close to their recipients, and as a result there are service workers to be found in every territory of the world. Some services, such as call centres and data entry, can occur anywhere as long as there are good channels of communication.

DISTRIBUTION OF MALE SERVICES WORKERS

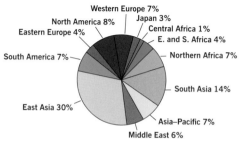

Western Europe 7%
Japan 3%
North America 8%
Central Africa 1%
Eastern Europe 4%
E. and S. Africa 4%
South America 7%
Northern Africa 7%
South Asia 14%
East Asia 30%
Asia–Pacific 7%
Middle East 6%

'These guys know accountancy, have computer skills, speak English and they are ready and willing – and that combination is a killer...'

Kiran Karnik, president of Nasscom, Indian software and services industry association, 2003

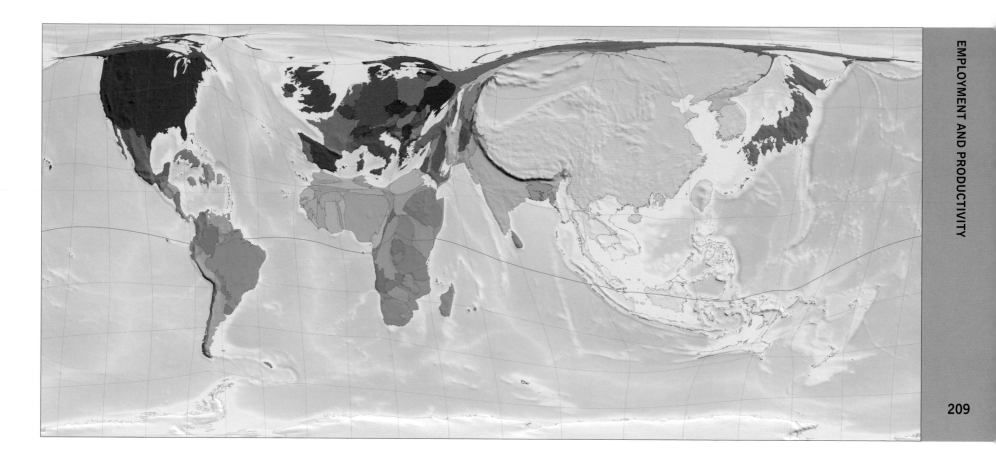

190 Service Sector Females

The size of each territory indicates the number of female service sector workers there. Sweden has a particularly high proportion of women working in services.

Worldwide, the largest number of females working in the service industries is found in China, where 260 million are thus employed. China accounts for 35% of all female services workers.

The territories with the highest proportions of females in the service industries are in Western Europe, North America and East Asia. The territories with the lowest proportions are in South Asia and the Middle East and Central Asia. In Sweden 23% of the population consists of females working in services, which is more than half of Sweden's economically active females.

DISTRIBUTION OF FEMALE SERVICES WORKERS

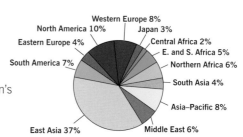

Western Europe 8%
North America 10%
Japan 3%
Eastern Europe 4%
Central Africa 2%
E. and S. Africa 5%
South America 7%
Northern Africa 6%
South Asia 4%
Asia–Pacific 8%
East Asia 37%
Middle East 6%

*'Girls are asking, "Do we get overtime? What are the benefits?"
Guangdong needs workers. Zhejiang and Shanghai need workers.
They have more choices. So it's difficult to find workers.'*
Kathy Deng, owner of a recruitment company in Guangzhou, China, 2006

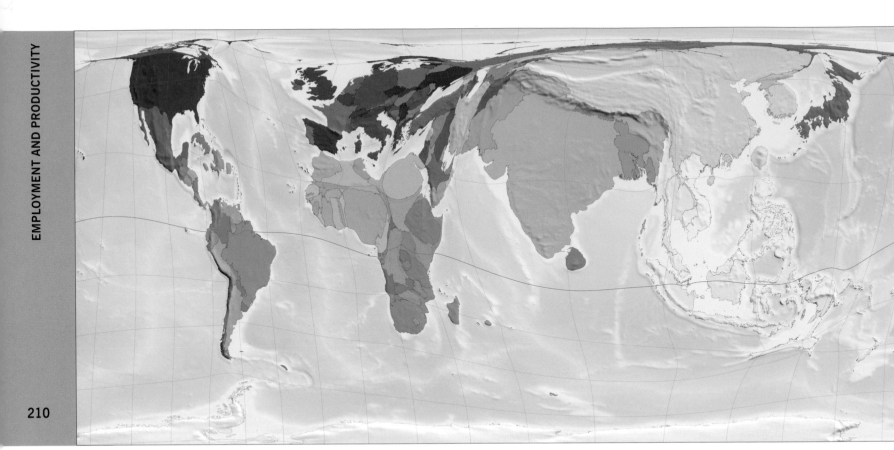

191 Female Domestic Labour

The size of each territory indicates the number of hours of unpaid or domestic labour done daily by females. Everywhere, girls and women do more of this work than men.

The average number of hours a day spent doing domestic or unpaid work is higher for females than for males in every region in the world. The biggest differences are found in Central Africa, where females spend an average of 7 hours and 6 minutes a day on this work and men only 1 hour and 30 minutes. (People of all ages are included in this calculation.)

Among the regions, numbers of hours of domestic labour performed by females are highest in Northern Africa, and lowest in Japan. In Northern Africa the average female there spends 7 hours and 12 minutes a day doing domestic work. In Japan the corresponding figure is just 3 hours and 43 minutes.

'Many of the women were holding babies in one arm and working over the stove with the arm that was left free.' Franz Kafka, *The Trial*, 1925

HOME HOURS WORKED BY WOMEN
average daily hours

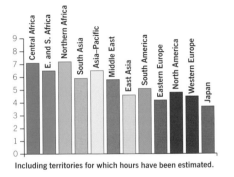

Including territories for which hours have been estimated.

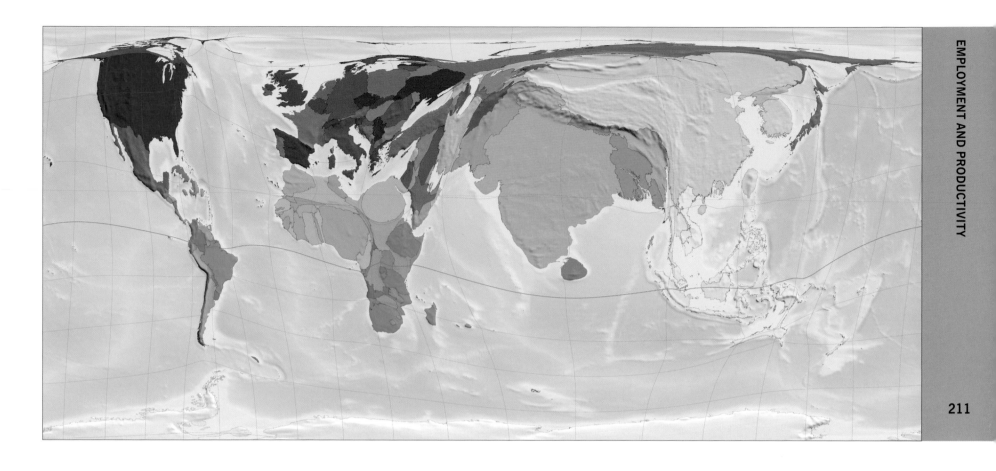

LONGEST AND SHORTEST HOME HOURS WORKED BY MEN

Rank	Territory	Hours[a]
1	Russia	2.94
2	Bulgaria	2.93
3	Georgia	2.88
4	Armenia	2.83
5	Slovenia	2.82
6	Romania	2.79
7	Croatia	2.77
8	Hungary	2.76
9	Ukraine	2.67
10	Israel	2.57
191	Laos	0.46
192	Rwanda	0.46
193	Dominican Republic	0.45
194	Solomon Islands	0.44
195	Paraguay	0.42
196	Nicaragua	0.41
197	Belize	0.37
198	Japan	0.36
199	Honduras	0.36
200	Guatemala	0.33

[a] daily hours worked by men averaged over all males, 2002

192 Male Domestic Labour

The size of each territory indicates the number of hours of unpaid or domestic labour done daily by males. Nowhere is this higher than the corresponding figure for females, though the differential varies.

Males in Japan and South America spend the least time every day doing domestic or unpaid labour. The average male in Japan spends 24 minutes a day doing domestic labour. (Men and boys of all ages are included in this calculation.) Males in the Middle East and Central Asia do the largest number of hours of domestic work: here the average is 3 hours and 24 minutes per day. (Note that Russia is included in the Middle East and Central Asia.) The highest average levels of male domestic labour among the regions of the world, however, are still lower than the lowest levels of female domestic labour.

HOME HOURS WORKED BY MEN
average daily hours

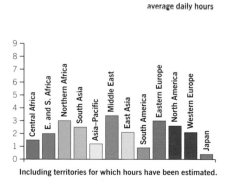

Including territories for which hours have been estimated.

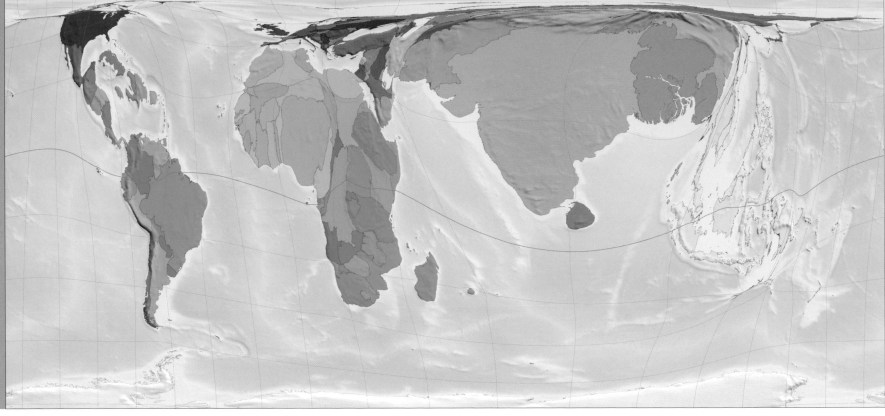

MOST AND FEWEST TEENAGE MOTHERS

Rank	Territory	Young mothers[a]
1	Niger	16.2
2	Mali	15.0
3	Mozambique	14.9
4	Chad	14.5
5	Guinea	13.8
6	Madagascar	13.4
7	Central African Republic	13.3
8	Bangladesh	13.0
9	Malawi	12.2
10	Gabon	12.1
191	Finland	0.26
192	France	0.26
193	Belgium	0.25
194	Netherlands	0.20
195	Switzerland	0.19
196	Spain	0.19
197	Sweden	0.19
198	Italy	0.19
199	Japan	0.13
200	South Korea	0.07

[a] teenage mothers per 1,000 people, 2003

193 Teenage Mothers

The size of each territory indicates the number of mothers giving birth in their teenage years (between 15 and 19). A third of them live in India.

For the purposes of this map, a teenage mother is a girl or woman aged between 15 and 19 (inclusive) who has had at least one child, though some girls have children when they are younger than 15.

There are three times as many teenage mothers living in South Asia as in any other region. The smallest (recorded) number of teenage mothers is in Japan.

Being a young parent is normal in some parts of the world, while in other parts teenage parenthood is a rarity. Where it is a rarity mothers tend to be considerably older – but internationally comparable data on older mothers do not yet exist.

TEENAGE MOTHERS
millions

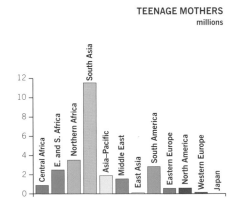

'Attitudes towards young mothers...shift in relation to prevailing moral values, and also to some extent reflect economic conditions.'

Debbie A. Lawlor, professor of epidemiology, University of Bristol, 2002

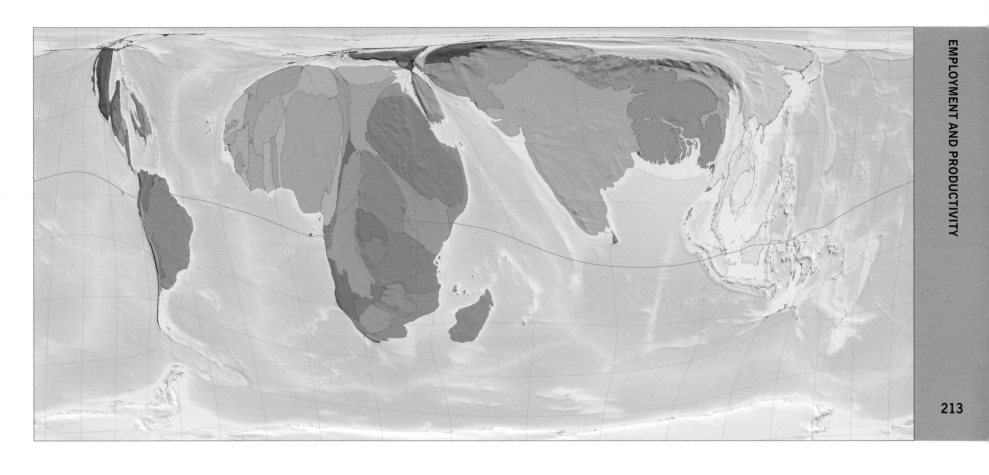

MOST CHILD LABOURERS

Rank	Territory	Children[a]
1	Mali	81
2	Burundi	75
3	Uganda	72
4	Niger	71
5	Bhutan	68
6	Burkina Faso	63
7	Rwanda	62
8	Ethiopia	62
9	Eritrea	57
10	Guinea-Bissau	56
11	Chad	55
12	Nepal	54
13	Tanzania	53
14	Kenya	53
15	Comoros	52
16	Madagascar	49
17	Mozambique	46
18	Timor-Leste	46
19	Malawi	45
20	Equatorial Guinea	45

[a] **children who work per 1,000 people, 2003**

194 Child Labour

The size of each territory indicates the number of children between the ages of 10 and 14 who are working, showing that most child labour occurs in African and South Asian territories.

This map shows only those children between the ages of 10 and 14, although younger and older children also work.

Of the 10 territories with the highest proportion of child labourers as a percentage of the population, 9 are in Africa; the 10th is Bhutan. The lowest proportion is in Italy, which also has the lowest proportion of children overall as a percentage of population.

India has the highest number of child labourers, twice as many as China, which has the second highest. Practically no children work in Japan; in Western Europe there are 13,000 child labourers.

'At our homes we had done a lot of ploughing, planting, weeding and harvesting; we had hewn wood and drawn water; we had tended sheep, goats and cattle; we had done one hundred and one odd jobs.' Ndabaninga Sithole, *African Nationalism*, 1959

CHILD LABOURERS
millions

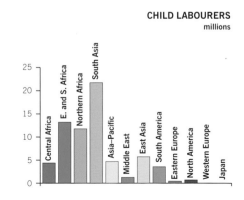

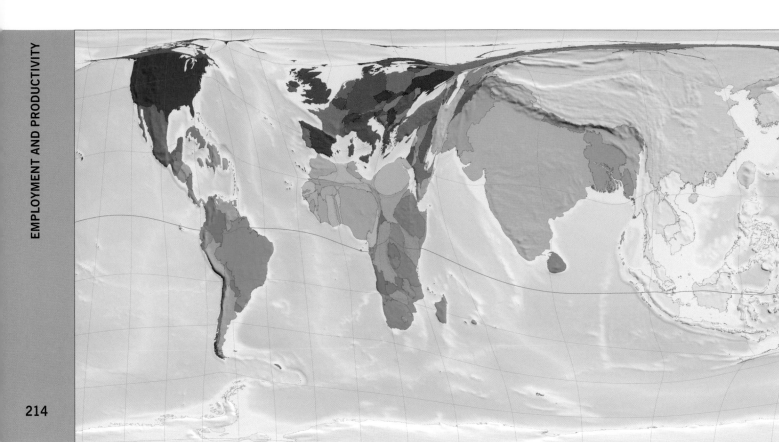

LONGEST AND SHORTEST MARKET HOURS WORKED BY MEN

Rank	Territory	Hours[a]
1	Hong Kong (China)	4.94
2	Thailand	4.91
3	Sao Tome and Principe	4.85
4	Rwanda	4.83
5	Japan	4.82
6	China	4.80
7	Taiwan	4.79
8	North Korea	4.79
9	Singapore	4.74
10	Equatorial Guinea	4.66
191	Turkmenistan	3.11
192	Kyrgyzstan	3.03
193	Uzbekistan	3.02
194	Oman	3.01
195	Jordan	2.96
196	Syria	2.96
197	Saudi Arabia	2.88
198	Tajikistan	2.82
199	Yemen	2.66
200	Gaza and West Bank	2.28

[a] **daily market hours worked by men averaged over all males, 2002**

195 Market Hours Worked by Males

The size of each territory indicates the total number of hours worked by males every day outside the home, either to earn money or for subsistence.

In Eastern Europe, Western Europe and North America the average male spends 4 hours and 30 minutes daily doing paid work. This is barely half the corresponding figure for Central Africa, where the average male does paid or other money-earning work for 8 hours and 6 minutes every day. Moreover, these averages include all males, not just those who actually work, so the number of hours spent working by males who do work is considerably more than the average.

Among the territories on the map, hours worked by males are lowest in Gaza and the West Bank, where many men cannot work. The 10 territories where males work fewest hours a day are all in the Middle East and Central Asia. The highest figures are recorded in Hong Kong (which is part of China) and Thailand.

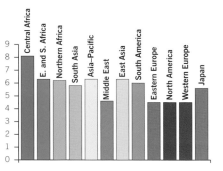

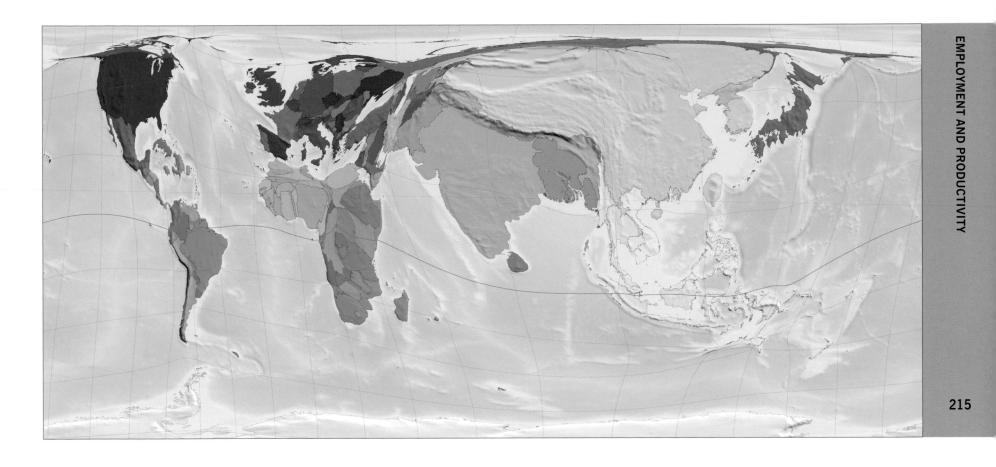

LONGEST AND SHORTEST MARKET HOURS WORKED BY WOMEN

Rank	Territory	Hoursª
1	China	3.43
2	Taiwan	3.40
3	North Korea	3.40
4	Mozambique	3.33
5	Tanzania	3.21
6	Russia	3.18
7	Armenia	3.16
8	Kenya	3.12
9	Mongolia	3.06
10	Malawi	3.01
191	Sudan	1.09
192	Guatemala	1.08
193	Malta	1.08
194	Algeria	1.05
195	Yemen	1.02
196	Libya	0.90
197	Belize	0.87
198	Saudi Arabia	0.86
199	Oman	0.81
200	Gaza and West Bank	0.33

ª daily market hours worked by women averaged over all females, 2002

196 Market Hours Worked by Females

The size of each territory indicates the total number of hours worked by females every day outside the home, either to earn money or for subsistence.

To enable accurate comparison, women of all ages, including girls, are included in this calculation because in different countries women begin and sometimes stop doing paid work at different ages. This means, however, that the average figures quoted are typically much lower than the actual number of hours worked by working women.

China has the largest total number of market hours worked by females, although this is partly because China has a very large population, including a large female workforce. However, it is also a result of long working hours: in East Asia, the region

in which China is located, females do market work on average for 4 hours 30 minutes daily across the whole female population, the second highest average of any region.

The region with the highest daily number of market hours worked by females is East and Southern Africa, where the average is 4 hours 32 minutes per day per female member of the population. The region with the smallest daily number of market hours worked by females is South America, with 2 hours and 12 minutes per day per female member of the population.

MARKET HOURS WORKED BY WOMEN
average daily hours

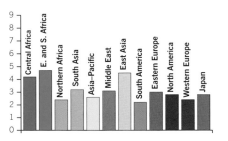

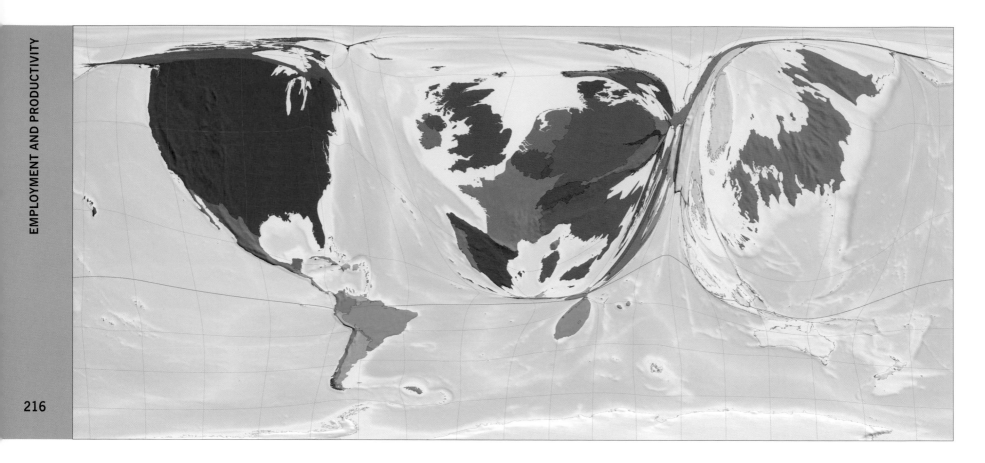

MOST MALE MANAGERS

Rank	Territory	%ᵃ
1	Luxembourg	52
2	Norway	32
3	Equatorial Guinea	28
4	Ireland	26
5	Denmark	23
6	Switzerland	22
7	Netherlands	20
8	Austria	20
9	United States	19
10	Canada	18
11	Japan	18
12	Italy	18
13	Belgium	17
14	Iceland	17
15	Hong Kong (China)	17
16	United Arab Emirates	17
17	France	17
18	Australia	17
19	United Kingdom	16
20	San Marino	16

ᵃ male managers as % of population, 2002

197 Male Managers

The size of each territory indicates the number of male managers there.
In Japan, management may become the most common career path for boys.

In Japan, 18% of the entire population are male managers and professionals. Since roughly half of the population is male, and many are retired and others too young to work, this means that boys living in Japan are now more likely to grow up to become managers than anything else. At least, this will be the case if past trends in Japan continue into the near future. However, given the rise in the number of females

working there and the increase in numbers of female managers elsewhere in the affluent world, our prediction is far from certain.

The total numbers of managers, both male and female, in North America and Western Europe are more than twice as large as in any other region and constitute a significant majority of all of the world's managers: these two regions are home to 73% of all managers in the world.

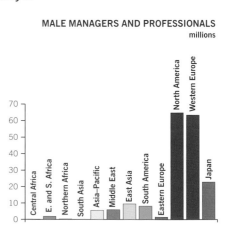

MALE MANAGERS AND PROFESSIONALS
millions

MOST FEMALE MANAGERS

Rank	Territory	%ᵃ
1	Luxembourg	32
2	Norway	20
3	United States	20
4	Ireland	17
5	Greenland	15
6	Canada	14
7	Australia	14
8	Denmark	13
9	Iceland	12
10	Switzerland	12
11	Austria	12
12	Netherlands	12
13	Germany	11
14	Belgium	11
15	France	10
16	United Kingdom	10
17	Finland	10
18	Sweden	10
19	New Zealand	9
20	Italy	9

ᵃ female managers as % of population, 2002

198 Female Managers

The size of each territory indicates the number of female managers there. Worldwide the gender gap is closing, but the variation from one region to another is huge.

In 2001 there were 130 million females employed in managerial and professional jobs worldwide, constituting 41% of all managers; male managers made up the remaining 59%. The largest numbers of female managers are found in North America and Western Europe, while there are very few in Northern Africa, Central Africa, East and Southern Africa, and South Asia.

Note that numbers of managers and professionals are not recorded directly for most territories, so the numbers used on this map and the preceding one are estimates based on income adjusted for local purchasing power.

However, even if data for actual numbers of managers were available, based on job titles, the information would not necessarily be more accurate, since job titles can be inflated and give an inaccurate description of the nature of a position.

FEMALE MANAGERS AND PROFESSIONALS
millions

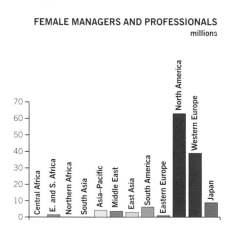

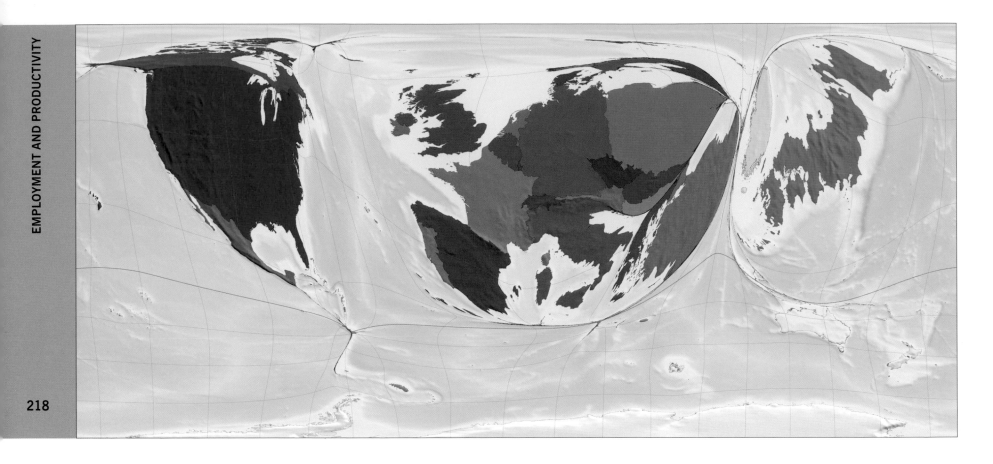

ᵃ **% of total population receiving unemployment
benefits, 2002 (OECD members only)**

199 Unemployed People

At any one time in any territory, some fraction of the population is unemployed.
The size of each territory indicates the number of unemployed people there.

These figures, and the data on which this map is based, relate only to members of the OECD, which tend to be the richer territories; most poorer, non-OECD territories do not pay unemployment benefits, and so do not record numbers of registered unemployed. There are, of course, unemployed people in these other territories as well. Unemployment tends to be high where there are few jobs and also where the unemployed receive support so that they are not compelled to take any work they can find just to survive.

At the time the data represented on this map were collated, around 4% of the total population in Western Europe were unemployed, 3% in Eastern Europe and Japan, and only 2% in North America. Added up, that comes to about 36 million unemployed people, which is just over half of 1 per cent of the total global population.

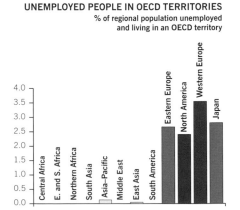

UNEMPLOYED PEOPLE IN OECD TERRITORIES
% of regional population unemployed
and living in an OECD territory

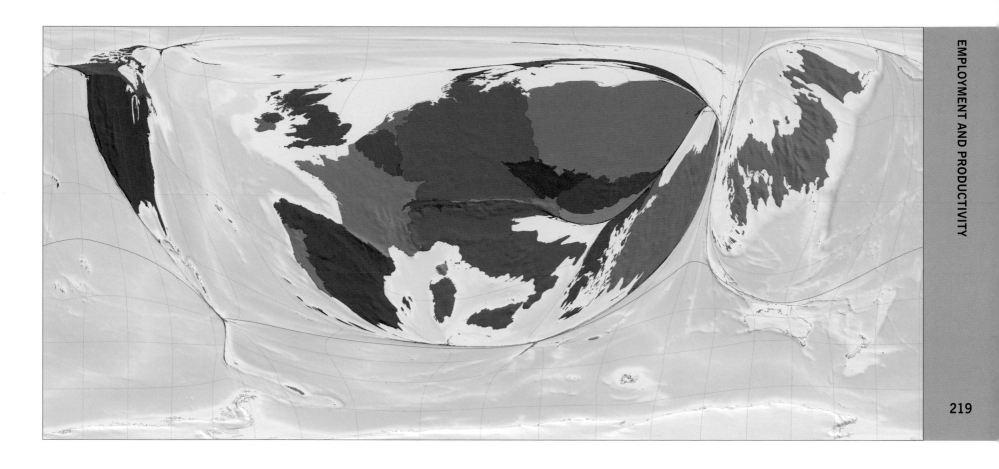

**HIGHEST AND LOWEST LONG-TERM
UNEMPLOYMENT RATES**

Rank	Territory	%ᵃ
1	Slovakia	5.4
2	Poland	4.3
3	Italy	2.2
4	Greece	2.0
5	Spain	2.0
6	Germany	2.0
7	Czech Republic	1.9
8	Belgium	1.6
9	France	1.4
10	Turkey	1.1
21	Switzerland	0.40
22	Luxembourg	0.40
23	Canada	0.39
24	New Zealand	0.38
25	Netherlands	0.28
26	United States	0.25
27	Iceland	0.20
28	Norway	0.13
29	South Korea	0.03
30	Mexico	0.00

ᵃ **% of total population registered as long-term
unemployed, 2002 (OECD members only)**

200 Long-Term Unemployed

The size of each territory indicates the number of people there who have
been out of work and looking for work for one year or longer. The highest
rates of long-term unemployment are found in European territories.

In 2002 almost 11 million long-term
unemployed people were living in OECD
territories. (The map reflects the position in
those territories that are part of the OECD;
most non-OECD territories do not pay
unemployment benefits, which means that
data on numbers of registered unemployed
are not available.)

In Slovakia 5.4% of the total population
are long-term unemployed; and since some
members of the population are too young,
too old or not able to work, the percentage
of long-term unemployed as a fraction of
the potential active workforce is even
higher – nearly 10%.

The lowest long-term unemployment
rate among the OECD countries is in
Mexico, which traditionally has offered
relatively little social support to
unemployed people.

**LONG-TERM UNEMPLOYED IN
OECD TERRITORIES**
% of regional poulation unemployed
and living in an OECD territory

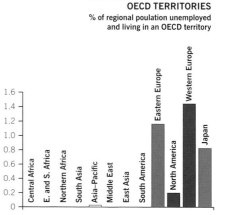

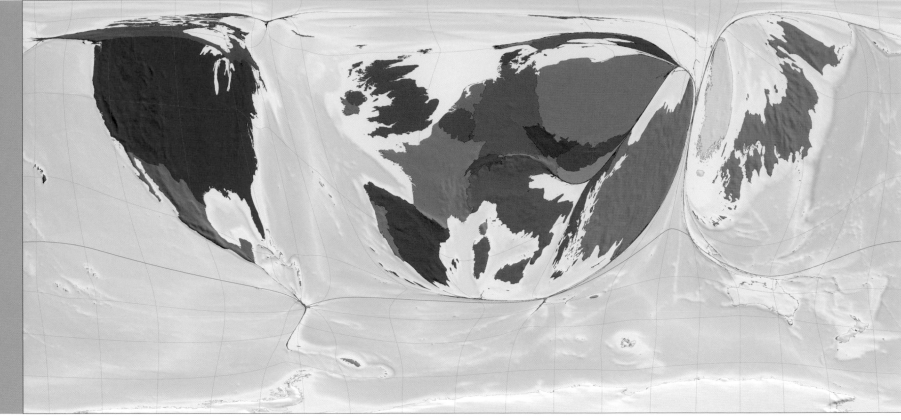

Rank	Territory	%ᵃ
1	Poland	2.9
2	Slovakia	2.7
3	Turkey	1.5
4	Italy	1.4
5	Finland	1.4
6	France	1.2
7	Greece	1.2
8	Spain	1.1
9	Canada	1.0
10	Belgium	1.0
21	South Korea	0.68
22	Iceland	0.67
23	Portugal	0.61
24	Ireland	0.60
25	Denmark	0.58
26	Switzerland	0.48
27	Austria	0.42
28	Netherlands	0.42
29	Luxembourg	0.35
30	Mexico	0.32

ᵃ male youth registered as unemployed as %
of total male population, 2002 (OECD
members only)

201 Male Youth Unemployed

The size of each territory indicates the number of young males (aged between 15 or 16 and 24) who are unemployed. Some Eastern European territories have particularly high rates.

The highest rates of unemployment among older boys and young men in OECD territories in 2002 were found in Poland, Slovakia and Turkey. The rates in Poland and Slovakia were almost twice the rate prevailing in any other OECD territory. The lowest rate of male youth unemployment was in Mexico, where state support for the unemployed is also weakest. (The map reflects the position in those territories that are part of the OECD; data for most of the remaining territories are not available as they do not pay unemployment benefits and so do not hold a register of unemployed people.)

The highest total number of unemployed young men and boys was in the United States: about 1.2 million males between 16 and 24 years of age. The smallest total was in Luxembourg, which had approximately 700 young men unemployed.

'If not addressed, in 10 years our unemployed young people will become unemployed middle-aged people...'

Nikos Koutsiaras, economist, University of Athens, 2006

UNEMPLOYED MALE YOUTH IN OECD TERRITORIES
male youth unemployed and living in OECD territories, thousands

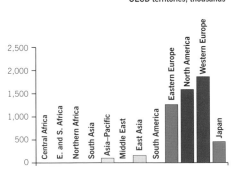

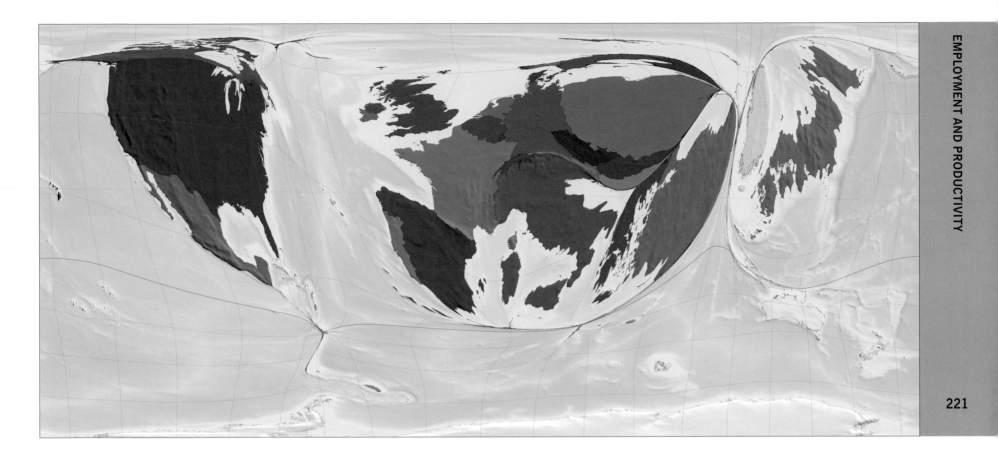

HIGHEST AND LOWEST FEMALE YOUTH UNEMPLOYMENT RATES

Rank	Territory	%[a]
1	Poland	3.0
2	Slovakia	2.4
3	Greece	2.1
4	Italy	1.9
5	Spain	1.7
6	France	1.4
7	Finland	1.3
8	Turkey	1.3
9	Czech Republic	1.1
10	Belgium	1.0
21	Japan	0.54
22	Germany	0.51
23	South Korea	0.47
24	Ireland	0.45
25	Mexico	0.40
26	Austria	0.40
27	Netherlands	0.36
28	Denmark	0.34
29	Iceland	0.31
30	Switzerland	0.26

[a] female youth registered as unemployed as % of total female population, 2002 (OECD members only)

202 Female Youth Unemployed

The size of each territory indicates the number of young females (aged between 15 or 16 and 24) who are unemployed. Western Europe has particularly high rates of unemployment in this group.

This map shows the numbers of unemployed older girls and young women in OECD territories. Most of these are in Western Europe, with the highest unemployment rates being found in Greece, Italy, Spain and France. (Most non-OECD territories do not pay unemployment benefits, which means that data on numbers of registered unemployed are not available.)

'In a majority of...[countries in] Europe...there is a higher youth unemployment rate for women than for men, but in many countries the differences are not large...' ECE, 2003

UNEMPLOYED FEMALE YOUTH IN OECD TERRITORIES
female youth unemployed and living in OECD territories, thousands

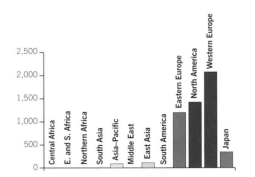

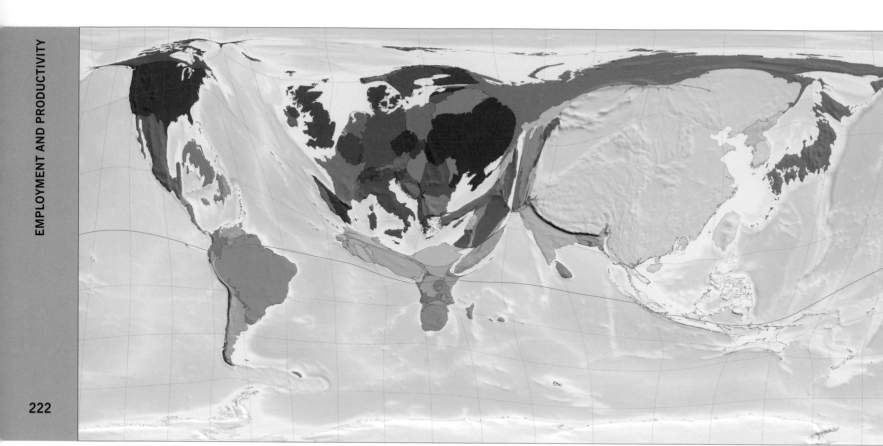

Rank	Territory	Members[a]
1	Ukraine	447
2	Sweden	419
3	Belarus	418
4	Finland	417
5	Iceland	403
6	Denmark	394
7	Norway	335
8	Russia	294
9	Belgium	262
10	Cuba	245
191	Thailand	6.7
192	India	6.1
193	Mauritania	5.4
194	Indonesia	4.6
195	Eritrea	4.5
196	Gabon	3.8
197	Uganda	2.5
198	Ethiopia	2.2
199	Pakistan	1.8
200	Guinea	1.5

[a] trade union members per 1,000
people, 2004

203 Trade Unions

The size of each territory shows the number of people who belong to
trade unions, combining to achieve better working conditions.

Trade unions are groups formed by
employees to bargain collectively with
employers on matters such as wages,
benefits and safety. They work on the
principle that people acting together can
achieve more than individuals acting alone.
   In 2004 there were 404 million trade
union members worldwide, a third of them
in China. Other territories that were home

to high numbers of unionized workers were
Russia, Ukraine and Brazil. Ukraine in
particular has a very high rate of union
membership: 45% of all people living there
belong to a union. Given that most children
and most elderly people in Ukraine do not
work, this figure suggests that a significant
majority of working people there are
unionized.

*'We are fighting for the right of the working people
to association and for the dignity of human labour.'*

Lech Walesa, Nobel Peace Prize winner, 1983

TRADE UNION MEMBERSHIP
trade union members per 1,000 people,
2004 or most recent data available

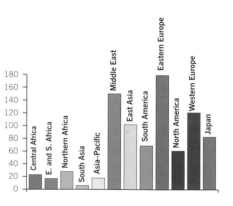

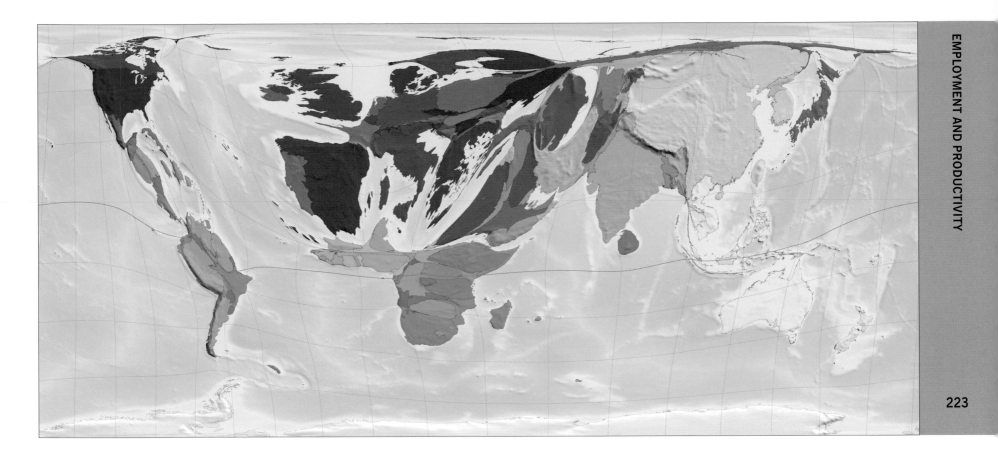

Rank	Territory	Participants[a]
1	Guyana	83
2	Israel	44
3	Azerbaijan	41
4	Finland	30
5	Australia	28
6	Greece	27
7	Cyprus	27
8	Denmark	22
9	Spain	19
10	New Zealand	16
191	Switzerland	0.21
192	Hong Kong (China)	0.18
193	Malawi	0.08
194	Thailand	0.07
195	Egypt	0.02
196	Burundi	0.00
197	Mali	0.00
198	Myanmar	0.00
199	Bahamas	0.00
200	Singapore	0.00

[a] strike and lockout participants per 1,000
people per year, 1980–1995

204 Strikes and Lockouts

The size of each territory indicates the number of people participating in strikes or lockouts
between 1980 and 1995. The right to strike is not, of course, universally enjoyed.

This map shows numbers of people
involved in labour disputes, either
employees withholding their labour
(strikes) or employers preventing
employees from working (lockouts). That
the halting of labour can function as an
effective form of protest for both employees
and employers is an indication of the
mutual dependence of the two groups.

Between 1980 and 1995, an average of
13.5 million people participated in strikes
and lockouts each year. The territories with
the largest numbers so involved were
China, India, Spain and Australia. As a
percentage of total population, the highest
rates of participation were in Guyana, Israel
and Azerbaijan.

In most territories there are now at least
some workers who have the right to strike,
although they sometimes suffer severely if
they do.

*'If you think you are too small to make a difference,
try sleeping with a mosquito.'* Anon., undated

PARTICIPATION IN STRIKES AND LOCKOUTS
participants per 1,000 peope, 1980–95

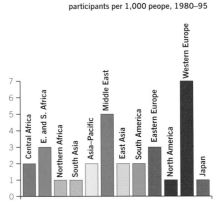

The Social World

Housing and Education

Communication and Media

Health and Illness

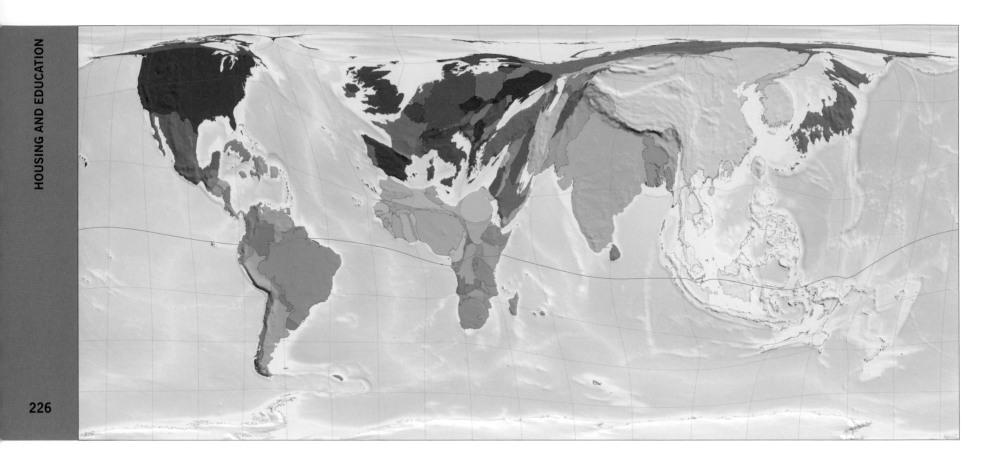

205 Urban Population

The size of each territory indicates the number of people living in cities and towns. In all, they amount to about half of all the people on Earth.

Cities and towns originally arose because their higher population densities made trade and services – and, later, industry – easier to conduct. The first known city was built by the Sumerians over 6,000 years ago in what is now Iraq.

In 2002, 48% of the world's population lived in urban areas; by the time you read this, that figure will almost certainly have passed 50%. In every territory on this map there are many people living in urban areas, but there are nonetheless great differences between territories, with some having far higher percentages of city-dwellers than others.

The most urbanized countries are not necessarily the richest. In Brazil 145 million people or 82% of the population live in towns and cities. In Bhutan the corresponding figure is just 180,000 or 8% of the population. In just 2 territories virtually the whole of the population lives in urban areas – Hong Kong (which is part of China) and Singapore.

POPULATION LIVING IN URBAN AREAS
% of total population

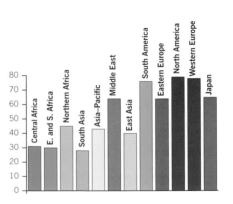

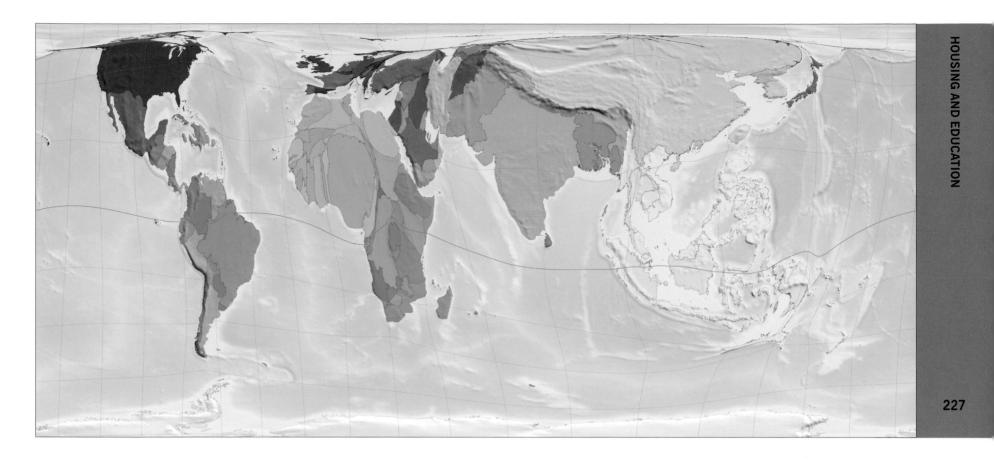

[a] people expected to be living in cities in 2015 who were not living in cities in 2002, as a % of 2002 population

206 Projected Increase in Urban Population

The size of each territory indicates the predicted increase in the number of people living in cities and towns from 2002 to 2015. This number is projected to increase in the vast majority of territories worldwide.

The number of people living in cities and towns is expected to increase in all but 14 of the 200 territories. Some of that increase will come from babies being born in cities, but more will come from people moving to urban from rural areas. There are many reasons for moving to a town or city, including increased opportunities for work, higher living standards or the disappearance of rural ways of life.

In the 186 territories in which an increase is predicted to occur there will be a combined total of 888 million more town- and city-dwellers by 2015. In the remaining 14 there will be a decrease of about 6.5 million.

'Massive urbanization means hundreds of already near-bankrupt cities trying to cope in 20 years with the kind of problems London or New York only managed to address with difficulty in 150 years.' John Vidal, *Guardian*, 2004

**URBAN POPULATION 2002 AND
GROWTH BY 2015**
urban population 2002 (bottom),
additional urban population by 2015 (top), millions

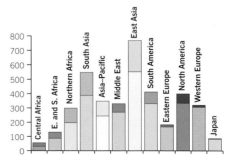

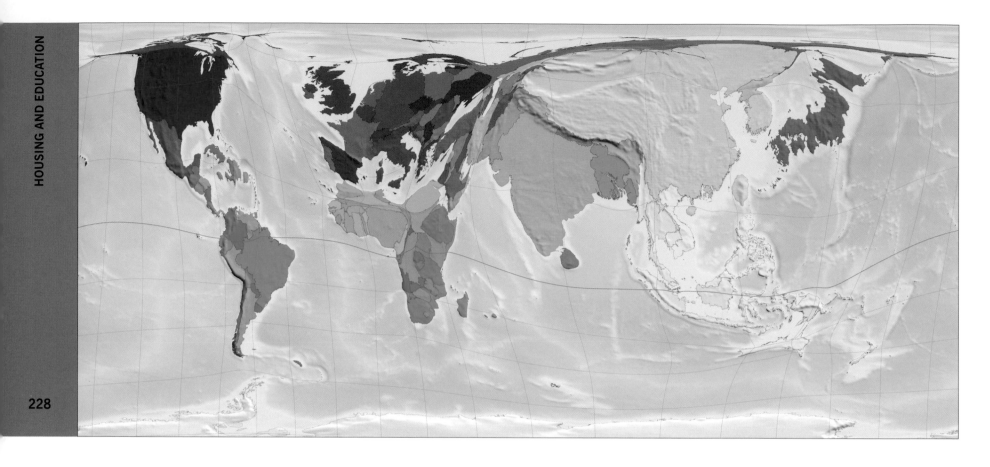

207 Number of Households

The size of each territory shows the number of households. The typical size of a household, in terms of its number of members, varies widely from one territory to another.

A household is defined as a social grouping, often a family, who eat together or share a living space. The number of households in a territory depends in part on the population of the territory, of course, but also on the usual household size, which varies widely. Among the territories the largest average household size is 7.7 people, in Iraq; among the regions the largest is 5, in Central Africa. Living in larger households tends to be cheaper per person, all other things being equal, because amenities are shared.

'*Having fewer people in more households means using more resources and putting more stress on the environment. Freedom and privacy come at a huge environmental cost.*'

Jianguo Lin, associate professor of fisheries and wildlife, Michigan State University, 2003

HOUSEHOLDS
per 100 people, 2002

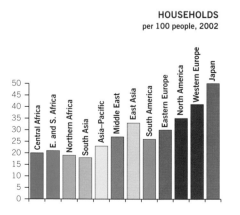

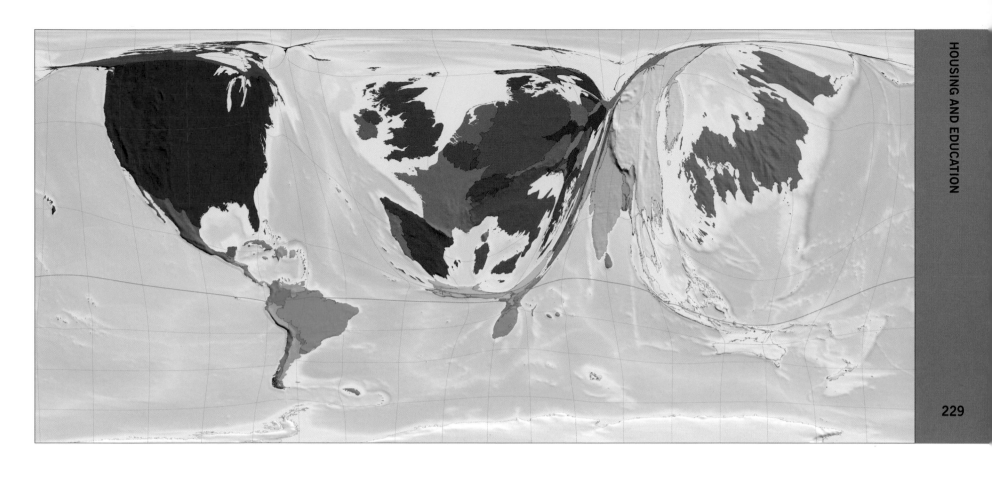

208 Housing Prices

The size of each territory shows the total value of all housing, adjusted for local purchasing power.

The most expensive housing in the world (measured per person living there) is in Europe. Even with the higher costs of housing in richer territories, however, the average household sizes tend to be smaller here.

The cheapest housing is found in the African regions and South Asia, even after allowing for the fact that money goes further here.

WORLD HOUSING PRICES
per person, PPP US$, thousands

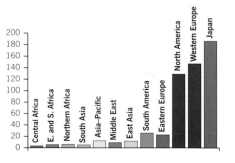

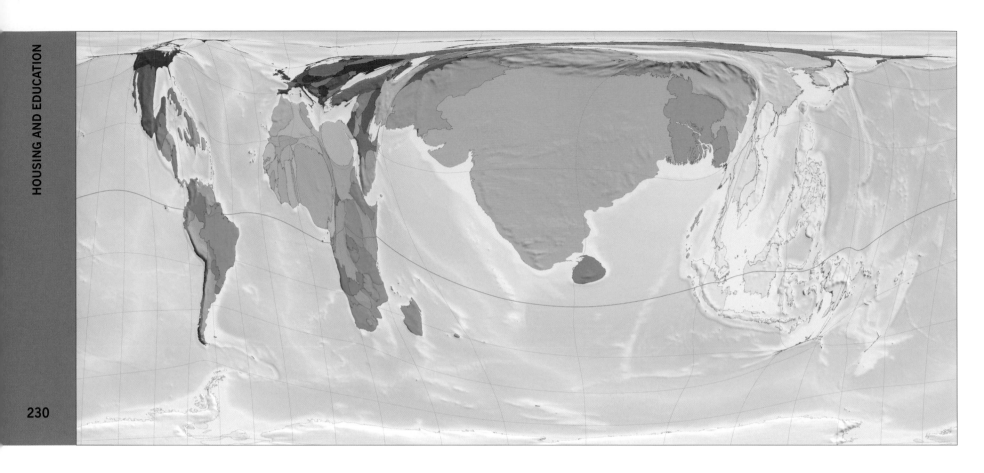

209 People Living in Overcrowded Homes

The size of each territory shows the number of people living in overcrowded homes, defined as those in which there are more than two people for each room in the house.

The level of crowding considered acceptable depends on social and cultural norms as well as wealth, but when given the choice most people prefer more space.

People in richer territories live in less crowded conditions than people in poorer territories. There are also large differences in rates of overcrowding within countries. In India 77% of the population live in conditions considered to be overcrowded, and there are high percentages in other South Asian territories as well: 72% in Pakistan, 61% in Bangladesh and 48% in Nepal. Australia, New Zealand, Japan, Western Europe and Canada, on the other hand, are barely visible on this map because they have very low levels of overcrowding.

OVERCROWDED HOMES
% of population living in overcrowded homes

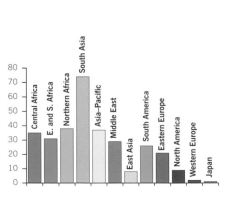

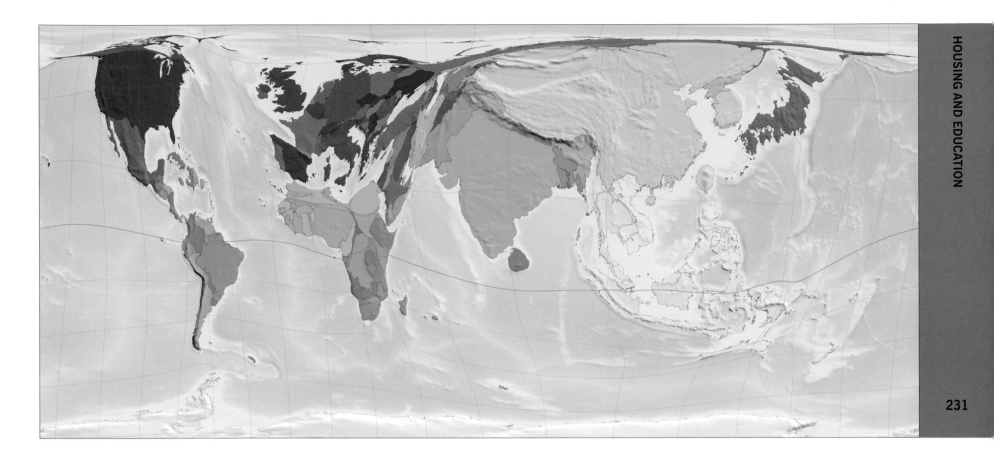

210 Durable Dwellings

The size of each territory shows the estimated numbers of permanent structures that people live in. Without durable dwellings, people's lives are inherently less secure.

Most people in most regions have access to durable dwellings, permanent structures designed to last for many years. However, in East and Southern Africa only 29% of the population lives in permanent structures. In Chad, where a sizeable proportion of the population are nomadic, not even 3% of people have such housing security.

Even those for whom a nomadic lifestyle is a matter of choice or tradition may suffer from lack of permanent housing. Their homes will be more likely to be damaged by bad weather, their legal right to live where they do may not be well established, and their habitations are likely to lack the facilities of more permanent structures.

DURABLE DWELLINGS
% of population living in permanent structures

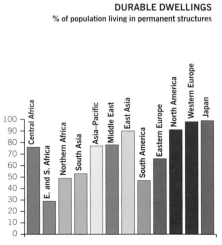

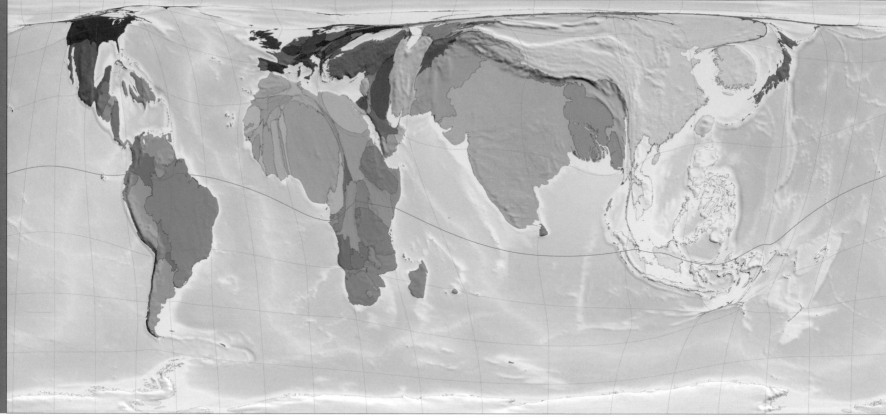

211 **Urban Slums**

The size of each territory shows the number of people living in slums – housing
legally identified as insecure by virtue of its poor condition.

In 2001, 927 million people – 15%
of the population of the world – lived
in slums. The inhabitants of slums tend
to be poor and have little in the way of
resources for improving their houses
and neighbourhoods. The legal status of
housing also affects the improvements that
inhabitants are willing to make; there is

less incentive to improve a dwelling
if you might be forced to leave it.

There are slums in almost all territories.
Among the regions, South America has
the largest proportion of its population
living in slums, at 26%, followed by
North Africa at 25%.

*'...there are two cities within one...[One] part...has all the benefits of
urban living, and the other part [is] the slums and squatter settlements,
where the poor often live under worse conditions than their rural
relatives.'* Anna Tibaijuka, executive director of UN-HABITAT, 2006

PEOPLE LIVING IN URBAN SLUMS
% of population living in urban slums, 2001

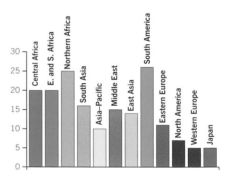

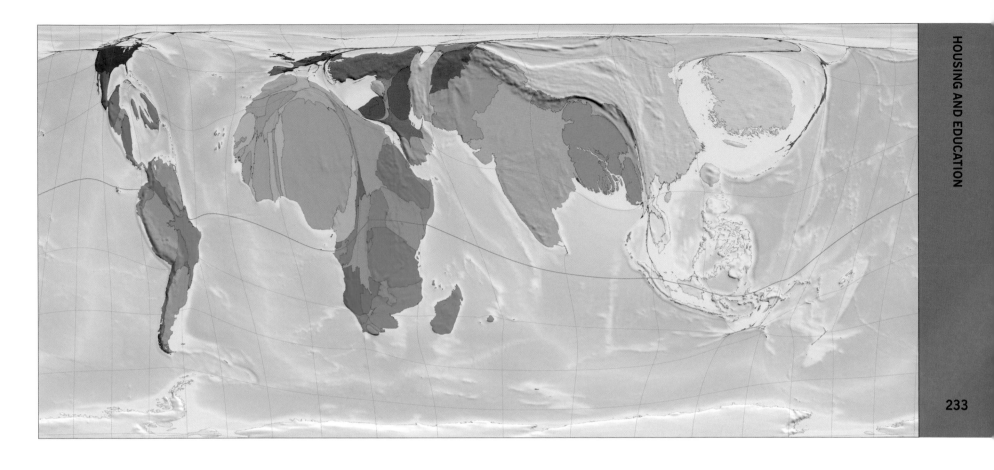

[a] additional people who started living in slums between 1990 and 2001, as % of total population

212 Slum Growth

Urban growth often partly includes the growth of slums. The size of each territory shows the increase in the number of people living in slums between 1990 and 2001.

Between 1990 and 2001 the number of people living in slums increased by 220 million. Among the regions of the world, the largest increases in slum populations were seen in South and East Asia, which already had the largest numbers of people in slums to begin with. At the territory level the largest increases were in China, India and Nigeria, all countries that also have large populations overall. Slum populations have grown more in poorer territories than in richer ones, but this decade saw an increase in the number of slum-dwellers in a large majority of all territories (85%).

SLUM POPULATIONS 1990 AND GROWTH BY 2001
millions of slum dwellers, 1990 (bottom)
millions of new slum dwellers, 1990–2001 (top)

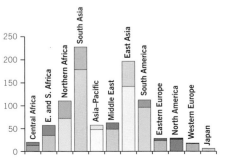

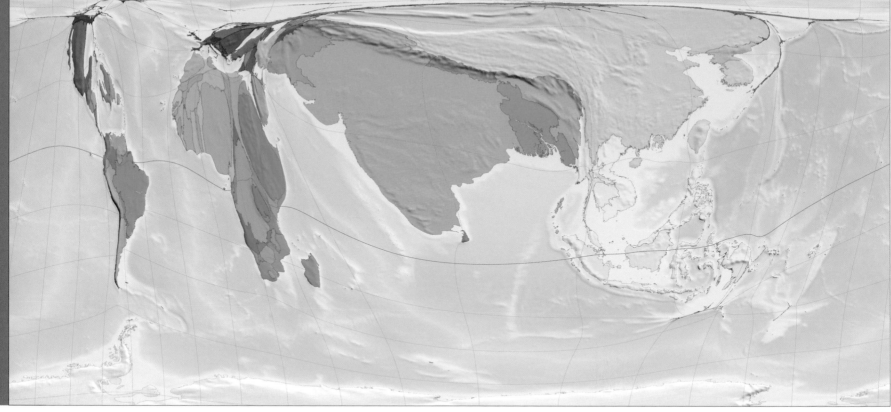

234

213 Minimal Sanitation

The size of each territory shows the number of people
with little or no access to sanitation facilities.

Effective sanitation is important for
preventing the spread of many diseases.
Worldwide, 2 people out of every 5 do not
have access to basic sanitation, meaning
latrines or toilets of any kind. In South Asia
and Central Africa, 65% of people live
without access to basic sanitation.

In Indonesia alone, 98 million people –
45% of the population – do not have
access to latrines. More people in
Indonesia live without access to basic
sanitation than in the whole of South
America, where 22% of the population
are without sanitation.

*'...none of us should tolerate a world in which over 1 million children
are, in a perversely literal sense, dying for a glass of water and a toilet.'*
Kevin Watkins, director of UN Human Development Report Office, 2006

POOR ACCESS TO BASIC SANITATION
% of population without sustainable
access to basic sanitation

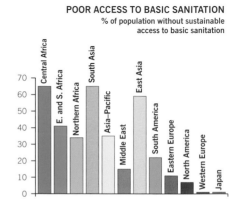

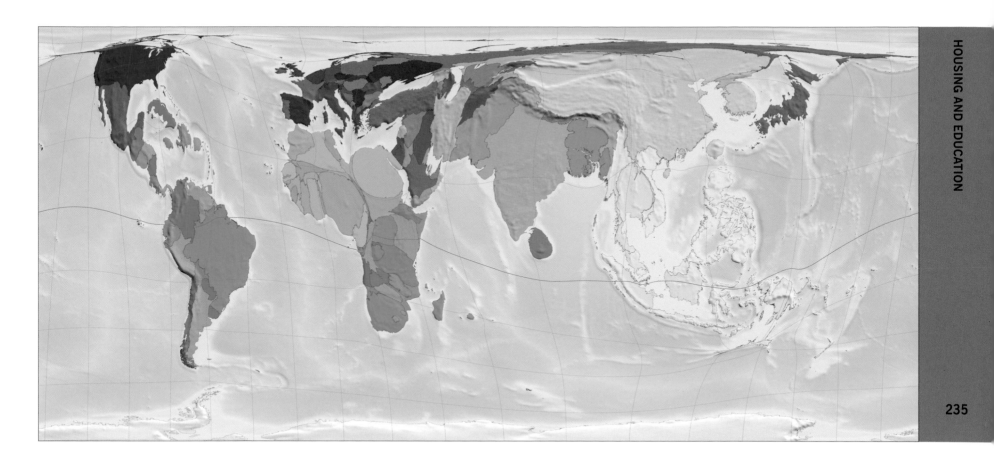

MOST PEOPLE RELYING ON BASIC (PRIVATE OR SHARED) SANITATION

Rank	Territory	%[a]
1	Comoros	97
2	Egypt	97
3	Cuba	97
4	Grenada	96
5	Libya	96
6	St Vincent and the Grenadines	95
7	Thailand	95
8	Chile	95
9	St Kitts–Nevis	95
10	Antigua and Barbuda	94
11	Sri Lanka	93
12	Paraguay	93
13	Uruguay	93
14	Suriname	92
15	Costa Rica	92
16	Greenland	92
17	Algeria	91
18	Oman	91
19	Panama	91
20	Djibouti	90

[a] % of population relying on improved sanitation facilities not connected to a sewage system, 2000

214 Basic Sanitation

The size of each territory shows the number of people with access to only basic sanitation facilities. Similar proportions can mask very different overall conditions.

Basic sanitation means pit latrines or toilets not linked to a sewerage system. Almost half (48%) of the world population uses this type of sanitation. The rest either have no sanitation facilities or are linked to a mains sewerage system. For instance, in both Rwanda and the United Kingdom 7% of the population uses only basic sanitation facilities with no sewerage. But in the UK the rest of the population has mains sewerage, while in Rwanda the rest of the population has no sanitation facilities at all.

'Are we to decide the importance of issues by asking how fashionable or glamorous they are? Or by asking how seriously they affect how many?'

Nelson Mandela, former president of South Africa, WSSD, Johannesburg, 2002

BASIC SANITATION
% of population with sustainable access to basic sanitation

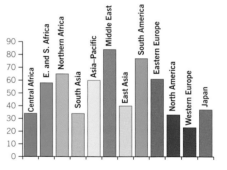

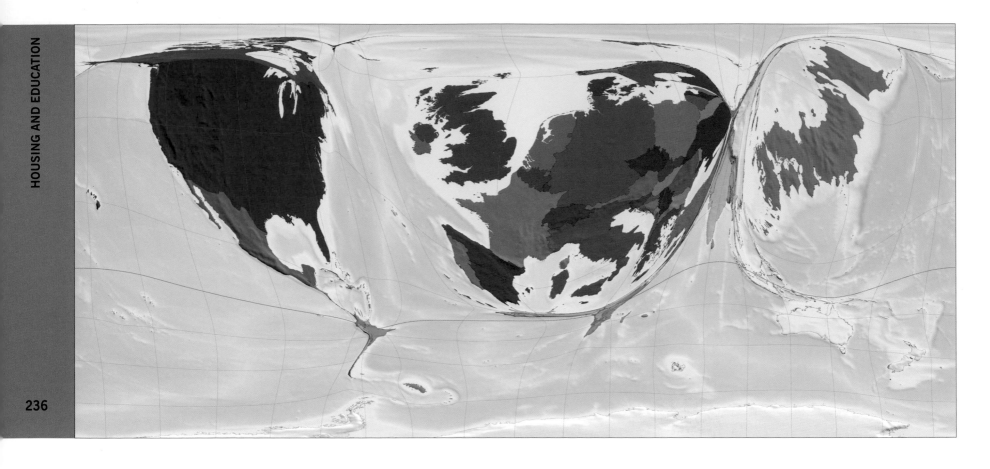

MOST PEOPLE WITH ACCESS TO SANITATION FACILITIES LINKED TO A SEWERAGE SYSTEM

Rank	Territory	%[a]
1	Netherlands	98
2	Switzerland	96
3	Luxembourg	95
4	Sweden	93
5	United Kingdom	92
6	Germany	91
7	Denmark	89
8	Austria	81
9	New Zealand	80
10	Finland	80
11	Australia	80
12	France	77
13	Canada	75
14	Norway	73
15	United States	71
16	Italy	63
17	Czech Republic	62
18	Slovenia	62
19	Japan	62
20	Ireland	61

[a] % of population with access to sanitation facilities linked to a sewerage system, 1999

215 Sewerage Sanitation

The size of each territory shows the number of people with access to toilets connected to public sewerage systems that remove waste from housing areas and treat it.

Sewerage systems help to reduce contamination of watercourses that supply water for drinking and washing. Pit latrines or cesspits do the same job, but rarely as well.

In more than three-quarters of territories toilets connected to sewers are available to 10% of the population or less. In only 6 territories are such toilets available to over 90% of people. The highest figure is 98% in the Netherlands.

In 8 of the 12 regions, less than 5% of people use toilets that are connected to sewerage systems.

'Rich societies have developed quite complicated and expensive systems for removing human wastes from houses and cities, usually by dumping them, treated to one degree or another, into subsoils or bodies of water.' Peter Bane, The Activist, 2006

CONNECTION TO PUBLIC SEWAGE TREATMENT
% of population with toilets connected to sewerage systems

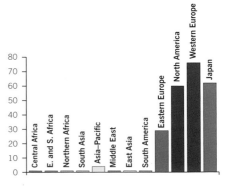

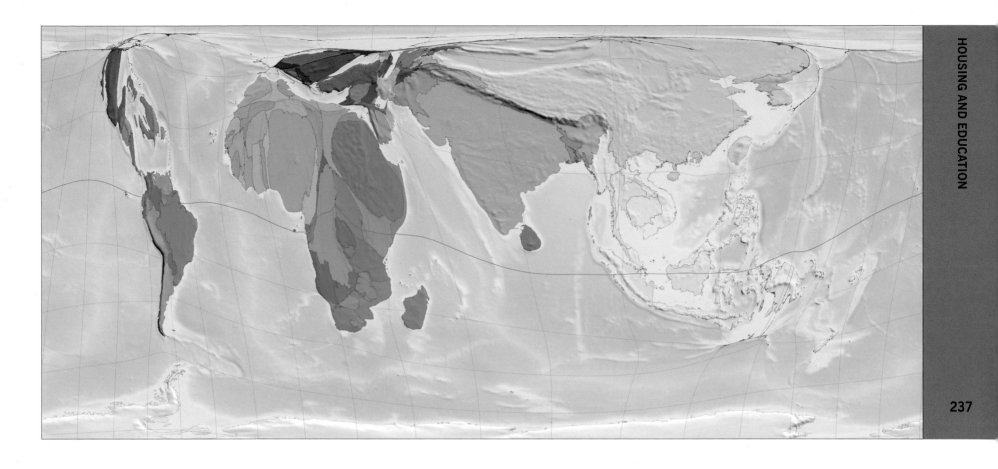

MOST PEOPLE WITH ACCESS ONLY TO POOR WATER

Rank	Territory	%[a]
1	Ethiopia	76
2	Chad	73
3	Cambodia	70
4	Mauritania	63
4	Laos	63
6	Angola	62
7	Oman	61
8	Rwanda	59
9	Burkina Faso	58
10	Papua New Guinea	58
11	Equatorial Guinea	56
12	DR Congo	55
13	Eritrea	54
14	Haiti	54
15	Madagascar	53
16	Fiji	53
17	Guinea	52
18	Sao Tome and Principe	50
19	Congo	49
20	Uganda	48

[a] % of population with access only to poor water, 2000

216 Poor Water

The size of each territory shows the number of people without reliable access to water that is safe to drink.

Drinking water is essential to life, but dirty drinking water is a major cause of disease. Safer water can be obtained from springs, from groundwater, or by treating less safe water to remove contaminants and kill microbes. In Western Europe, most people have good access to safe water, but only 50% of people in Central Africa do.

Worldwide, 18% of people do not have safe drinking water. The territory with the largest number of these is China, where 324 million people, or 25% of the population, lack access to safe water.

POOR ACCESS TO WATER
% of population without sustainable access to safe water

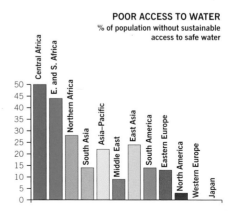

'It is the cause of debilitating diseases for the majority of children. That is how serious a lack of sanitation and clean water is.'

Hans Spruijt, head of water and environmental sanitation for UNICEF, Ethiopia, 2004

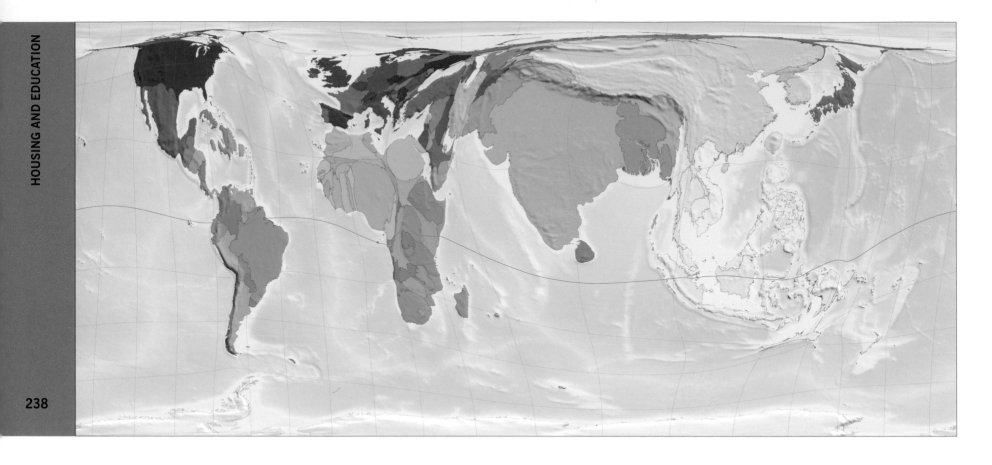

217 Children in Primary Education

The size of each territory indicates the number of children enrolled in primary education, and thus with at least a hypothetical chance of beginning a school career.

'Everyone has the right to education.' So says the 1948 Universal Declaration of Human Rights. The second of the UN's list of eight Millennium Development Goals (discussed in conjunction with Map 153) is to achieve universal primary education. In 2002, 5 out of every 6 children in the appropriate age range across the world were enrolled in primary education. This figure, however, although it sounds promising, is also somewhat misleading, since it hides a wide variation between territories. In some parts of the world enrolment is impressively high – indeed, it can exceed 100% if children stay in primary education for longer than is expected. Argentina, for example, has an impressive 108% enrolment rate. Meanwhile, on the other side of the Atlantic, just 30% of children in Angola are enrolled in primary school. (It is also important to bear in mind that enrolment does not guarantee that children will actually attend school, or that, if they do, they will complete their primary schooling.)

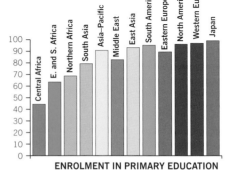

ENROLMENT IN PRIMARY EDUCATION
% projected uptake of 5 years of primary education by children of relevant age group

'If you are planning for a year, sow rice; if you are planning for a decade, plant trees; if you are planning for a lifetime, educate people.' Chinese proverb

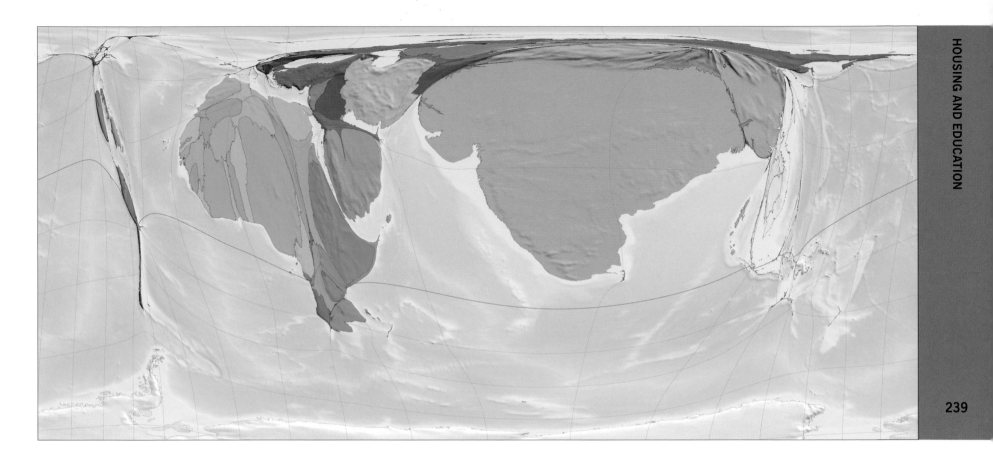

GREATEST EXCESS OF MALE OVER FEMALE PRIMARY SCHOOL ENROLMENT

Rank	Territory	%[a]
1	Yemen	20
2	Benin	13
3	Chad	11
4	Côte d'Ivoire	10
5	Iran	9
6	Togo	9
7	Guinea-Bissau	7
8	India	7
9	Guinea	7
10	Equatorial Guinea	7
11	Iraq	6
12	Afghanistan	6
13	Russia	6
14	Uzbekistan	6
15	Turkmenistan	6
16	Mali	6
17	Burkina Faso	6
18	Niger	6
19	Ethiopia	5
20	Burundi	5

[a] % of relevant age group by which boys' enrolment in primary education exceeds girls'

218 Gender Balance in Primary Education

The size of each territory shows the number of boys enrolled in primary education minus the number of girls enrolled. The widest gender gaps tend to be found in poorer countries.

This map shows how many more boys there are than girls enrolled in primary education in each territory. Territories in which there are fewer boys enrolled than girls have size zero on this map and so do not appear.

The territory with the widest gap between boys and girls is India, where there are about 8 million fewer girls than boys in the first 5 years of education: 10 times the number in any other territory. Part of the reason for this difference is that India has a large population to begin with, but this is not the whole explanation, since in other countries with large populations,

such as China, there is little or no difference between the enrolment of boys and of girls.

When viewed in terms of percentage enrolment rates, the biggest gaps between boys and girls are found in Yemen, Benin, Chad and Côte d'Ivoire. As a region, Northern Africa has the widest gap, with 2.5 million fewer girls than boys enrolled in primary education.

102 territories – more than half – have roughly equal numbers of boys and girls enrolled in primary education or, in some cases, slightly more girls than boys.

GENDER GAP IN PRIMARY EDUCATION
excess of boys over girls enrolled in primary education, as % of relevant age group

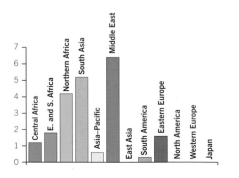

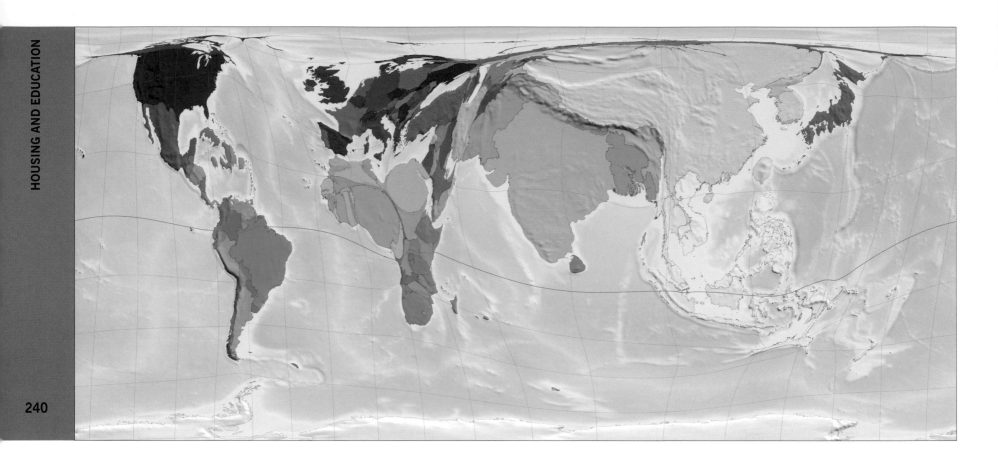

219 Children in Secondary Education

The size of each territory indicates the number of children enrolled in secondary education. In India and much of Africa, under half of children attend school at secondary level.

Worldwide, an average of 73 million children, out of a possible 122 million, are enrolled in each year of secondary education. Even assuming that all those who enrol actually attend, that still means only 60% of children are receiving a secondary education.

In China 89% of children receive a secondary education. In India only 49% do, and the figures for Africa are even lower: 45% in Northern Africa, 25%

in East and Southern Africa, and 13% in Central Africa. The lowest enrolment in secondary school education of any territory is 5%, in Niger.

Secondary education is compulsory in some territories but a rarity in others. The largest population of secondary school students is in China, which is home to a quarter of all the secondary school students in the world.

ENROLMENT IN SECONDARY EDUCATION
as % of relevant age group

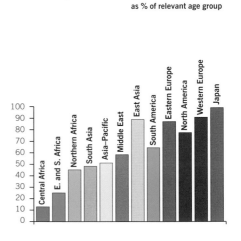

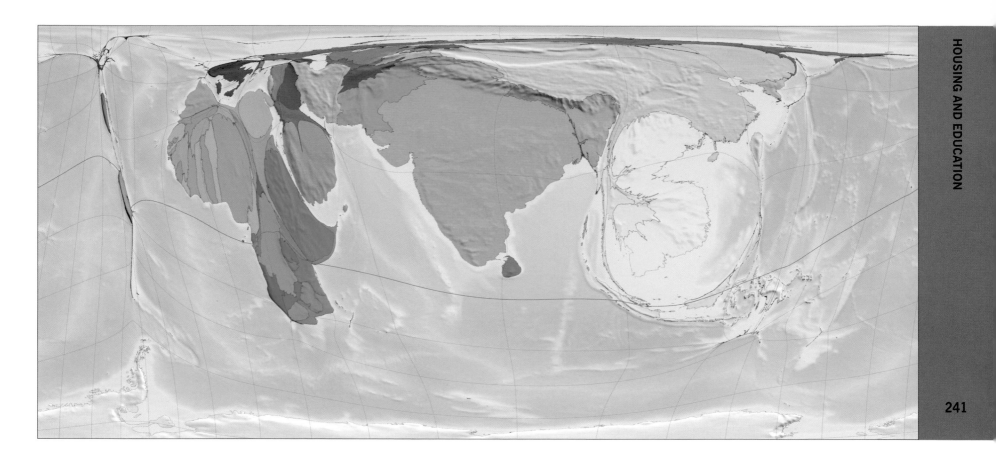

220 Gender Balance in Secondary Education

The size of each territory shows the number of boys enrolled in secondary education minus the number of girls enrolled. Trends established at primary level tend to be reinforced later.

Territories with a big difference between the percentages of boys and girls enrolled in primary education often have an even larger difference at the secondary level. In 104 territories boys and girls are equally enrolled, or girls' enrolment slightly exceeds that of boys. (Territories in which there are fewer boys enrolled than girls have size

zero on this map and so do not appear.) Most of these territories are found in South America, North America, Eastern Europe and Western Europe, although there are anomalies in these regions where, overall, more boys go to secondary school than girls, such as Peru, Guatemala, Bulgaria, Italy and Switzerland.

GENDER GAP IN SECONDARY EDUCATION
excess of boys over girls enrolled in secondary education, as % of relevant age group

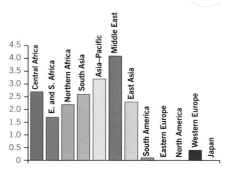

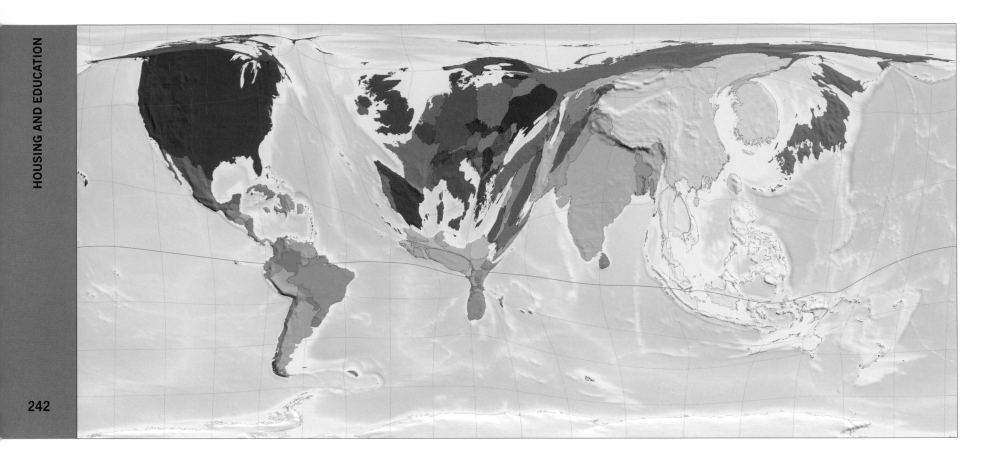

221 Tertiary Education

The size of each territory shows the number of people enrolled in tertiary education, including university study and vocational training after secondary school.

In 2002, about 105 million students worldwide were enrolled in tertiary education. The highest enrolment as a fraction of population in the 15–24 age bracket was in Finland, where the enrolment rate is currently 3.6 times the world average. A young person between the ages of 15 and 24 is 140 times more likely to receive tertiary education in Finland than in Mozambique.

Most territories have at least some people in tertiary education and hence are visible on this map. Of the territories with such low rates of tertiary education as to be invisible, most are in Central Africa.

ENROLMENT IN TERTIARY EDUCATION
as % of relevant age group

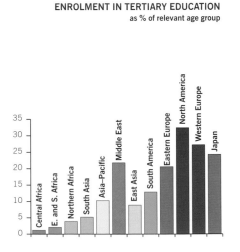

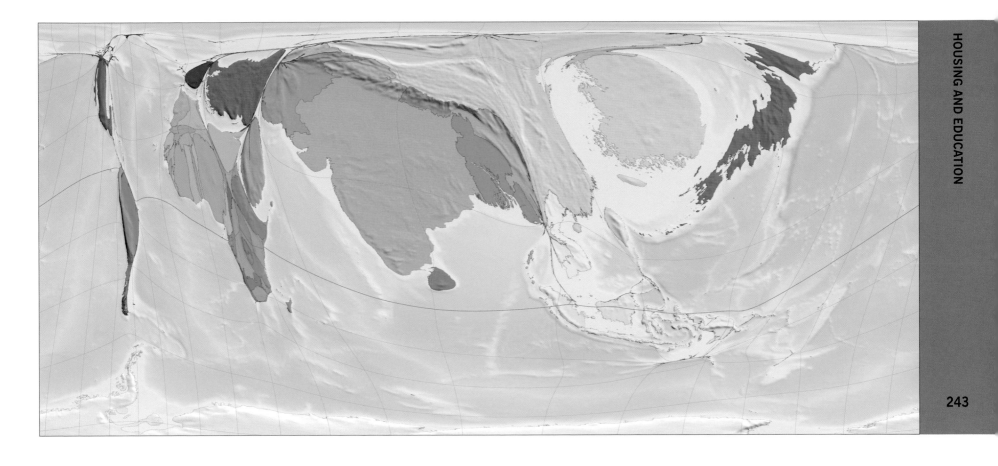

GREATEST AND LEAST EXCESS OF MALE OVER FEMALE TERTIARY ENROLMENT

Rank	Territory	%ᵃ
1	South Korea	10
2	Bolivia	5
3	Tajikistan	4
4	Yemen	3
5	Switzerland	3
6	Turkey	2
7	Japan	2
8	Côte d'Ivoire	2
9	Nepal	1
10	Central African Republic	1
69	Peru	0.16
70	Gaza and West Bank	0.15
71	Angola	0.15
72	Cape Verde	0.14
73	Sudan	0.13
74	Madagascar	0.10
75	Comoros	0.09
76	Hong Kong (China)	0.07
77	Djibouti	0.06
78	Mozambique	0.05

ᵃ % of relevant age group by which male exceeds female enrolment in tertiary education (female enrolment equals of exceeds male in 122 territories)

222 Gender Balance in Tertiary Education

The size of each territory shows the number of males enrolled in tertiary education minus the number of females. Worldwide, the gender gap is virtually non-existent, but this masks wide local variations.

In most territories where it is common for girls to receive a secondary education, more young women than men subsequently enrol in tertiary education. In 122 territories the proportion of women enrolling in tertiary education is the same as or greater than that of men. Exceptions are Japan, where 46% of tertiary students are female, and South Korea, with 34%.

Conversely, in places where fewer girls than boys receive a secondary education, the gap between females and males is usually even wider at tertiary level; and the situation is still more pronounced in places where even primary education is uncommon for girls. In Central Africa, Northern Africa, South Asia, and East and Southern Africa, for instance, gender differences in education begin in primary school.

Territories in which there are fewer boys enrolled than girls have size zero on this map and so do not appear.

Worldwide, the total numbers of male and female students are in fact approximately equal, the territories in which there are more female than male students roughly balancing those in which there are more male than female.

GENDER GAP IN TERTIARY EDUCATION
excess of males over females in tertiary education, as % of relevant age group

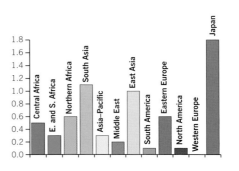

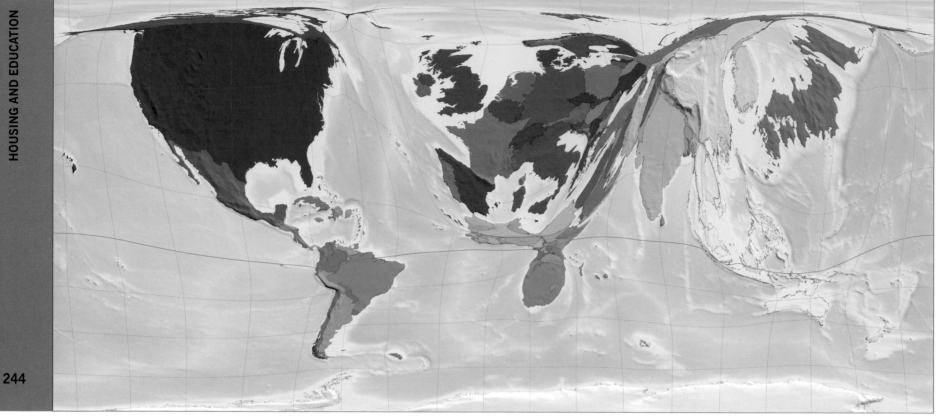

244

Rank	Territory	PPP US$[a]
1	Norway	6,151
2	Luxembourg	4,346
3	Denmark	4,074
4	Switzerland	3,777
5	Sweden	3,621
6	United States	3,460
7	Italy	3,119
8	Belgium	3,074
9	Canada	2,915
10	Austria	2,864
191	Malawi	23
192	Chad	22
193	Burundi	22
194	Madagascar	21
195	Mali	21
196	Niger	20
197	DR Congo	19
198	Zambia	17
199	Myanmar	17
200	Guinea-Bissau	14

[a] spending per child of primary
school age per year, 2001

223 Primary Education Spending

The size of each territory shows spending on primary education
in 2001, in US dollars, adjusted for local purchasing power.

In 2001 US$784 billion was spent
worldwide on primary education, after
adjusting for local purchasing power.
The largest amount was spent in the
United States, which accounted for
28% of all the spending in the world.
By contrast, only 0.28% of the world
total was spent in Nigeria.

Spending per child also varies widely.
Among the regions, Japan, North America
and Western Europe spend the most,
often more than three times the level that
prevails elsewhere. Central Africa, which
has the lowest rate of primary school
enrolment, also spends less than any
other region.

PRIMARY EDUCATION SPENDING
PPP US$ spent per child of primary school age, 2001

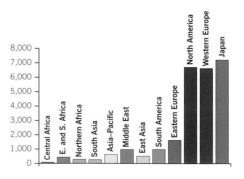

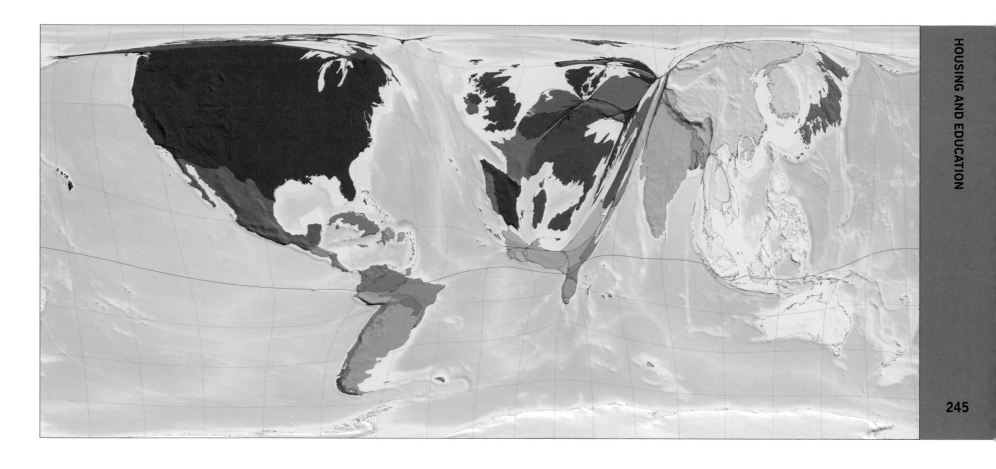

HIGHEST AND LOWEST GROWTH IN SPENDING ON PRIMARY EDUCATION

Rank	Territory	PPP US$[a]
1	Australia	2,101
2	Denmark	2,029
3	Luxembourg	2,028
4	Italy	1,712
5	Norway	1,509
6	United States	1,312
7	Belgium	1,245
8	Estonia	1,104
9	Singapore	1,039
10	Greenland	961
121	Benin	15
122	Ethiopia	14
123	Guatemala	14
124	Somalia	14
125	Tanzania	12
126	Swaziland	12
127	Togo	12
128	Solomon Islands	6
129	Malawi	4
130	Madagascar	4

[a] increase from 1990 to 2001, per child of primary school age per year (spending was constant or fell in 70 territories)

224 Growth in Primary Education Spending

The size of each territory shows the increase in spending on primary education between 1990 and 2001. About two-thirds of territories registered at least some increase.

Between 1990 and 2001 spending on primary education increased in 130 of the 200 territories on the map. The largest increase in spending per child was in Australia; Denmark was next, followed by Luxembourg.

'You can't broaden access to primary education unless you have enough teachers trained to teach primary school. You can't train these teachers...without raising the educational...levels of the entire society.'

Adriana Puiggros, professor at the University of Buenos Aires and representative of the Confederation of Educational Workers of the Argentine Republic, 1996

PRIMARY EDUCATION SPENDING GROWTH
increase per child of primary school age, PPP US$ 1990–2001

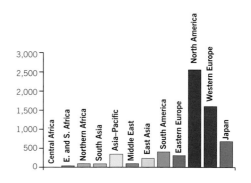

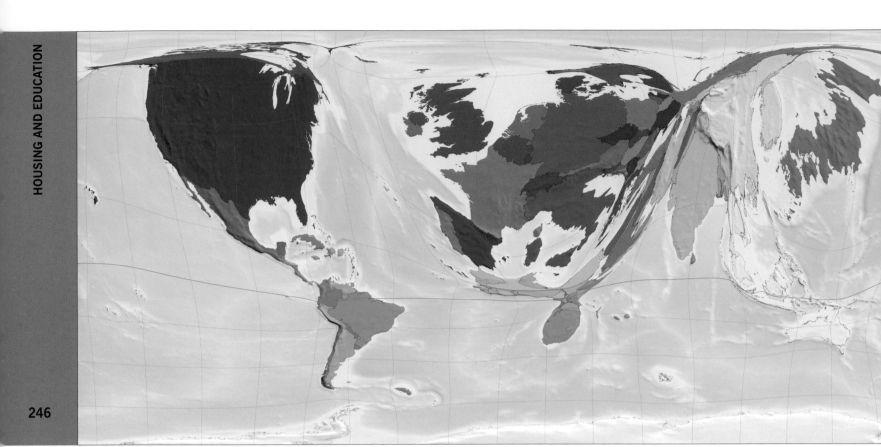

225 Secondary Education Spending

The size of each territory shows spending on secondary education
in 2001, in US dollars, adjusted for local purchasing power.

Although it seems incredible on first glance,
it is in fact true that only a sixth of the
children in the world receive above the
average spending on their secondary
education. The education of that sixth,
however, is so well funded, even after
adjusting for variations in local purchasing
power, that the average spending is high
enough for all the other children in the
world to fall below it. In Western Europe,
North America and Japan over 5 times
the world average is spent per child on
secondary education. At the other end
of the scale, the territories that spend little
on secondary education do so principally
because most children get none at all, but
typically there is also very little spent even
on those who do receive it.

SECONDARY EDUCATION SPENDING
PPP US$ per child of secondary school age, 2001

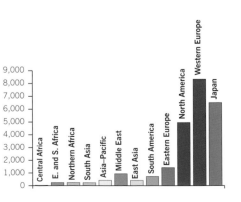

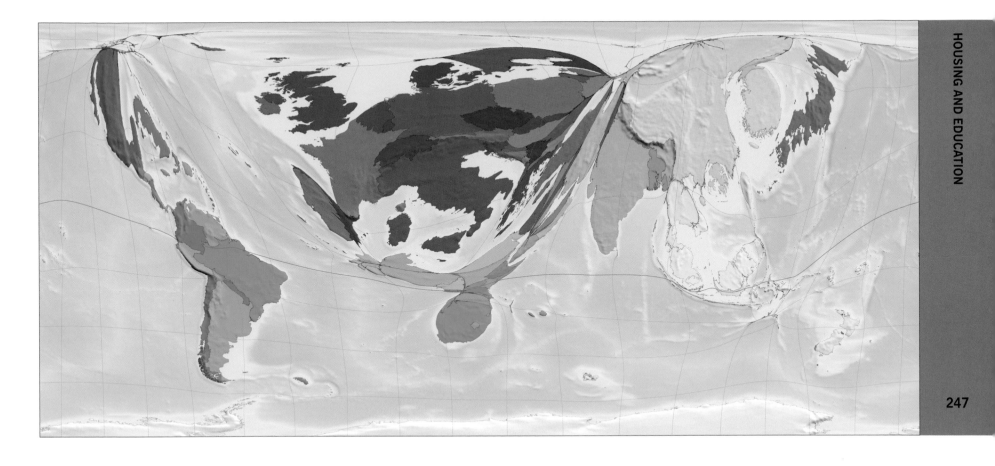

HIGHEST AND LOWEST GROWTH IN SPENDING ON SECONDARY EDUCATION

Rank	Territory	PPP US$ᵃ
1	Luxembourg	3,073
2	Denmark	2,158
3	Sweden	1,966
4	Iceland	1,951
5	Italy	1,800
6	Switzerland	1,756
7	Portugal	1,545
8	Czech Republic	1,471
9	Estonia	1,393
10	New Zealand	1,345
137	Tanzania	14
138	Kenya	14
139	El Salvador	14
140	Senegal	13
141	Ethiopia	13
142	Malawi	10
143	Gambia	9
144	Mozambique	6
145	Central African Republic	4
146	Madagascar	2

ᵃ **increase from 1990 to 2001 per child of secondary school age per year (spending was constant or fell in 54 territories)**

226 Growth in Secondary Education Spending

The size of each territory shows the increase in spending on secondary education between 1990 and 2001. Few territories did not increase spending at all.

Spending on secondary education increased in most territories between 1990 and 2001. The biggest overall increases in spending were in China, Italy and India. The increase in China was 146%, in Italy 67%, and in India 37%. India and China now spend similar amounts per child, but Italy spends 30 times as much as either.

Of the 32 territories in South America, 28 increased their spending on secondary education during this period. The largest increases were in Brazil, Argentina and Colombia, where the largest populations live.

'When salaries are regularly late (in some cases by up to 3 weeks), teachers are understandably left frustrated, demoralized and in financial hardship.'

Sutapa Choudhury, *Teacher Professionalism in Punjab: Raising Teachers' Voices*, 2005

SECONDARY EDUCATION SPENDING GROWTH
increase per child, PPP US$, 1990–2001 of secondary school age

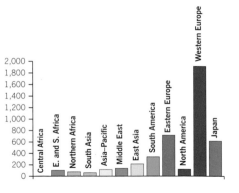

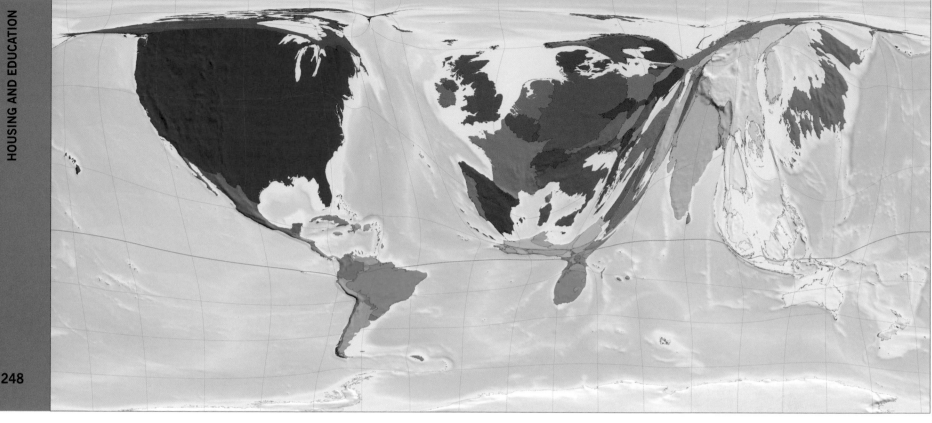

227 Tertiary Education Spending

The size of each territory shows spending on tertiary education in US dollars, adjusted for local purchasing power.

61% of all spending on tertiary education, when measured in US dollars adjusted for local purchasing power, occurs in North America and Western Europe. At the other extreme, the total tertiary education spending in all of Central, Eastern and Southern Africa was just 1.5% of the worldwide total. Tertiary education includes the training of doctors, engineers and scientists, and holds substantial benefits for territories that are able to invest in it.

'With regard to the issue of equity in social expenditure, investment in primary education reaches a much broader cross section of society than does funding to higher education...' Michael C. Gonzales, *Africa Today*, 1999

TERTIARY EDUCATION SPENDING
PPP US$ per person, 2001

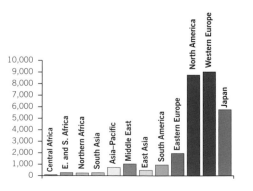

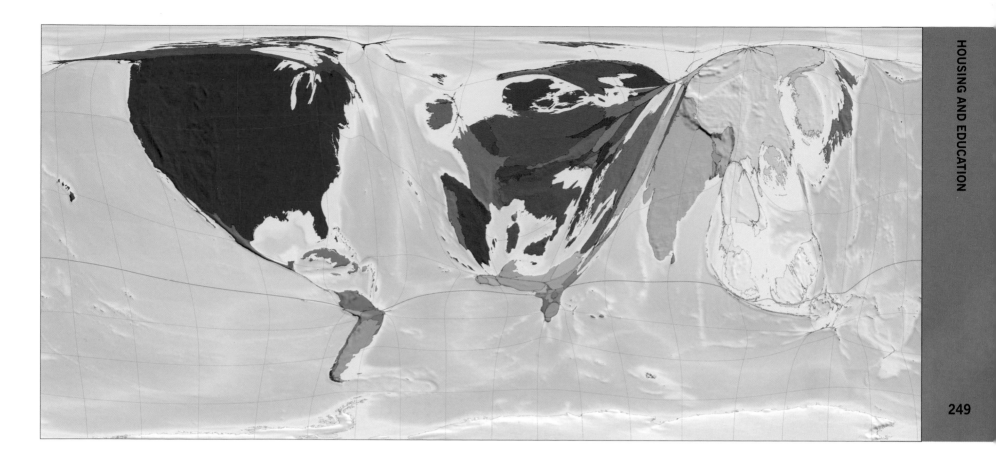

228 Growth in Tertiary Education Spending

The size of each territory shows the increase in spending on tertiary education between 1990 and 2001. Again, most territories registered some increase, but amounts vary very widely.

Between 1990 and 2001 there were increases in spending on tertiary education in 135 of the 200 territories. North America and South Asia are the only regions in which every territory has increased spending, although in East Asia there has been a spending increase in every territory except for Mongolia and in Central Africa there has been an increase in every territory except for Burundi.

The size of these increases in spending, however, varies hugely. The increases in spending per person in North America and Western Europe over this decade were over three times those in any other region.

'...if Europe wants to retain its competitive edge at the top of the global value-added chain, the education system must be made more flexible, more effective and more easily accessible to a wider range of people.' Andreas Schleicher, member of the OECD Education Directorate, 2006

TERTIARY EDUCATION SPENDING GROWTH
increase per person, PPP US$, 1990–2001

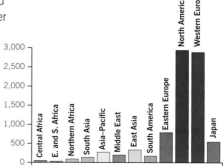

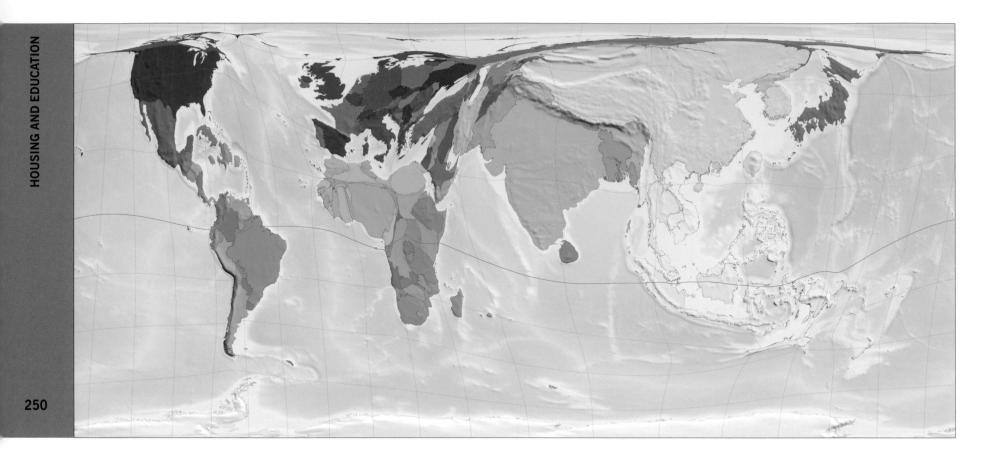

229 Youth Literacy

The size of each territory indicates the number of literate young people between the ages of 15 and 24. In a few African territories, less than half of young people can read and write.

The minimal definition of literacy used here is the ability to read, write and understand a short, simple statement about one's everyday life. Of all the 15- to 24-year-olds living in the world, 88% are literate. Over half of these live in Asia.

In most territories most of the young people can read and write. Only 5 territories have youth literacy rates lower than 50%; of these, 4 are in Northern Africa.

8 of the 12 regions have youth literacy rates over 95%. The highest rate is found in Japan.

YOUTH LITERACY
million literate people aged 15–24, 2002

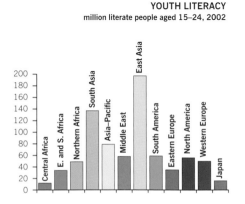

'*The freedom promised by literacy is both freedom from – from ignorance, oppression, poverty – and freedom to – to do new things, to make choices, to learn.*' Koïchiro Matsuura, director general of UNESCO, 2001

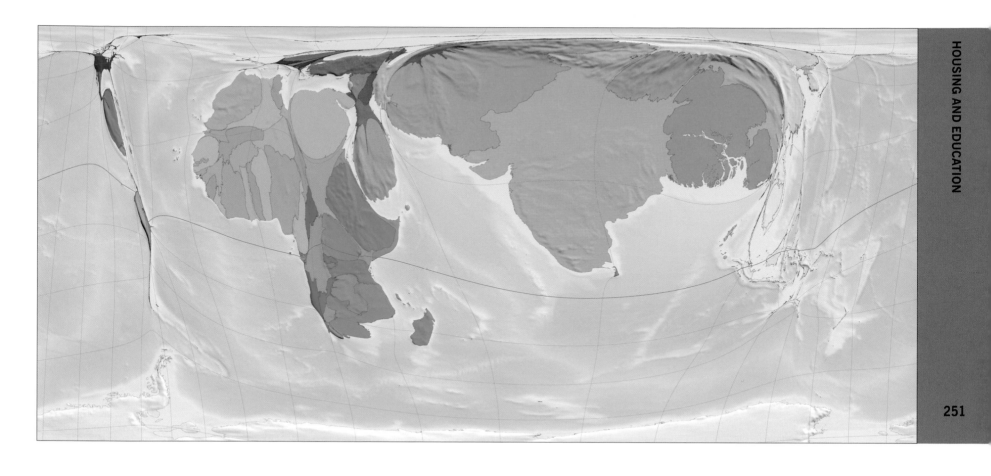

230 Gender Balance of Illiterate Youth

The size of each territory indicates the number of illiterate females aged 15 to 24 years minus the number of illiterate males in the same age range.

This map shows how many more illiterate girls and young women there are in each territory than illiterate boys and young men. Put another way, it shows how many of the former would have to be taught to read and write to reach the same literacy rates as the latter.

Territories in which there are fewer illiterate females than males have size zero on this map and so do not appear.

The largest gaps between male and female literacy are in South Asia, Northern Africa, and East and Southern Africa. In Pakistan, for instance, there are 2.6 million more girls who cannot read than boys; 24% of the girls and young women living in Pakistan are illiterate. Region by region, the territories with the biggest literacy gaps between male and female youth are Yemen in the Middle East and Central Asia, Turkey in Eastern Europe, Indonesia in the Asia–Pacific and Australasia, Guatemala in South America and the United States in North America.

GREATER FEMALE THAN MALE YOUTH ILLITERACY
extra women aged 15–24 who are illiterate compared with men, as a % of the youth population

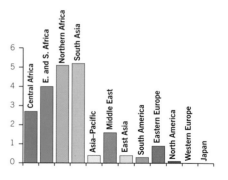

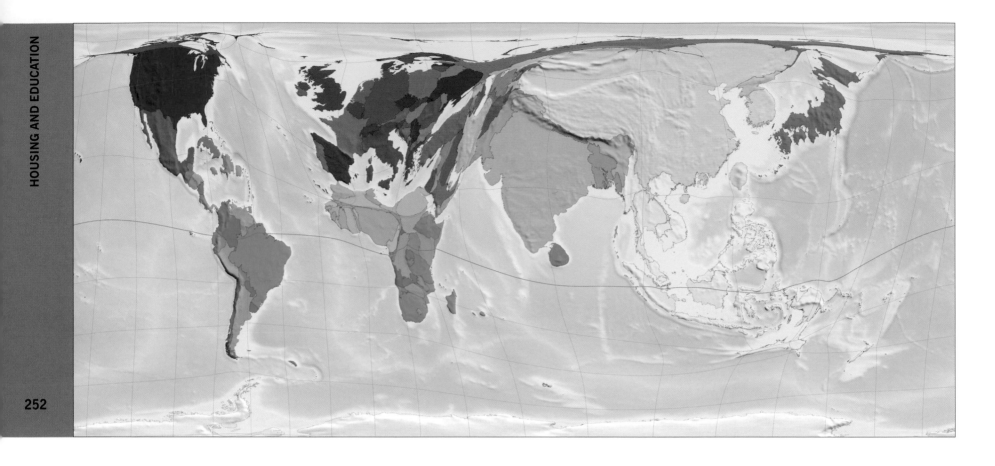

231 Adult Literacy

The size of each territory shows the number of literate people aged 15 years and older. Younger people are overtaking their elders in reading and writing.

Worldwide, there are 3.6 billion literate adults, where adults are defined as people of 15 years of age or older. 82% of the adult population of the world is able to read, write and understand simple statements.

The proportion of literate adults is, however, lower than the proportion of literate young people in every region of the world. The widest gaps between youth and adult literacy rates are in Northern Africa, Central Africa, and East and Southern Africa. The narrowest gap is in Japan. The largest populations of literate adults live in China, India and the United States. India has a literacy rate of 61%; in both China and the United States, by contrast, 91% of people aged 15 and over are literate.

'I am somehow less interested in the weight and convolutions of Einstein's brain than in the near certainty that people of equal talent have lived and died in cotton fields and sweatshops.'

Stephen Jay Gould, palaeontologist, evolutionary biologist and historian of science, 1980

DISTRIBUTION OF ADULT LITERACY
millions of literate people aged 15 and over, 2002

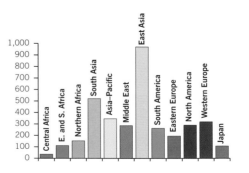

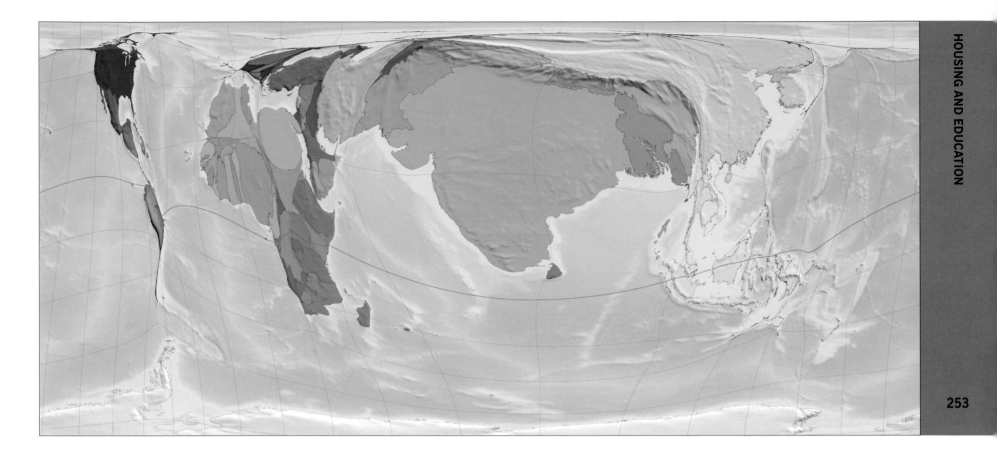

GREATEST EXCESS OF FEMALE OVER MALE ADULT ILLITERACY

Rank	Territory	%[a]
1	Yemen	21
2	Nepal	18
3	Mozambique	15
4	Central African Republic	15
5	Benin	14
6	Togo	14
7	India	14
8	Malawi	13
9	Pakistan	13
10	Bhutan	13
11	Morocco	12
12	Egypt	12
13	Libya	11
14	Laos	11
15	Sudan	11
16	Cambodia	10
17	Tunisia	10
18	Mauritania	10
19	Bangladesh	10
20	Uganda	10

[a] % of total population aged over 15 by which female exceeds male illiteracy

232 Gender Balance of Illiterate Adults

The size of each territory indicates the number of illiterate women aged 15 years or older minus the number of illiterate men in the same age bracket.

In Western Europe, Japan and South America, rates of literacy between men and women are similar, and so these regions appear small (in some cases invisible) on this map. Elsewhere, however, particularly in India, China, Pakistan and Iran, the numbers of women who cannot read or write are much larger than the numbers of men. When we look at percentage literacy rates rather than total numbers, the biggest gaps are in Yemen, where male and female literacy rates are 69% and 28% respectively, Nepal (62% and 26%), Mozambique (62% and 31%) and the Central African Republic (64% and 34%).

Territories in which there are fewer illiterate females than males have size zero on this map and so do not appear.

'...illiteracy is essentially a manifestation of social inequality, the unequal distribution of power and resources in society.'
Bharati Silawal-Giri, UNDP, Nepal, 2003

GENDER GAP IN ADULT LITERACY
extra women aged 15+ who are illiterate compared with men aged 15+, as % of the adult population

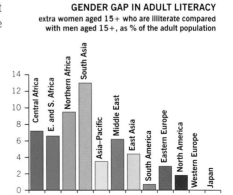

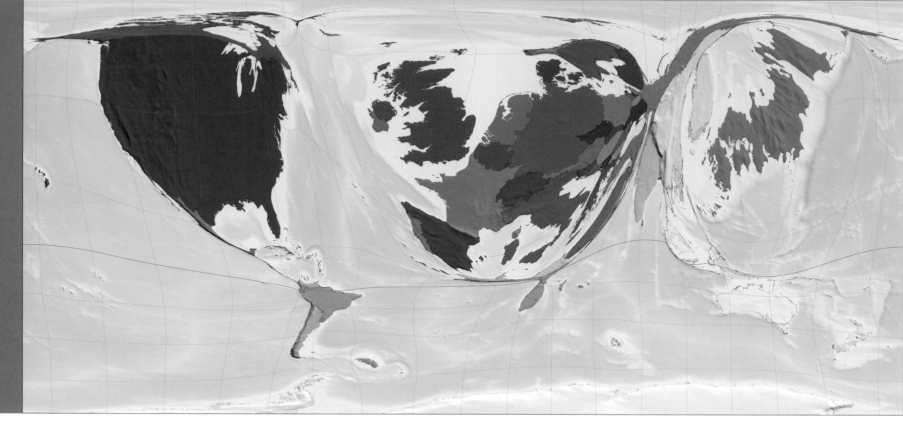

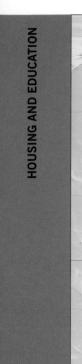

233 Science Research

The size of each territory indicates the number of scientific papers published by researchers resident there.

Articles published in scientific journals are the central means by which scientists announce and describe their discoveries to the world and to other scientists. In this context, 'science' is interpreted to cover physics, biology, chemistry, mathematics, clinical medicine, biomedical research, engineering, computer science, technology, and the earth and space sciences.

Researchers in the United States publish more than three times as many papers per year than their counterparts in any other territory. Japan is next. More scientific research tends to be carried out, and therefore more results published, in richer territories. Thus over four times as many scientific papers are published per person in Western Europe, North America and Japan as in any of the other regions.

SCIENTIFIC PUBLICATIONS
papers published per million people per year

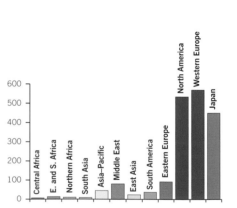

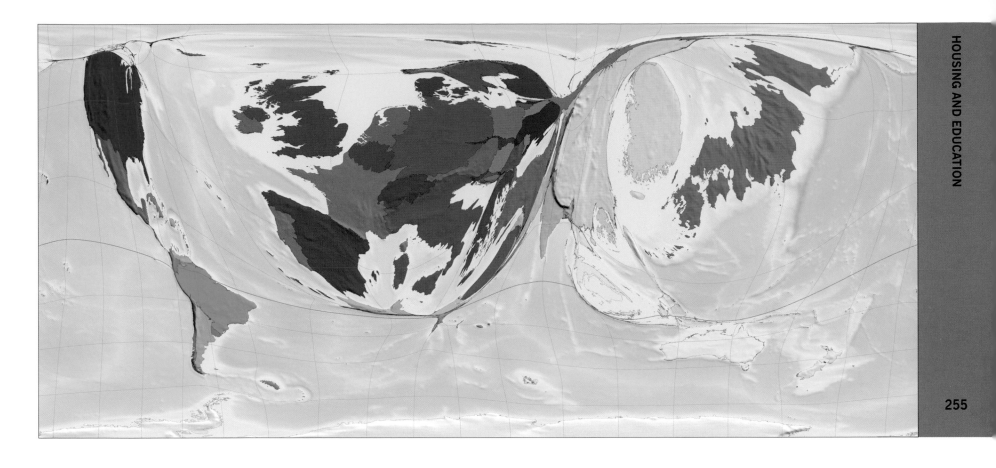

234 Growth of Science Research

The size of each territory shows the increase in the number of scientific papers published by researchers resident there between 1990 and 2001.

This map shows the growth, if any, in scientific research in each territory between 1990 and 2001, measured in terms of the number of scientific papers published. In 1990, 80 scientific papers were published per million people living in the world. By 2001 this figure had increased to 106 per million, the increase being concentrated primarily in territories with strong scientific research programmes to begin with. However, the United States, which has the largest scientific research output by far, experienced a smaller total increase than several other countries, including Japan, China, Germany and South Korea. When measured per person, the territory with the largest increase in scientific publications was Singapore.

'Singapore is engaging robustly in the materials science research, as we position ourselves for the global, knowledge-driven economy, and for our next phase of development as a society.'

Tharman Shanmugaratnam, acting minister for education, Singapore, 2003

GROWTH IN SCIENTIFIC PUBLICATIONS, 1990–2001
additional papers published per million people

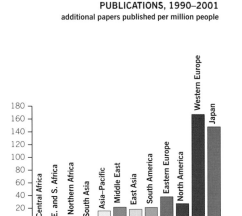

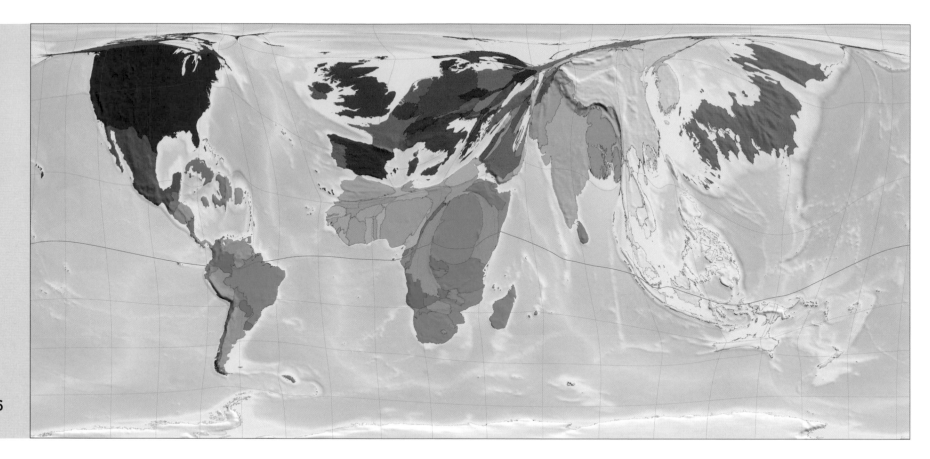

235 Cost of Universal Telephone Service

The size of each territory indicates the total cost of providing
every person with a landline telephone service.

This is a hypothetical map: it shows how
much it would cost, in aggregate, for
everyone in every one of the 200 territories
to have a landline telephone. If telephone
service cost the same everywhere, this map
would be indistinguishable from a simple
map of population. But it doesn't. In fact,
landline telephones tend to cost most in
areas that are either poor or rich and least
in areas of medium wealth. The world
average cost of a landline telephone

service is US$630 per year, but regional
averages range from US$265 in South Asia
to US$2,805 in Japan.

In reality, of course, nowhere near
the entire population of the planet has
a telephone. Overall, there are just 12
landline telephones for every 100 people
on Earth. In Western Europe there are
57 per 100 people. In Central Africa
there are 3 per 100.

REVENUE FROM ONE TELEPHONE
revenue to provider from one telephone main line, US$, 2002

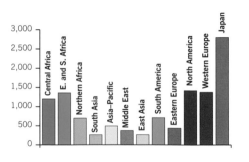

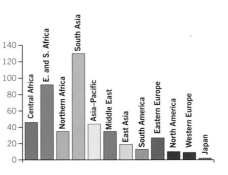

MOST AND FEWEST TELEPHONE FAULTS

Rank	Territory	Faults[a]
1	Tonga	761
2	Zimbabwe	223
3	Kenya	221
4	Bangladesh	208
5	Mali	178
6	Myanmar	169
7	Swaziland	160
8	Liberia	144
9	Bhutan	130
10	India	126
191	Oman	1.8
192	Brazil	1.7
193	Japan	1.7
194	Spain	1.5
195	South Korea	1.5
196	Hong Kong (China)	1.3
197	Egypt	1.0
198	Gabon	0.5
199	Netherlands	0.5
200	United Arab Emirates	0.3

[a] faults per 100 main line telephones per year, 2002

236 Telephone Faults

The size of each territory indicates the number of people who experience faults with their landline telephone service each year.

Telephone faults are problems with telephone lines that prevent communication. This map shows in which territories people are most likely to experience such faults. By measuring faults in terms of numbers of people affected rather than number of telephones, we allow for the fact that a single telephone can serve many people. In some cases an entire village may be served by one line, in which case a fault on that line affects far more people than a fault on a line used by just a single person. Less reliable telephone service often coincides with widespread sharing of telephone lines, so that in areas with large numbers of faults each fault may also affect many people.

The reliability of telephone lines varies greatly between regions. South Asia, for instance, has 19% of all telephone lines but 58% of all telephone faults.

RATE OF TELEPHONE FAULTS
telephone faults reported per 100 main lines per year, 2002

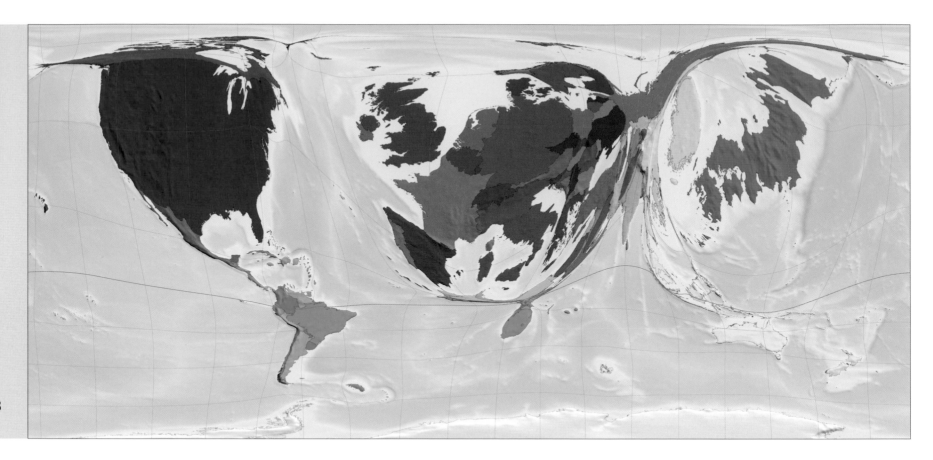

MOST AND FEWEST LANDLINE TELEPHONES IN 1990

Rank	Territory	Landlines[a]
1	Sweden	681
2	Switzerland	574
3	Denmark	567
4	Canada	565
5	United States	547
6	Finland	534
7	Iceland	510
8	Norway	502
9	France	495
10	Luxembourg	481
191	Uganda	2
192	Myanmar	2
193	Laos	2
194	Niger	1
195	Mali	1
196	DR Congo	1
197	Burundi	1
198	Chad	1
199	Viet Nam	1
200	Cambodia	0.5

[a] landline telephones per 1,000 people, 1990

237 Telephone Lines in 1990

The size of each territory indicates the number of landline telephones in 1990. In some wealthier regions, most households had one; in poorer areas, whole communities had no telephone service.

In 1990 there were 516 million landline telephones worldwide. 63% were in Western Europe and North America; along with Japan, these regions were the only parts of the world where there were more than 4 landlines for every 10 people, or about one landline per family. In Central Africa and South Asia, by contrast, there was less than 1 landline for every 100 people.

The territories with the largest total numbers of landline telephones in 1990 were the United States, at 141 million, Japan at 53 million and Germany at 36 million.

'...universal access to basic communication and information services is a fundamental human right.'

Pekka Tarjanne, ITU secretary general, 1997

TELEPHONE LANDLINES
number per 1,000 people, 1990

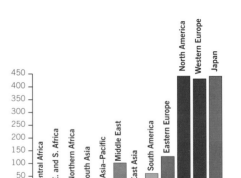

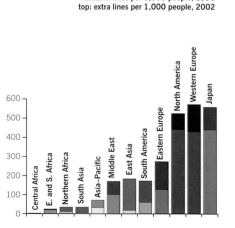

259

238 Telephone Lines in 2002

The size of each territory indicates the number of landline telephones in 2002. The world total had nearly doubled in less than a decade.

In 2002 there were just over 1 billion landline telephones worldwide, almost twice the number in 1990. Interestingly, the 6 territories with the largest numbers of landlines were all in different regions. They were, in order, China, the United States, Japan, Germany, India and Brazil. The territories with the smallest numbers of landlines were Andorra and Niue, both of which have very small populations to begin with.

The regions with the largest numbers of landlines per person in 2002 were North America, Western Europe and Japan, just as in 1990. The largest increase in landlines, however, was in East Asia, where there were roughly 10 times as many landlines (both in total and per person) in 2002 as in 1990.

INCREASE IN LANDLINES
bottom: lines per 1,000 people, 1990
top: extra lines per 1,000 people, 2002

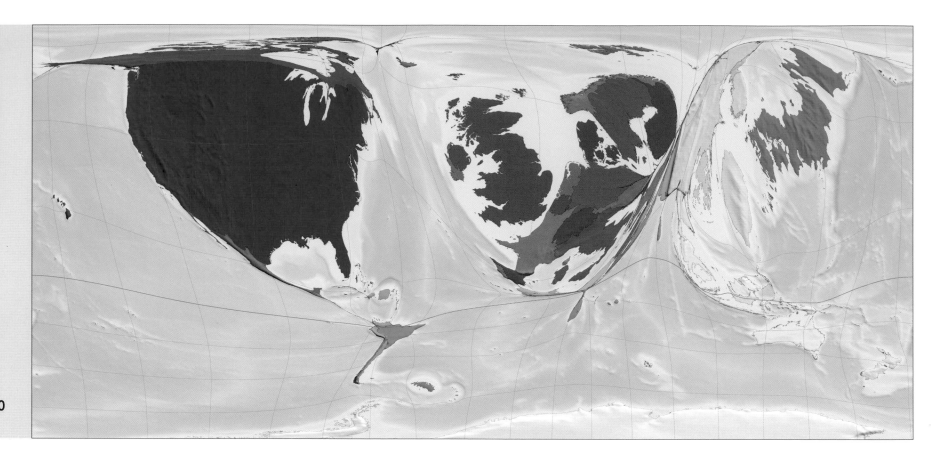

Rank	Territory	Mobile phones[a]
1	Sweden	54
2	Finland	52
3	Norway	46
4	Iceland	39
5	Denmark	29
6	Hong Kong (China)	24
7	Canada	22
8	United States	21
9	United Arab Emirates	19
10	United Kingdom	19
11	Switzerland	18
12	Singapore	17
13	Taiwan	12
14	Kuwait	12
15	Australia	11
16	Bahrain	10
17	Austria	10
18	Bahamas	8
19	Brunei	7
20	Japan	7

[a] mobile phone subscriptions per 1,000 people, 1990

239 Mobile Phones in 1990

The size of each territory indicates the number of mobile (cellular) telephone subscriptions, including prepaid subscriptions, in 1990, when these phones were still cumbersome and costly.

Worldwide in 1990 just 12 million people subscribed to a mobile telephone service. Put another way, there were 2.4 mobile phone subscriptions per 1,000 people. At that time mobile phones were large and expensive, and worked only in relatively limited areas. The United States had the largest number of mobile phone subscriptions at 5 million, followed by the United Kingdom at 1 million, Japan at 800,000, and Canada and China with 600,000 subscriptions each.

Many territories had no mobile phone network at all in 1990. 138 of the 200 territories had fewer than 1,000 mobile phone subscribers each; consequently, large areas of the Middle East and Central Asia, South Asia and Central Africa are not visible at all on this map.

'When they were new, rich people flaunted them to show they were connected.'
Anthony Zwane, sociologist at the University of Swaziland, 2004

MOBILE PHONE SUBSCRIBERS
millions, 1990

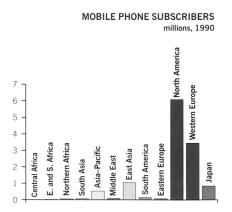

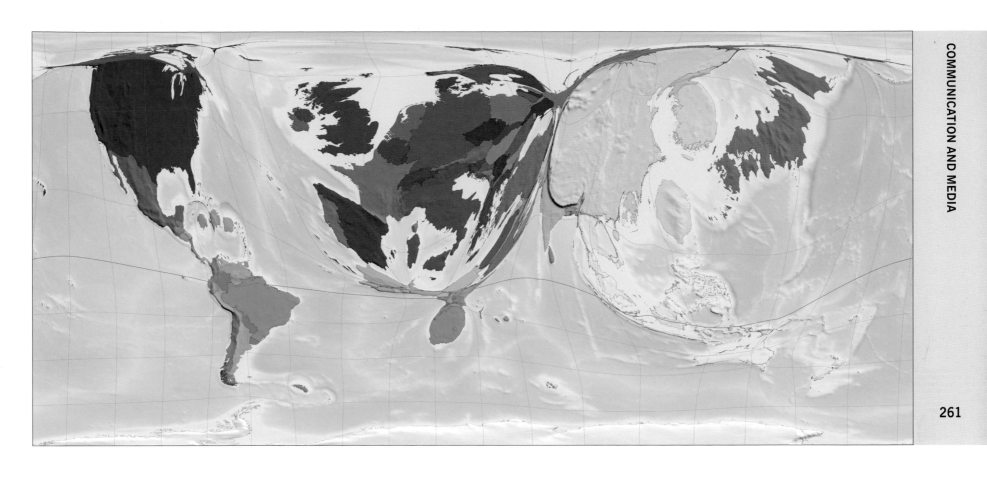

MOST AND FEWEST MOBILE PHONES IN USE IN 2002

Rank	Territory	Mobile phones[a]
1	Taiwan	1,083
2	Luxembourg	1,061
3	Israel	955
4	Hong Kong (China)	942
5	Italy	939
6	Iceland	906
7	Sweden	889
8	Finland	867
9	Czech Republic	849
10	Greece	845
191	Niger	1.0
192	Nepal	1.0
193	Myanmar	1.0
194	Micronesia	0.9
195	Iraq	0.8
196	Liberia	0.6
197	Guinea-Bissau	0
198	Eritrea	0
199	Comoros	0
200	Bhutan	0

[a] mobile phone subscriptions per 1,000 people, 2002

240 Mobile Phones in 2002

The size of each territory indicates the number of mobile (cellular) telephone subscriptions, including prepaid subscriptions, in 2002, showing how rapidly and how far usage has increased.

The worldwide number of mobile phone subscriptions increased 100-fold between 1990 and 2002, a period during which many territories were installing the infrastructure needed to support the service. On this map of subscriptions in 2002 there are many territories of moderate or large size that are virtually invisible on the map for 1990 opposite, among them Bangladesh, Cameroon, El Salvador and Georgia. Worldwide, by 2002, there were 188 mobile phone subscriptions per 1,000 people. Despite the spread of mobile phone technology, however, the territories with the largest numbers of mobile phones in 1990 still had the largest in 2002.

The territories with the largest numbers of mobile phones per person are Taiwan and Luxembourg, where there are now more mobiles than people. Some territories that had limited landline telephone networks to begin with have moved almost entirely to cellular telephone systems, in preference to improving their landline networks.

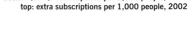

INCREASE IN MOBILE PHONE USE
bottom (dark): subscriptions per 1,000 people, 1990
top: extra subscriptions per 1,000 people, 2002

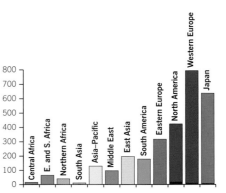

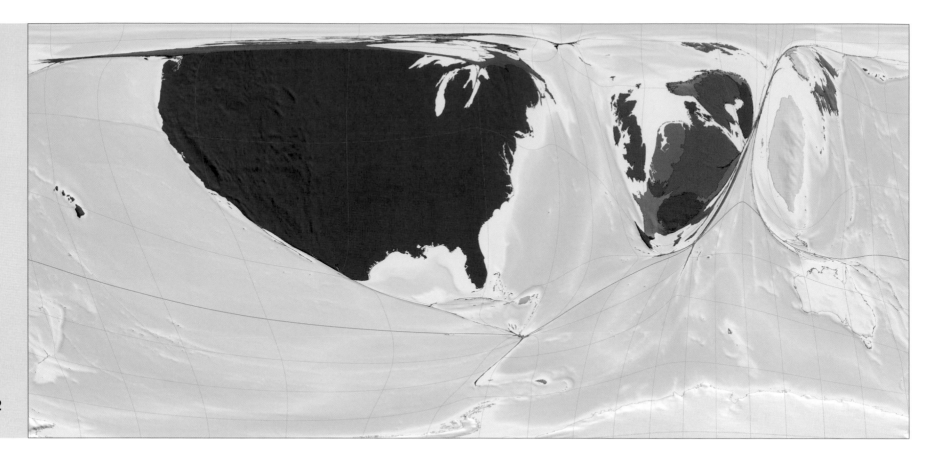

MOST INTERNET USE IN 1990

Rank	Territory	Internet users[a]
1	United States	8.0
2	Norway	7.1
3	Australia	5.9
4	Switzerland	5.8
5	Sweden	5.8
6	Finland	4.0
7	Canada	3.7
8	Netherlands	3.3
9	Germany	1.4
10	Austria	1.3
11	Israel	1.1
12	Denmark	1.0
13	Andorra	0.9
14	United Kingdom	0.5
15	Belgium	0.5
16	France	0.2
17	South Korea	0.2
18	Japan	0.2
19	Italy	0.2
20	Spain	0.1

[a] internet users per 1,000 people, 1990
(only 20 countries reported use)

241 Internet Users in 1990

The size of each territory indicates the number of people using the infant internet in 1990.

In 1990 the internet had existed in its modern form for only about ten years. Just 3 million people worldwide had access to it at that time: 73% of these were in the United States and 15% in Western Europe, with just a few in other places such as Canada, Australia, Japan, South Korea and Israel and practically no access anywhere else.

'I sit down comfortably in the university's modern computer lab and take advantage of the technology available to enter the digital world. The digital world!' Gabriela Tôrres Barbosa, who lives in a favela (slum) in Rio de Janeiro, 2006

INTERNET USERS IN 1990
thousands

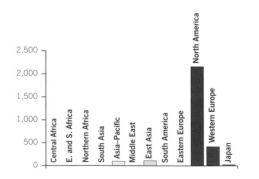

263

[a] internet users per 1,000 people, 2002

242 Internet Users in 2002

The size of each territory indicates the number of internet users in 2002, by which time its reach and sophistication had increased dramatically.

The number of people using the internet saw a 200-fold increase between 1990 and 2002. By 2002 there were 631 million internet users worldwide. The distribution of users also changed radically. In 1990 the vast majority of internet users were in the United States, Western Europe, Australia, Japan and Taiwan. By 2002 there were substantial numbers of internet users also in the Asia–Pacific and Australasia region, South Asia, South America, Eastern Europe and China, and small but significant numbers in Northern Africa, East and Southern Africa, and the Middle East and Central Asia.

'We [strive] to achieve a "warm-hearted digital world" where everybody, including the elderly, the young, men and women, enjoys ubiquitous access to communications technologies for the greater good...'
Yeon Gi Son, president and CEO of Korea Agency for Digital Opportunity and Promotion, 2006

INTERNET USERS IN 2002
users per 1,000 people, 2002

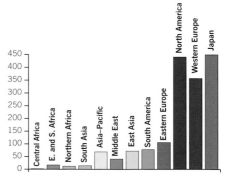

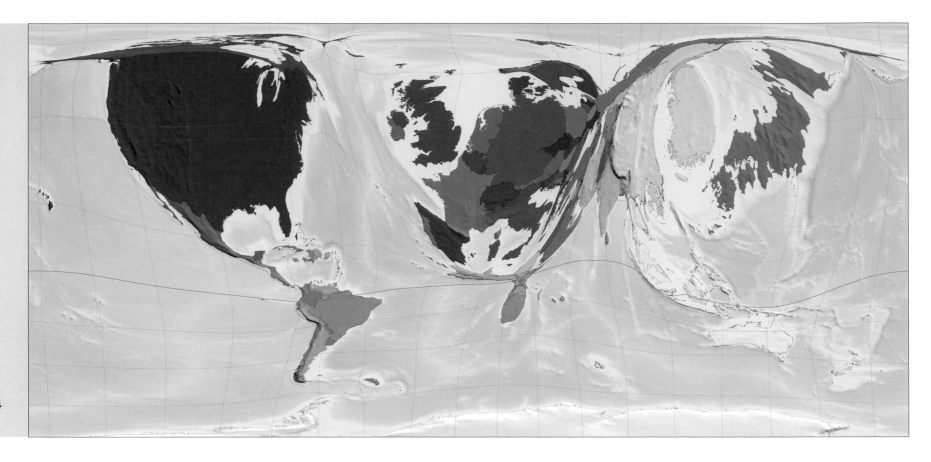

MOST AND FEWEST PERSONAL COMPUTERS

Rank	Territory	Computersᵃ
1	San Marino	760
2	Switzerland	709
3	United States	659
4	Singapore	622
5	Sweden	621
6	Luxembourg	594
7	Denmark	577
8	Australia	565
9	South Korea	556
10	Norway	528
191	Central African Republic	2.0
192	Cambodia	2.0
193	Angola	1.9
194	Chad	1.7
195	Burkina Faso	1.6
196	Ethiopia	1.5
197	Mali	1.4
198	Malawi	1.3
199	Burundi	0.7
200	Niger	0.6

ᵃ computers per 1,000 people, 2002

243 Personal Computers

The size of each territory indicates the number of personal computers used both in homes and in workplaces.

Personal computers are computers designed to be used by a single individual. This includes laptop and desktop computers, computers used at work and those used at home. In 2002 there were almost 600 million such computers in use worldwide, or about 1 computer for every 10 people.

The largest numbers of computers are found in the United States, Japan, China and Germany. The 12 territories with the largest numbers of computers are home to 75% of all the computers in the world. The remaining 25% are distributed (very unevenly) among the remaining 188 territories.

PERSONAL COMPUTERS
number per 1,000 people, 2002

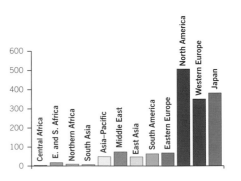

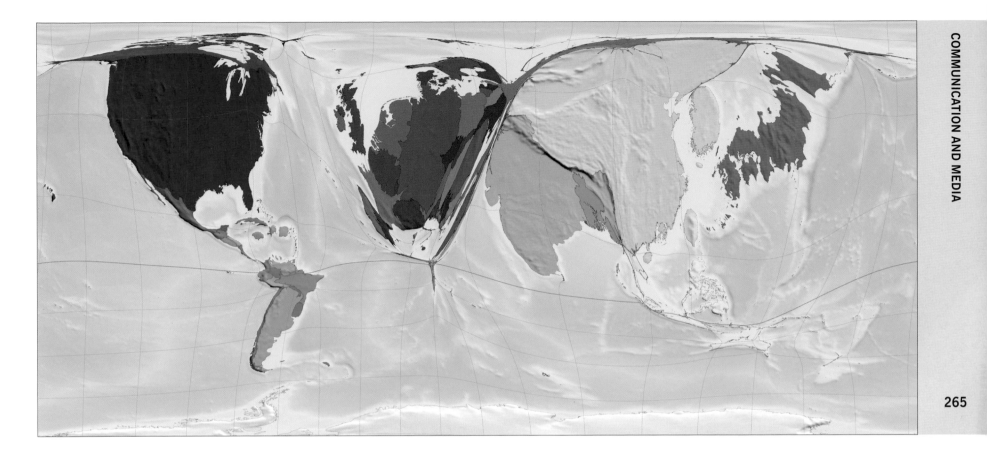

265

244 Cable Television

The size of each territory indicates the number of subscribers to cable television networks.

This map shows the distribution of subscriptions to multichannel television services delivered via fixed-line connection, commonly called cable television. In 2002 there were 355 million households that subscribed to cable television services, of which 27% were in China, 21% in the United States and 12% in India.

In contrast, 36 territories had no cable television at all. Of these, 12 were in Northern Africa, 7 in the Asia–Pacific and Australasia, 5 in the Middle East and Central Asia, 5 in East and Southern Africa, and the remaining 7 in other regions. Cable subscription rates as a percentage of population were highest in the territories of Monaco, Liechtenstein and the Netherlands.

'In the US, pay-TV began because the rural and outlying areas could not get good over-the-air reception...new content alternatives increase[d] the value-add of cable systems, spawning their growth into the denser city areas.' StarHub, cable television service in Singapore, 2007

CABLE SUBSCRIBERS
number per 1,000 people, 2002

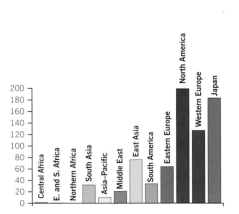

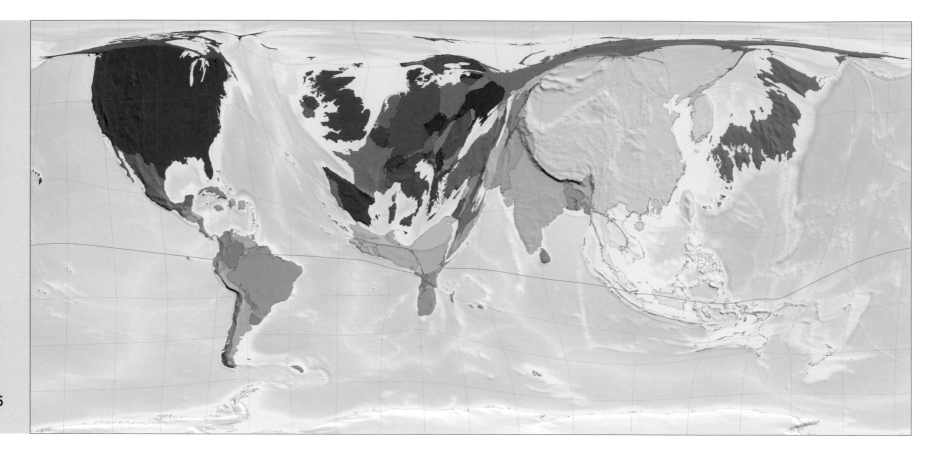

MOST AND FEWEST TELEVISIONS IN USE

Rank	Territory	Televisions[a]
1	Sweden	965
2	United Kingdom	950
3	United States	938
4	Norway	884
5	Qatar	869
6	San Marino	863
7	Denmark	859
8	Latvia	850
9	Japan	785
10	Monaco	761
191	Myanmar	7.6
192	Cambodia	7.6
193	Haiti	6.0
194	Central African Republic	5.8
195	Ethiopia	5.7
196	Malawi	3.9
197	Comoros	3.7
198	DR Congo	1.9
199	Chad	1.9
200	Rwanda	0.1

[a] televisions per 1,000 people, 2002

245 Televisions in Use

The size of each territory indicates the number of television sets in use. Over half the world's population have access to, on average, at least one television per household.

In 2002, less than 100 years after the invention of television, there were almost 2 billion television sets in use worldwide, or 1 television for every 3 people. Televisions are primarily seen as a source of entertainment, but they are also a key channel for providing information about the world and a primary means of advertising goods and services.

The territory with the largest number of television sets is China, where there are

454 million. This is equivalent to 35 televisions for every 100 people. In the United States, which has the second highest total number of television sets (273 million), there are 94 televisions for every 100 people. Some 52% of the world's population live in territories with 1 or more television for every 4 people.

TELEVISIONS IN USE
number per 1,000 people, 2002

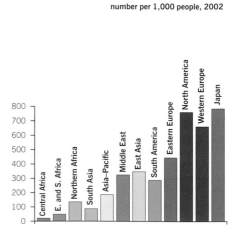

MOST AND FEWEST RADIOS IN USE

Rank	Territory	Radios[a]
1	Norway	3,324
2	Sweden	2,811
3	United States	2,109
4	Australia	1,996
5	Finland	1,624
6	United Kingdom	1,445
7	Denmark	1,400
8	Estonia	1,136
9	Iceland	1,081
10	Samoa	1,063
191	Lesotho	61
192	Somalia	60
193	Guinea	52
194	Bhutan	50
195	Mongolia	50
196	Bangladesh	49
197	Mozambique	44
198	Nepal	39
199	Azerbaijan	22
200	Haiti	18

[a] radios per 1,000 people, 2002

246 Radios in Use

The size of each territory indicates the number of radios in use.
Where literacy is low, communication by sound has importance.

Radio broadcasting can be a very effective means of widespread communication. It can often reach people, especially those in remote places, faster than written or printed communications such as newspapers. Also, while many people find reading difficult, a much smaller number cannot hear.

Radios first came into widespread use among the general public in the 1920s. Today there are 2.6 billion radios in use worldwide. In 13 territories there are more radios in use than there are people living there. In another 55 territories there is more than 1 radio for every 2 people, and in 126 territories there is more than 1 per 4 people. (The world average household size is 4 people, so this corresponds to roughly one radio per household.) The largest number of radios per person is found in Norway, where there are more than 3 radios for every member of the population. By contrast, there is only 1 radio in use for every 50 people in Haiti.

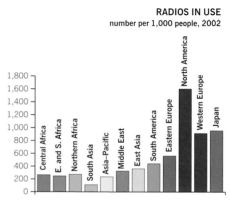

RADIOS IN USE
number per 1,000 people, 2002

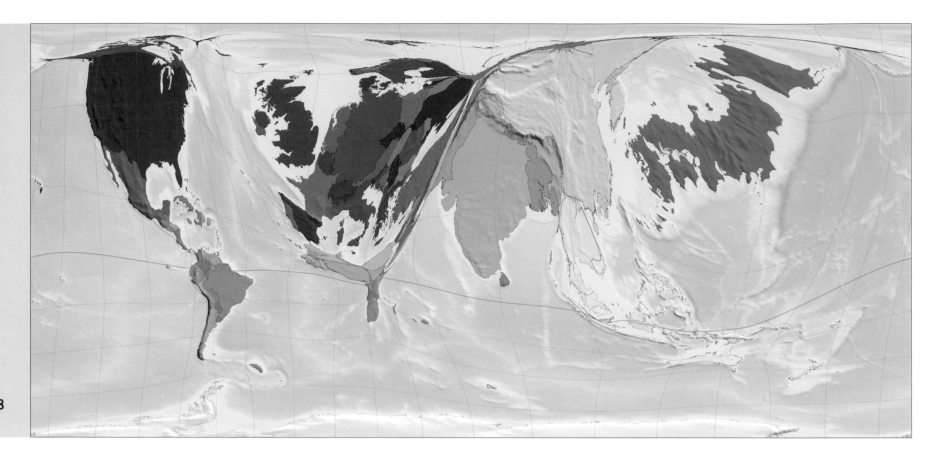

HIGHEST AND LOWEST CIRCULATION OF DAILY NEWSPAPERS

Rank	Territory	Copies[a]
1	Norway	569
2	Japan	566
3	Finland	445
4	Sweden	410
5	Switzerland	372
6	United Kingdom	326
7	Iceland	322
8	Austria	309
9	Germany	291
10	Denmark	283
191	Burundi	2.5
192	Togo	2.2
193	Central African Republic	1.7
194	Gambia	1.7
195	Burkina Faso	1.3
196	Mali	1.1
197	Rwanda	0.9
198	Ethiopia	0.4
199	Chad	0.2
200	Niger	0.2

[a] copies of daily newspapers in circulation per 1,000 people, 2002

247 Daily Newspaper Circulation

There are daily newspapers in every territory in the world. The size of each territory indicates the number of copies of daily newspapers distributed.

In 2002 a total of 507 million copies of newspapers were distributed every day, including both newspapers that are free and those that are sold. China distributes the largest number of daily newspapers, followed by Japan and then India. In Japan 566 newspapers are distributed for every 1,000 inhabitants. Per-person circulation in China and India is a tenth of this, but total circulation is still high because of the large populations of those countries.

The world average daily newspaper circulation is 81 newspapers per 1,000 people.

'I think the duty of doctors is to give health to their patients, the duty of the singer to sing. The duty of [the] journalist [is] to write what this journalist sees in the reality. It's only one duty.'

Anna Politkovskaya, Russian journalist (assassinated in 2006), 2004

CIRCULATION OF DAILY NEWSPAPERS
millions of copies, 2000

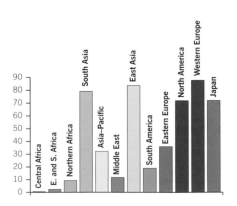

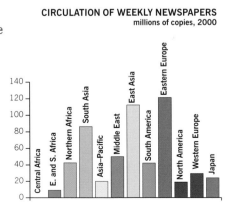

269

[a] copies of non-daily newspapers in circulation per 1,000 people, 2002

248 Weekly Newspaper Circulation

The size of each territory indicates the numbers of copies distributed of newspapers that have fewer than 4 issues a week. Most are weeklies.

This map shows the distribution of copies of non-daily newspapers, including those sold directly, those sold by subscription and those given away for free.

Newspaper circulation tends to be highest in regions that have intermediate levels of wealth and education: Eastern Europe, East Asia and South Asia, Northern Africa and South America. Togo and Belarus, for example, both print more weekly newspapers than they have people to read them, meaning that a lot of people must be reading more than one weekly paper.

Circulation of newspapers is lower in poorer areas where people cannot afford to buy them and may not be able to read, but also in richer areas, where television has supplanted the newspaper as the leading medium for the distribution of news.

'The smart way to keep people passive and obedient is to strictly limit the spectrum of acceptable opinion, but allow very lively debate within that spectrum...That gives people the sense that there's free thinking...' Noam Chomsky, professor of linguistics (emeritus) at MIT, undated

CIRCULATION OF WEEKLY NEWSPAPERS
millions of copies, 2000

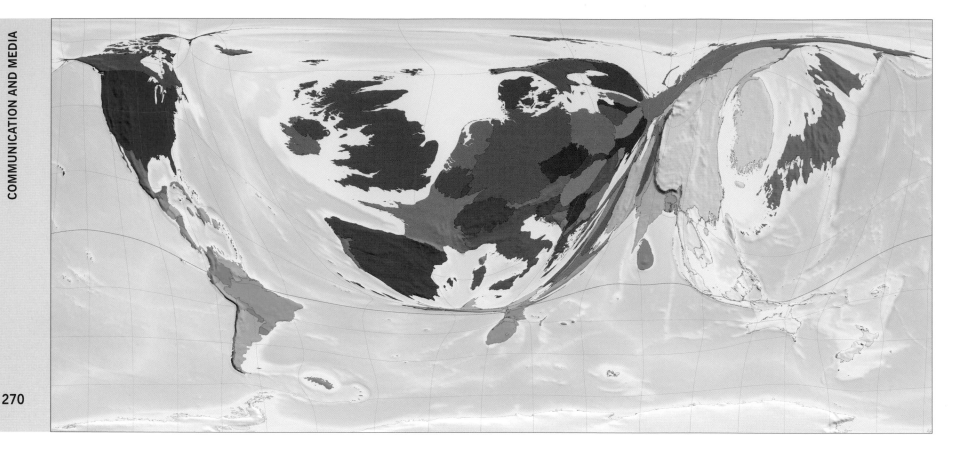

249 Books Published

The size of each territory shows the number of new book titles published each year.

This map shows the numbers of new book and pamphlet titles published in each territory in 1999. Note that this is different from the total number of copies of books sold: each new book published is counted only once on this map, regardless of how many copies it sells. Books and pamphlets differ only in how many pages they have: a book is defined as having at least 50 pages; a pamphlet has 5 to 49 pages. Publications with fewer than 5 pages are not shown on this map.

Worldwide, about a million new book titles were published in 1999, with the largest numbers published in the United Kingdom, China and Germany. Overall the map is dominated by Western Europe, which is home to a large number of well-established publishing houses.

The highest rate of publication of new titles per person is in the Vatican City, where 1 new title was produced for every 5 residents in 1999. By comparison, the worldwide average figure was 1 new title for every 6,000 people.

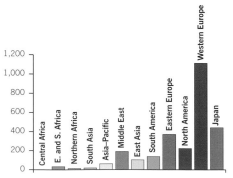

BOOK TITLES PUBLISHED
number per million people, 1999

250 Books Borrowed

The size of each territory indicates the number of loans of books from public libraries.

Public libraries are libraries that lend books to members of the public free or for a nominal charge. The map shows the total number of instances in which a book – any book – is borrowed from a library in the territory in question. In other words, a single book that gets borrowed many times gets counted many times.

The largest number of book loans occurs in Russia, although there are also high rates of borrowing in Western Europe, Japan and Eastern Europe. Most territories

in these regions report at least some book borrowing. Elsewhere, book borrowing is less frequent and many territories report very little. Certainly some of the variation is attributable to differences in literacy rates – people who cannot read rarely borrow books – and poverty also plays a role, with extensive libraries being less common in less wealthy countries. Thus, in places where many people cannot afford to buy books, they often cannot borrow them either.

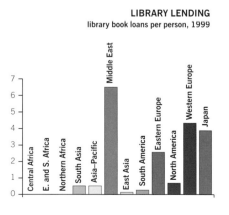

LIBRARY LENDING
library book loans per person, 1999

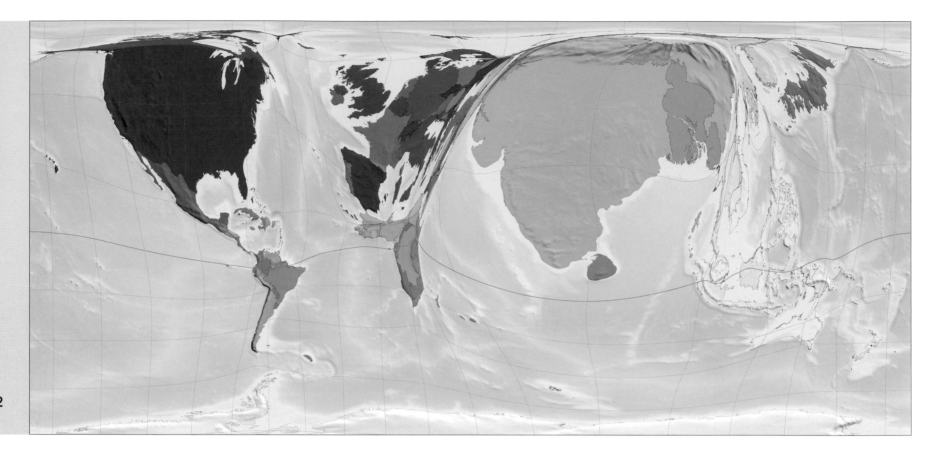

MOST AND FEWEST CINEMA VISITS PER YEAR

Rank	Territory	Visits[a]
1	Iceland	5.1
2	United States	5.0
3	Australia	4.5
4	New Zealand	4.4
5	Singapore	4.0
6	Greenland	4.0
7	Bahamas	4.0
8	Hong Kong (China)	4.0
9	Canada	3.6
10	Georgia	3.6
191	Tajikistan	0.07
192	Pakistan	0.06
193	Kazakhstan	0.06
194	Kyrgyzstan	0.06
195	Benin	0.04
196	Kenya	0.03
197	Moldova	0.02
198	Algeria	0.02
199	Azerbaijan	0.02
200	Malaysia	0.01

[a] average cinema trips per person, 1999

251 Films Watched

The size of each territory indicates the number of film viewings in cinemas.

This map shows where people watch the most films (not including short films) at commercial screenings. Of the 7.6 billion film viewings a year represented on this map, almost 3 billion are in India.

In some territories it is now more popular to watch films at home than in cinemas. Even in these places, however, cinemas persist because many recently released films can be viewed only at the cinema.

The largest number of viewings per person is in Iceland, where the average person watches 5 films in cinemas every year. In Malaysia, by contrast, only 1 person in every 100 sees even one film at a cinema during the average year.

'Currently, there are around 12,000 movie theaters in India, ranging from small ramshackle neighborhood screening rooms to overly extravagant movie halls.' Phurba Gyalzen, *Green Cine*, 2005

CINEMA TRIPS PER YEAR
visits to a cinema per person per year, most recent data 1995–9

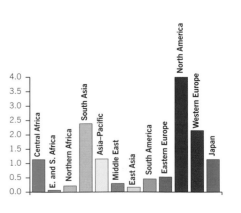

252 Electricity Access

The size of each territory indicates the number of people with electrical power in their homes. How much they use, and how many powered facilities they have at their disposal, of course vary hugely.

This map shows the number of people in each territory who have at least some electrical power in their homes, including power from a public electricity grid and locally generated power, such as that from solar, wind or hydroelectric sources. Electricity in homes can be used to provide power for lighting, heating, cooking, radios, televisions, computers, washing machines and other appliances. This map shows only the numbers of people with electrical power and not the quantity of power they use.

The proportion of people with electricity in their homes is over 97% in East Asia, Eastern Europe, North America, Western Europe and Japan. 7 of the 10 territories where the lowest proportions of people have electricity are in East and Southern Africa.

ELECTRICITY ACCESS
% of people with access to electricity, 2000

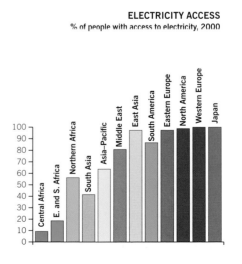

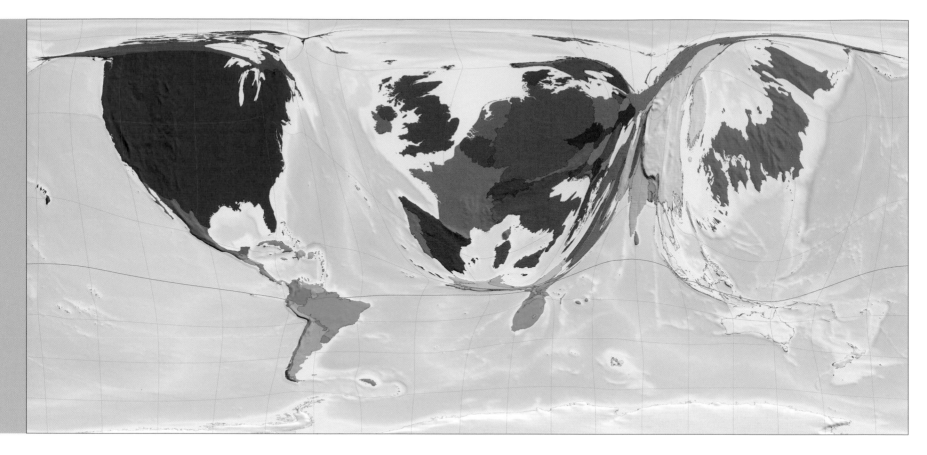

HIGHEST AND LOWEST PUBLIC SPENDING ON HEALTHCARE

Rank	Territory	PPP US$[a]
1	Luxembourg	3,304
2	Norway	2,525
3	Iceland	2,261
4	United States	2,217
5	Germany	2,195
6	Denmark	2,166
7	Canada	2,005
8	France	1,965
9	Sweden	1,954
10	Switzerland	1,891
191	Burundi	13
192	Sudan	13
193	Tanzania	12
194	Niger	11
195	Ethiopia	11
196	Tajikistan	10
197	DR Congo	10
198	Madagascar	10
199	Nigeria	7
200	Myanmar	4

[a] public spending on healthcare per person, 2001

253 Public Health Spending

The size of each territory shows public spending on healthcare, adjusted for local purchasing power. Public health spending reduces, or in some cases eliminates, the direct cost of healthcare to an individual.

As defined here, public health spending includes all spending by governments on healthcare, along with any money from grants, social insurance or non-governmental organizations, but not direct payments by individuals or corporations.

The highest rates of public spending on healthcare per person are in the regions of Western Europe, North America and Japan. The territories with the highest spending per person are Luxembourg, Norway and Iceland.

The figures on which this map is based are adjusted for purchasing power parity (PPP), which measures spending in terms of what the money would buy in the territory in question. It is important to realize, however, that many medicines are not manufactured within the territories where they are consumed, but are brought in from other regions of the world, so that their prices sometimes do not vary as the local prices of other goods do. For this reason PPP-adjusted spending figures may be less useful in this example than in some others.

PUBLIC HEALTHCARE SPENDING
PPP US$ per person, 2001

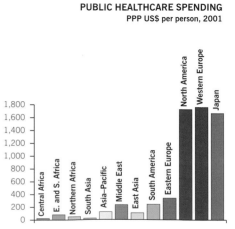

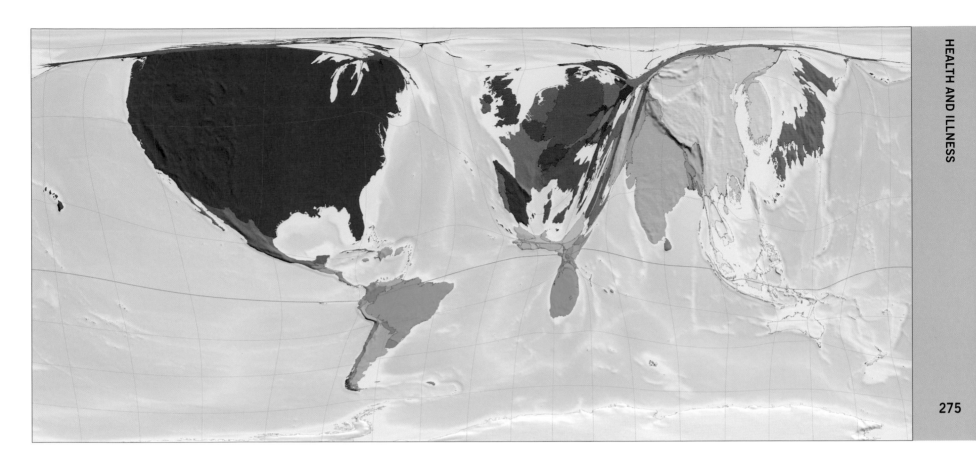

254 Private Health Spending

The size of each territory shows private spending on
healthcare, adjusted for local purchasing power.

Private health spending means direct
spending on healthcare services or goods
by individuals or corporations. The level of
private spending depends not only on the
need for healthcare but also on the ability
to pay. Among the 12 regions on the map,
private spending on healthcare was
greatest in North America and least in
Central Africa. Total spending in North
America was more than 100 times that
in Central Africa.

*'The health industry focuses on people with the greatest
ability to pay rather than the greatest need for care.'*

Phineas Baxandall, *Dollars and Sense*, 2001

PRIVATE HEALTHCARE SPENDING
PPP US$ per person, 2001

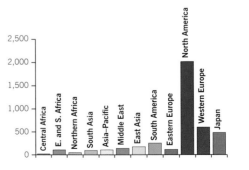

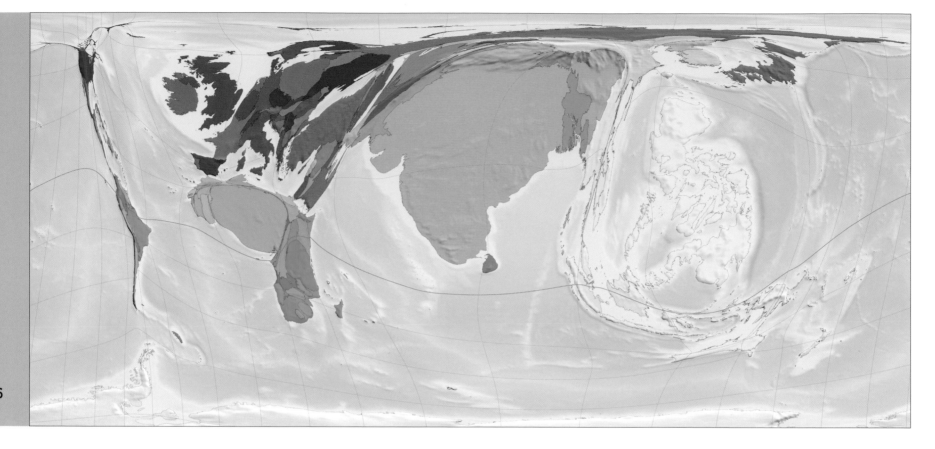

MOST AND FEWEST WORKING MIDWIVES

Rank	Territory	Midwivesᵃ
1	Seychelles	395
2	Maldives	185
3	Philippines	179
4	Azerbaijan	122
5	Brunei	121
6	Namibia	117
7	Tuvalu	96
8	Niue	95
9	Turkmenistan	83
10	Uzbekistan	82
191	China	3.5
192	Uganda	3.4
193	Mali	2.4
194	Eritrea	2.2
194	Equatorial Guinea	2.2
196	Chad	2.0
197	Ethiopia	1.7
198	Samoa	1.6
199	Cameroon	0.5
200	Rwanda	0.1

ᵃ number per 100,000 people, 2002

255 Working Midwives

The size of each territory shows the number of working midwives.
Variations reflect culture and custom as well as wealth and training.

Midwives assist women in giving birth. The prevalence of midwives is not solely related to wealth or the availability of good healthcare. Some territories have few midwives because they never adopted the custom or the custom has ceased. Others prefer births to be attended by doctors. The value of midwives has only recently been realized in North America, for example, so that the number of working midwives is still low there.

The territory with the largest number of midwives is India, where there are estimated to be almost half a million. The Philippines comes second. The number of midwives per person is highest in the island territories of the Seychelles and the Maldives, and lowest in Cameroon and Rwanda.

'The lives and health of many millions more would be saved with greater investments in midwives.'

Thoraya Ahmed Obaid, executive director, UNFPA, 2006

WORKING MIDWIVES
midwives per 100,000 people, 2004

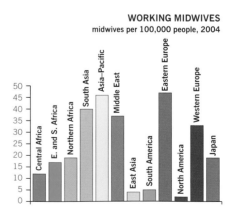

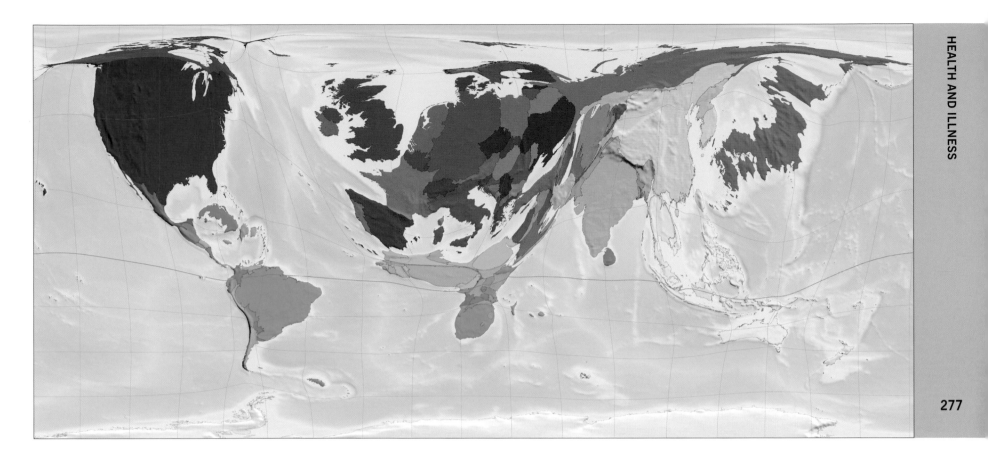

MOST AND FEWEST WORKING NURSES

Rank	Territory	Nursesᵃ
1	Finland	2,171
2	Norway	2,065
3	Ireland	1,662
4	Monaco	1,430
5	Netherlands	1,334
6	Belarus	1,234
7	Belgium	1,074
8	Canada	1,010
9	Uzbekistan	997
10	Sweden	977
191	Togo	17
192	Eritrea	16
193	Chad	15
194	Bangladesh	13
195	Mali	13
196	Gambia	13
197	Haiti	11
198	Central African Republic	9
199	Liberia	6
200	Uganda	5

ᵃ **number per 100,000 people, 2002**

256 Working Nurses

The size of each territory indicates the number of working nurses, now a highly exportable resource.

The United States, China and Russia have the highest total numbers of working nurses, but the highest ratios of nurses to members of the population are in Western European territories such as Finland and Norway. The lowest ratios are in the Central African Republic, Liberia and Uganda, all territories where there is a great need for nurses.

'[Nurses] are also being poached by industrialized countries. There are more nurses from Malawi in Manchester than in Malawi...' Glenys Kinnock, EU co-president, 2005

WORKING NURSES
nurses per 100,000 people, 2004

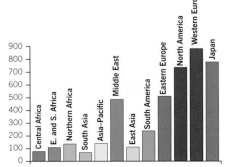

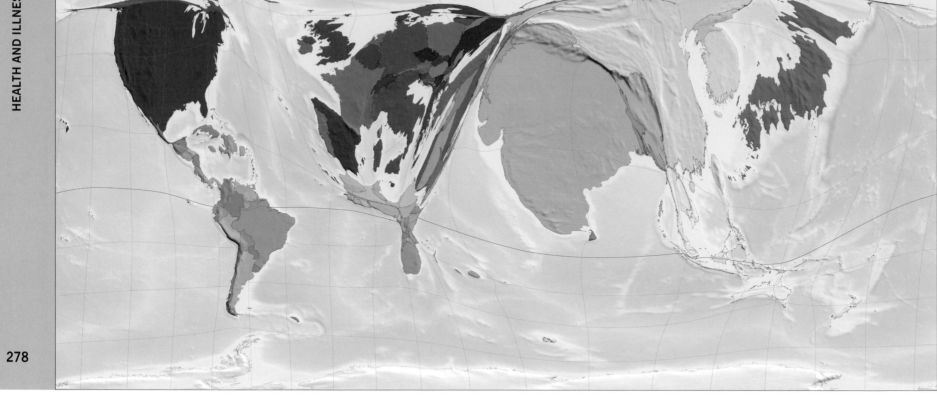

257 Working Pharmacists

The size of each territory indicates the number of working pharmacists,
on whom many people rely for supplies of ameliorative and life-saving medication.

A pharmacist prepares and dispenses medicinal drugs. Those drugs can range all the way from simple painkillers and rehydration salts to complicated prescriptions such as anti-retroviral drugs for the treatment of HIV.

The largest numbers of working pharmacists are found in the territories with the biggest populations, India and China. 42% of all pharmacists work in these two territories. The territories with the fewest pharmacists are in the regions of the Asia–Pacific and Australasia, Central Africa, Northern Africa, and East and Southern Africa.

'When we write out prescriptions we have to take into account the financial situation of our patients. If we prescribe medicine that is too expensive, then they just won't buy it…'

Alphonsine, nurse working in Togo, 2006

WORKING PHARMACISTS
pharmacists per 100,000 people, 2004

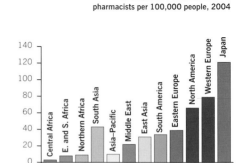

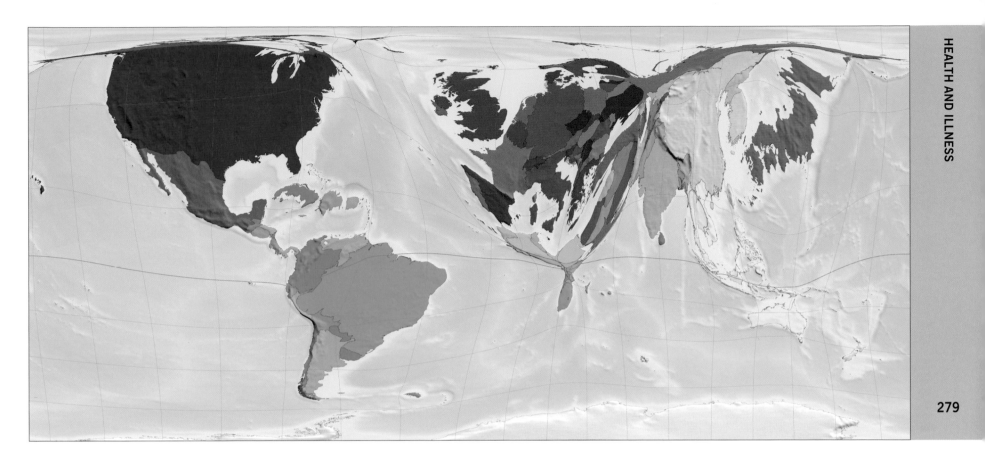

MOST AND FEWEST WORKING DENTISTS

Rank	Territory	Dentists[a]
1	United States	163
2	Jordan	129
3	Estonia	128
4	Finland	128
5	Lebanon	121
6	Israel	117
7	Seychelles	117
8	Uruguay	116
9	Greece	113
10	Brazil	111
191	DR Congo	0.29
192	Rwanda	0.25
193	Burundi	0.20
194	Somalia	0.19
195	Benin	0.17
196	Chad	0.17
197	Ethiopia	0.13
198	Niger	0.12
199	Sierra Leone	0.10
200	Angola	0.02

[a] **number per 100,000 people, 2002**

258 Working Dentists

The size of each territory indicates the number of working dentists.

In 2004 there were 1.8 million working dentists in the world, or 29 dentists per 100,000 people.

The three territories with the largest numbers of dentists are the United States, Brazil and China, though there were 10 times as many dentists per person in Brazil as in China.

There are almost twice as many dentists per person in North America as in any other region. In many African territories there is less than 1 dentist per 100,000 people. The smallest number of dentists per person is in Angola.

WORKING DENTISTS
dentists per 100,000 people, 2004

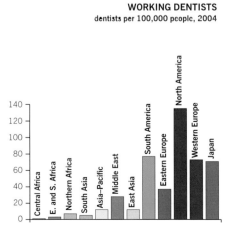

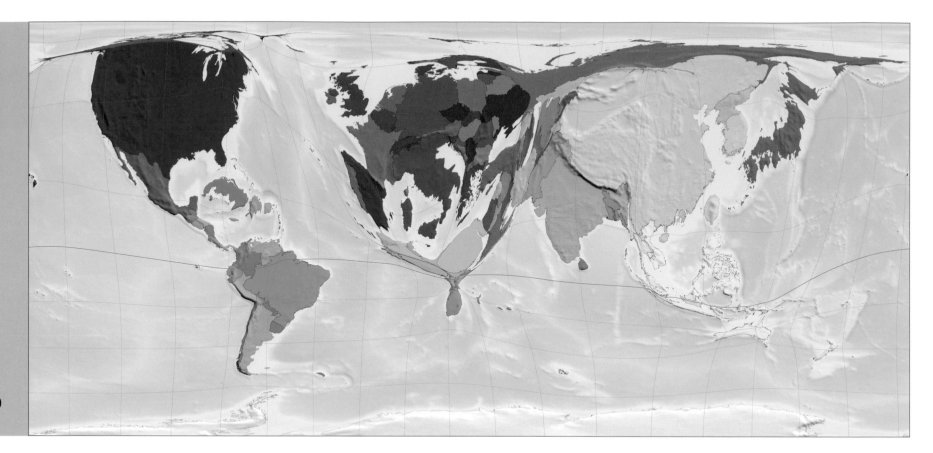

MOST AND FEWEST WORKING PHYSICIANS

Rank	Territory	Physiciansª
1	Italy	606
2	Cuba	591
3	Monaco	586
4	United States	549
5	St Lucia	518
6	Belarus	450
7	Greece	440
8	Greenland	433
9	Belgium	418
10	Russia	417
191	Gambia	4
192	Niger	3
193	Eritrea	3
194	Ethiopia	3
195	Chad	3
196	Mozambique	2
197	Liberia	2
198	Tanzania	2
199	Rwanda	2
200	Malawi	1

ª number per 100,000 people, 2002

259 Working Physicians

The size of each territory indicates the number
of working physicians (medical doctors).

In 2004 there were 7.7 million physicians
working around the world. The territory
with the largest number of physicians
was China, which also has the largest total
population. If physicians were distributed
evenly, there would be 124 physicians
for every 100,000 people in the world; but
they are not. Half of all working physicians
are practising in the territories that are
home to the richest fifth of the world
population in terms of physician care.
The poorest fifth have just 2%.

Italy and the Caribbean island of Cuba
have the largest numbers of working
physicians per resident; the African
territories of Malawi and Rwanda have
the smallest.

WORKING PHYSICIANS
physicians per 100,000 people, 2002

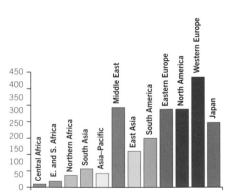

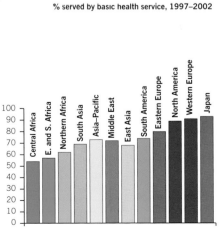

281

HIGHEST AND LOWEST HEALTH SERVICE QUALITY

Rank	Territory	Score[a]
1	Japan	93
2	Switzerland	92
3	Norway	92
4	Sweden	92
5	Luxembourg	92
6	France	92
7	Canada	92
8	Netherlands	92
9	United Kingdom	92
10	Austria	91
191	Malawi	52
192	Afghanistan	52
193	Nigeria	52
194	Mozambique	51
195	Ethiopia	51
196	Liberia	50
197	Niger	50
198	Somalia	49
199	Central African Republic	46
200	Sierra Leone	36

[a] **WHO Health Service Quality score 1997 (0 = low, 100 = high) for providing good basic healthcare**

260 Basic Healthcare Provision

The size of each territory shows the number of people receiving good basic healthcare.

This map shows the prevalence of good basic healthcare in each territory, calculated by multiplying the size of the population by the Health Service Quality Score. The Health Service Quality Score is an index used by the WHO to measure the quality of healthcare in a territory. It is equal to the estimated fraction of the population with access to basic healthcare. The scores used to make this map were from 1997, the most recent year for which data were available for most territories.

The world average Health Service Quality Score is 72 out of 100, meaning that 72% of the world population, or 4.5 billion people, have access to good basic healthcare. The populations with the poorest healthcare availability are in Sierra Leone and the Central African Republic. The Sierra Leonean health system scored 36 out of 100, half the world average score. Note that only the most basic care is measured here.

HEALTH SERVICE QUALITY
% served by basic health service, 1997–2002

Central Africa | E. and S. Africa | Northern Africa | South Asia | Asia–Pacific | Middle East | East Asia | South America | Eastern Europe | North America | Western Europe | Japan

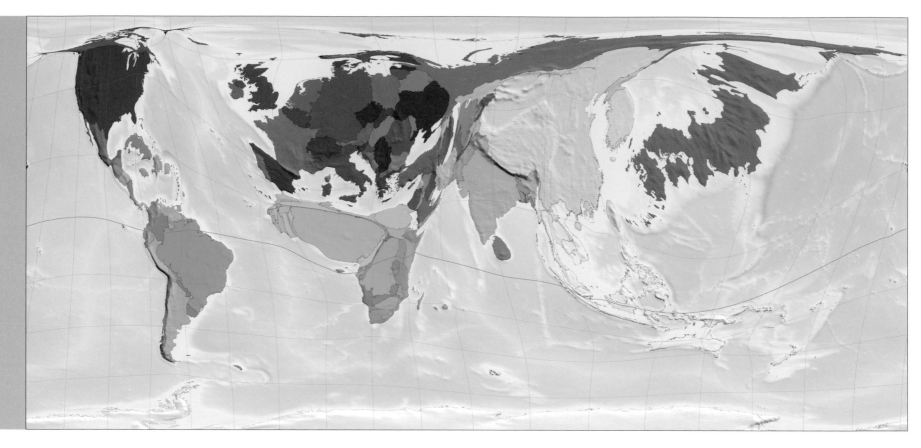

261 Hospital Beds

The size of each territory indicates the number of hospital beds.

In 2002 there were an estimated 19.6 million hospital beds in the world, the largest numbers being in China, Japan and Russia. The largest numbers of beds per person are in Sao Tome and Principe and Monaco. There is one hospital bed for every 25 people living in Sao Tome and Principe, and one for every 46 people in Monaco. In Niger, on the other hand, there is only about one hospital bed for every 10,000 people.

HOSPITAL BEDS
beds per 10,000 people, 2002

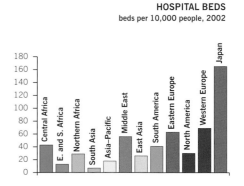

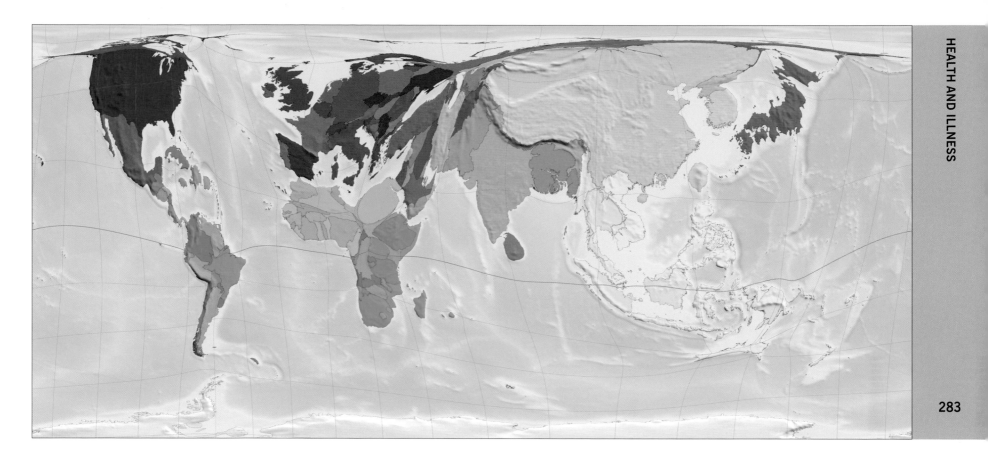

262 **Affordable Drugs**

The size of each territory indicates the numbers of people with access to essential drugs at affordable prices.

The World Health Organization defines essential medicines as those that satisfy the highest-priority healthcare needs of the population and says that these should be available in adequate amounts, with assured quality and accompanying information, at locally affordable prices. Of the total world population, 69% have access of this kind to essential drugs. The remainder, 1.9 billion people, do not.

'An Indian company...has challenged global orthodoxy and conscience by offering AIDS treatment at US$350 a year – the same cocktail costs US$15,000 a year in the developed world.'

Salil Tripathi, *Index on Censorship*, special issue on AIDS, 2004

AFFORDABLE DRUGS
% population with sustainable access to affordable drugs

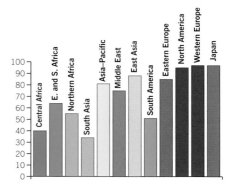

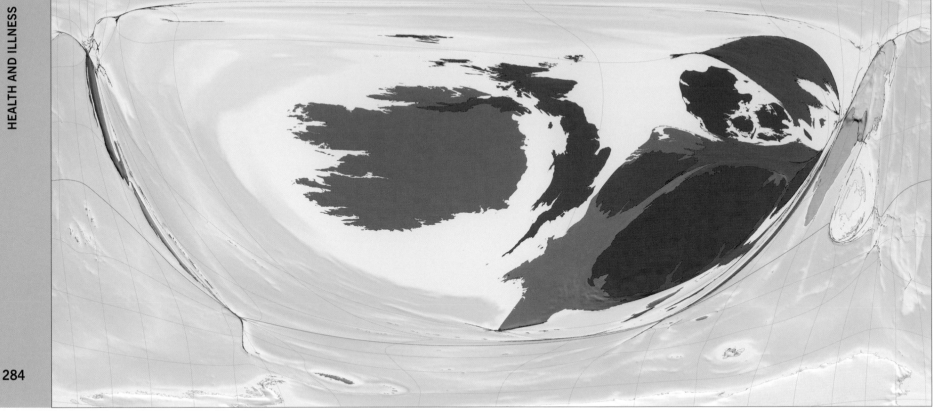

263 Medical Exports

The size of each territory indicates the dollar value of its net exports of medicines and medical equipment. Three-quarters of this revenue goes to Western Europe.

Medical exports make up 3.2% of all export earnings worldwide. 74% of those worldwide earnings go to Western European territories. Ireland has the highest net export earnings in terms of dollar value, although a large fraction of that comes from the re-export of imports.

Non-European net exporters include China, India, Mexico and Singapore. India is a major source of medicines, but Indian medicines are typically sold more cheaply than European medicines, so India's earnings are lower in dollar value, making India smaller on the map.

'...in the pharmaceutical sector the winners will be the large northern-based transnational companies...'

John Sulston, co-founder of the Human Genome Project, 2001

NET EXPORTS OF MEDICAL PRODUCTS
annual earnings, US$ billion

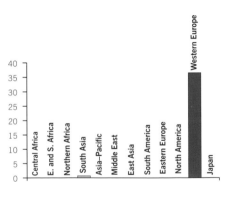

264 Medical Imports

The size of each territory indicates the dollar value of its net imports
of medicines and medical equipment.

Nearly 90% of the territories on the map
are net importers of medicine and medical
equipment. Across territories, however,
there is huge variation in spending on
medical imports per person. The highest
figure is for Luxembourg, where US$406
is spent per person per year.

*'The World Trade Organization's rules on patents restricts
the supply of medicines and drives up the prices.'*

Phil Bloomer, policy director of Oxfam, 2005

NET IMPORTS OF MEDICAL PRODUCTS
annual spending, US$ billion

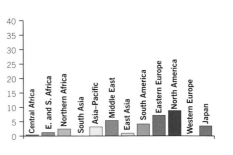

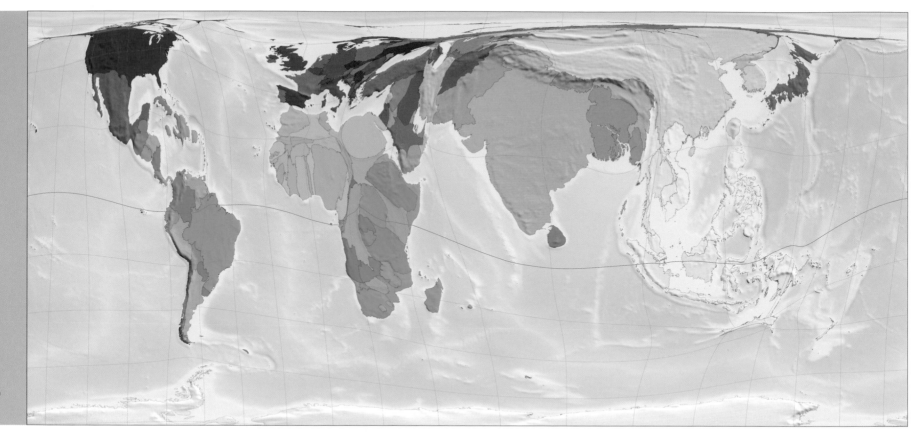

LOWEST RATES OF IMMUNIZATION AGAINST MEASLES

Rank	Territory	%ᵃ
1	Mali	33
2	Central African Republic	35
3	Congo	37
4	Nigeria	40
5	Vanuatu	44
6	DR Congo	45
7	Burkina Faso	46
8	Guinea-Bissau	47
9	Timor-Leste	47
10	Niger	48
11	Sudan	49
12	Equatorial Guinea	51
13	Ethiopia	52
14	Cambodia	52
15	Haiti	53
16	Guinea	54
17	Senegal	54
18	Chad	55
19	Laos	55
20	Gabon	55

ᵃ % of infants immunized against measles, 2002

265 Measles Immunization

The size of each territory indicates the number of children vaccinated against measles, a disease that can range from a mildly unpleasant infection to a fatal malady.

Effective vaccines against measles have been in existence for 40 years. In 2002 an estimated 75% of one-year-olds worldwide were vaccinated against the disease. The largest numbers of them were in India and China; however, the largest numbers of unvaccinated infants were also in India and China. The 10 highest death rates from measles were all in African countries. The 5 highest were in Guinea-Bissau, Somalia, Niger, Sierra Leone and Guinea.

MEASLES IMMUNIZATION
% infants vaccinated, 2002

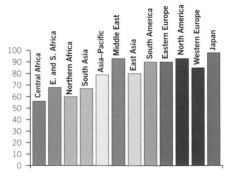

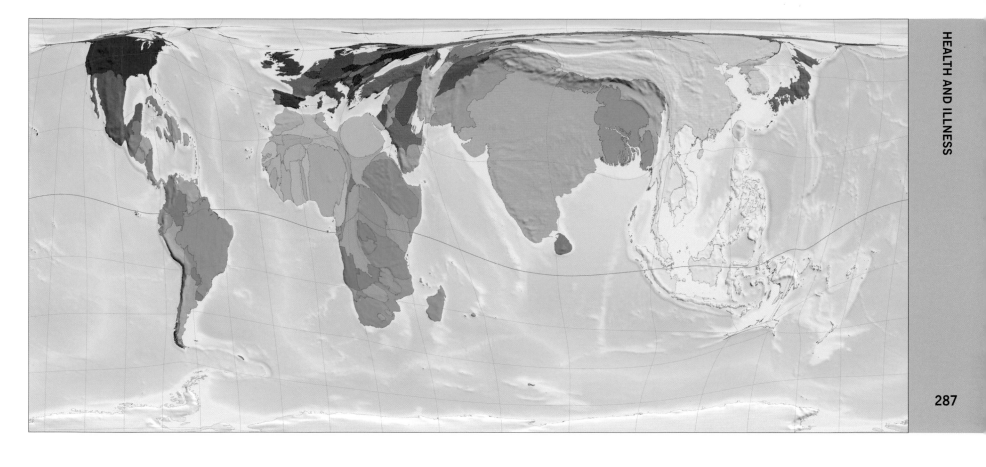

287

266 Infants not at Risk from TB

The size of each territory shows numbers of infants protected
from tuberculosis, either by immunization or by lack of contact.

In some territories there is little or no
tuberculosis (TB), so vaccination is
unnecessary. In other territories the disease
is more common and vaccination is needed
to protect children from infection.

Of all infants in the world, 81% are
protected from tuberculosis, but that means
that almost 1 in every 5 is not. The biggest
populations of protected infants live in
India and China. In 36 territories 99%
or more of infants are protected from
tuberculosis, and in all but two territories
over 50% of infants have some protection
from tuberculosis. The two remaining
territories are Sudan and Niger.

*'I contracted tuberculosis at the age of 14 and was
hospitalized for 20 months. I'm here to witness that
tuberculosis is a curable and preventable disease.'*

Desmond Tutu, archbishop of Cape Town, 2006

TUBERCULOSIS IMMUNIZATION
% infants protected

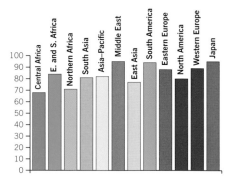

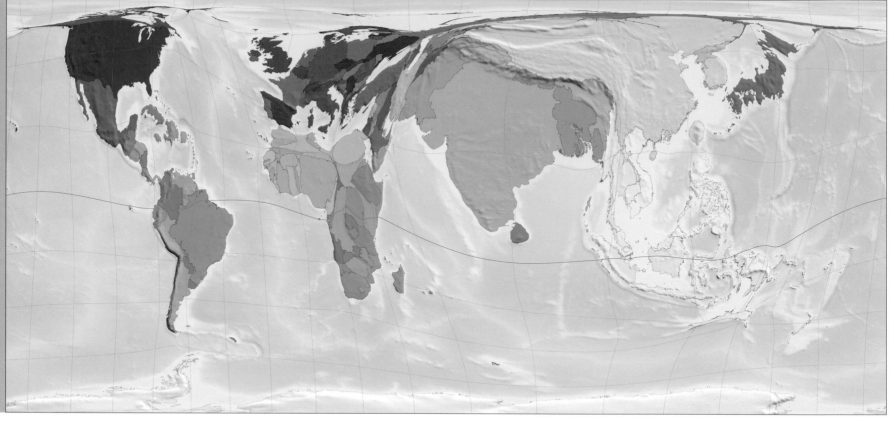

Rank	Territory	%ª
1	Botswana	88
2	Canada	72
3	Zimbabwe	69
4	Lebanon	69
5	Latvia	69
6	United States	65
6	Greenland	63
8	Bahamas	63
8	Uganda	62
10	Brazil	59
191	Tanzania	31
192	Cameroon	31
193	Ethiopia	30
194	Mali	30
195	Haiti	30
196	Kazakhstan	28
196	Mauritius	26
198	Bolivia	22
199	Slovenia	17
200	Chad	2

ª % aged 15–24 using condoms on most
recent instance of high-risk sex, 2002

267 Condom Use by Men

The size of each territory shows the numbers of boys and men aged 15
to 24 years old who used a condom last time they had high-risk sex.

Condoms are among the few
contraceptives that also provide users
with protection from sexually transmitted
infections (STIs). It is wise for someone to
use a condom if they have sex with a
person who they are not certain is free from
infection. (Many people are themselves
unaware whether they have an STI.)

47% of young men worldwide report

using a condom the last time they had
high-risk sex, defined as sex with someone
who is neither a spouse nor a cohabitant.
(This is a loose definition: marriage, of
course, does not itself provide protection
from STIs.) The region with the highest rate
of condom use is North America, at
63.5%. The region with the lowest is the
Middle East and Central Asia, at 36.8%.

*'When condoms are expensive, the costs of sexual freedom
can be very high.'* Dan Smith, *The State of the World Atlas*, 2003

CONDOM USE BY MEN
% of men aged 15–24 using condoms, 2002

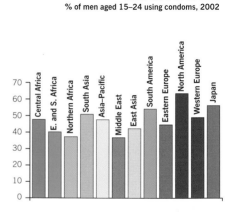

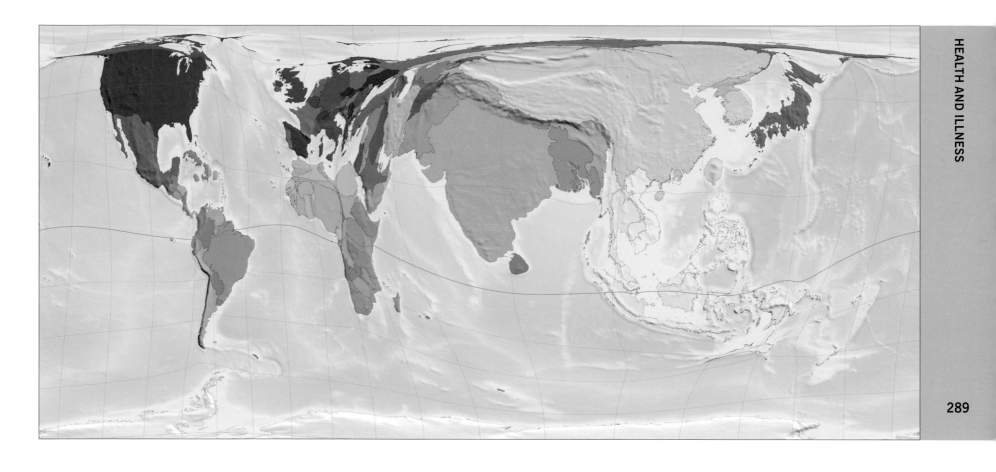

ᵃ % aged 15–24 using condoms on most recent instance of high-risk sex, 2002

268 Condom Use by Women

The size of each territory shows the numbers of girls and women aged 15 to 24 years old who used a condom the last time they had high-risk sex.

This map shows the numbers of young women who used a condom last time they had high-risk sex. Not everyone has access to condoms, can demand their use, or wants to use them. Overall, only 40% of young women around the world used condoms the last time they had risky sex.

Figures for condom use by males and females do not agree exactly, for several reasons. Homosexual sex changes the numbers for men and women separately.

Also, many men may sleep with the same woman (or vice versa), while the figures on which this map and the previous one are based correspond only to the most recent instance of high-risk sex. This is particularly an issue where female prostitution is concerned, which is widespread and a significant risk factor for the transmission of disease. Also, condom use and high-risk sex may be reported differently by men and women.

'In many societies, it is taboo even at home to speak about sexual matters, sexual choices and sexual diseases...'

Clive Wing, UNESCO, Bangkok, 2005

CONDOM USE BY WOMEN
% women aged 15–24 using condoms, 2002

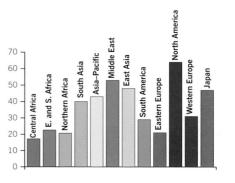

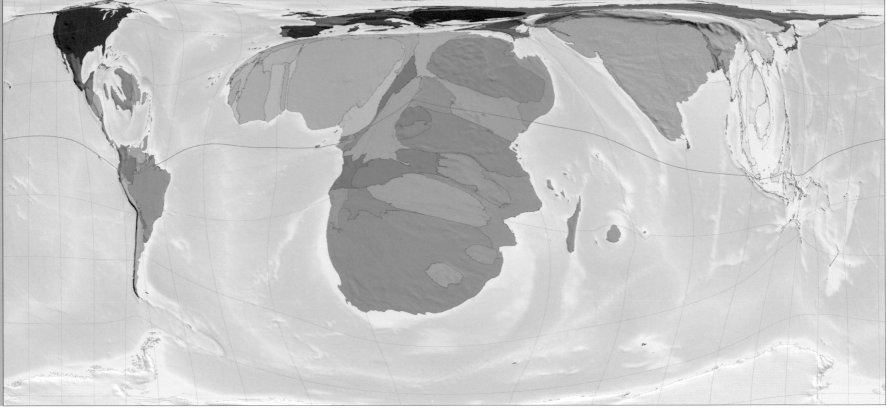

269 # HIV Prevalence

The size of each territory shows the numbers of people aged 15 to 49 infected with HIV, the virus that causes AIDS.

HIV, the human immunodeficiency virus, attacks the immune system. It eventually causes AIDS or acquired immune deficiency syndrome. AIDS cases were first recognized in the United States in 1981, although they were almost certainly occurring quite widely for many years before that. AIDS destroys the body's ability to fight infections and tumours, and is eventually fatal unless treated.

HIV can be transmitted through sex, through infected hypodermic needles,

in the womb and during childbirth. Other routes of transmission include breastfeeding and the transfusion of contaminated blood and blood products. Infected children, however, are not included on this map.

In 2003 the highest HIV prevalence was recorded in Swaziland, where 38% of people aged between 15 and 49, almost 4 in every 10, were HIV-positive (that is, known to carry the virus). The 10 territories with the highest prevalence of HIV are all in Central or East and Southern Africa.

HIGHEST RATES OF HIV INFECTION

Rank	Territory	%ᵃ
1	Swaziland	38
2	Botswana	37
3	Lesotho	28
4	Zimbabwe	24
5	South Africa	21
6	Namibia	21
7	Zambia	16
8	Malawi	14
9	Central African Republic	13
10	Mozambique	12
11	Tanzania	9
12	Gabon	8
13	Côte d'Ivoire	7
14	Cameroon	7
15	Kenya	7
16	Ethiopia	6
17	Burundi	6
18	Haiti	6
19	Nigeria	5
20	Rwanda	5

ᵃ % of people aged 15–49 living with HIV, 2003

HIV PREVALENCE
% of people aged 15–49 with HIV, 2003

[a] cases per 100,000 people, 2003

270 Tuberculosis Cases

The size of each territory shows the number of cases of tuberculosis, a disease that is spread through the air when an infected person sneezes or coughs.

The WHO reports that someone with the 'open' form of tuberculosis (the most infectious form) can infect 10 to 15 people a year. This makes it very hard to stop the spread of the disease once a certain number of people are infected. In the past 50 years drugs have been developed to treat tuberculosis, but the disease has since evolved strains that are resistant to this medication.

In 2003 tuberculosis affected 8.7 million people. Most of these people were in Asia and Africa; only a small proportion were residents of Europe and the Americas.

TUBERCULOSIS CASES
cases per 100,000 people

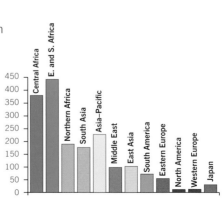

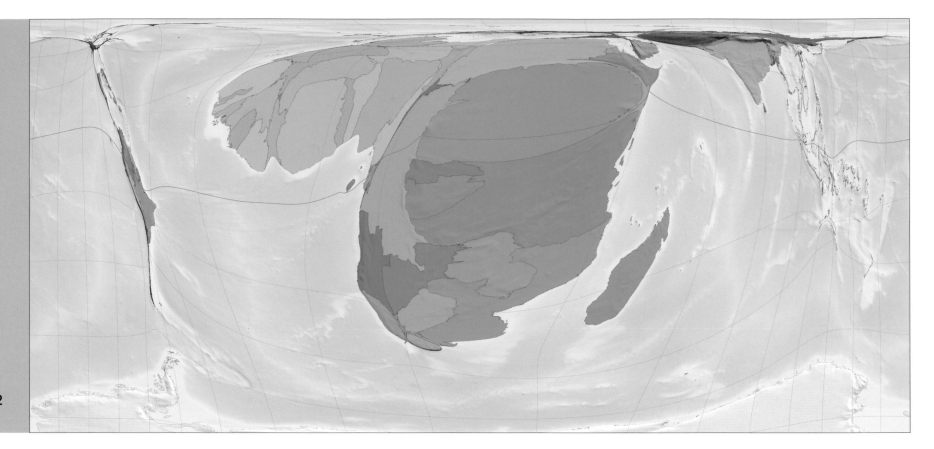

271 Malaria Cases

The size of each territory shows the number of people with malaria, a parasitic infection spread by mosquitoes.

Symptoms of malaria include fever and vomiting. The incidence of the disease can be reduced by mosquito control or by the preventive use of anti-malarial medications. The term 'malaria' comes from the medieval Italian *mala aria*, meaning 'bad air'. The term was coined at a time before the mosquito had been identified as the carrier of the disease.

Malaria is curable, but there are nonetheless many people living with the disease in chronic form. In 2003 an estimated 72 million people had malaria, of whom 92% were in African territories. Almost half the people living in Uganda suffer from malaria.

Most territories outside Africa are barely visible on this map as the prevalence of malaria is so low elsewhere.

'...malaria remains the biggest cause of death for children under five in Africa.' Jong-Wook Lee, director general of WHO, 2006

MALARIA CASES
cases per 100 people, 2003

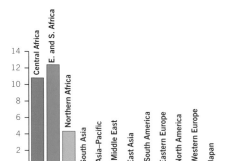

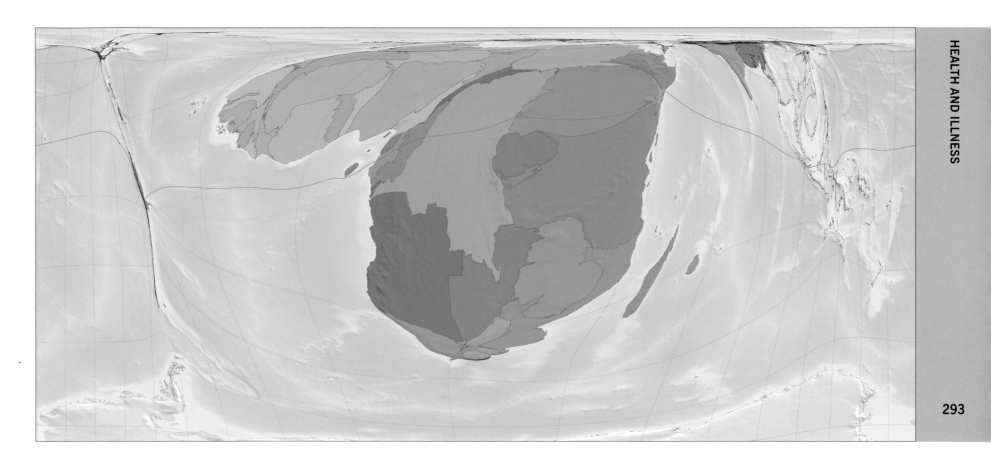

MOST MALARIA DEATHS

Rank	Territory	Deaths[a]
1	Angola	86
2	Sao Tome and Principe	77
3	Malawi	59
4	Guinea-Bissau	56
5	Namibia	55
6	Zambia	54
7	Tanzania	39
8	Burkina Faso	35
9	Uganda	34
10	Rwanda	32
11	DR Congo	32
12	Mozambique	19
13	Togo	16
14	Ghana	16
15	Solomon Islands	14
16	Swaziland	14
17	Senegal	14
18	Central African Republic	13
19	Chad	12
20	Mali	10

[a] deaths caused by malaria per 100,000 people, 2003

272 Deaths from Malaria

The size of each territory shows the number of deaths from malaria.

In 2003, 94% of all deaths from malaria occurred in African territories. The other 6% occurred mainly in the regions of South Asia and the Asia–Pacific and Australasia.

Malaria does not kill its victims quickly, and many people live with the disease for years. As a result, deaths from malaria are typically very much fewer in number than malaria cases. In 2003, for example, malaria deaths recorded were only 0.15% of the total number of cases. However, while most malaria cases occurred in East and Southern Africa, most deaths caused by the disease were in Central Africa.

MALARIA DEATHS
deaths from malaria, 2003

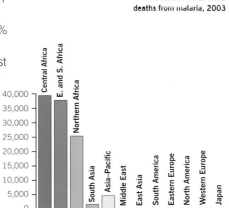

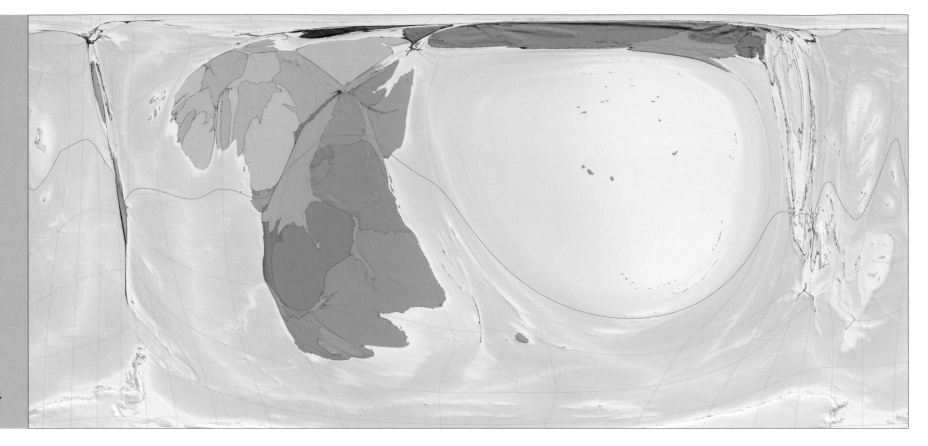

273 Cholera Cases

The size of each territory indicates the number of cases of cholera, an infection of the intestine that causes diarrhoea, vomiting and leg cramps.

Although cholera is not intrinsically fatal, without proper care an infected person can quickly become dehydrated and die. Cholera is transmitted by contaminated food or water and is most common where sanitation is poor and/or access to clean water is limited.

The data used to create this map are from 2004. In that year 70% of all cholera cases identified were in the regions of East and Southern Africa, Northern Africa and Central Africa. Mozambique had the highest number of cholera cases, at 20,000.

An epidemic in 1978 in the small population of the Maldives affected 3.8% of the population, but the Maldives are barely visible on this 2004 map.

'Cholera is a disease of the poor, born of too many people living cheek by jowl without the infrastructure of a sewer system or clean drinking water.' Pascale Harter, BBC News website, 2005

PREVALENCE OF CHOLERA
cases per 100,000 people per year

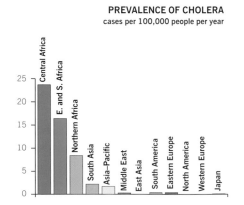

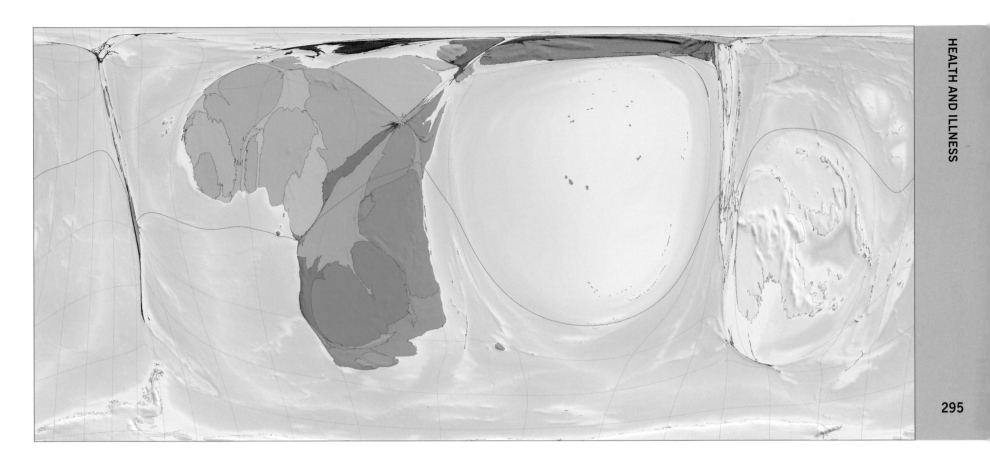

274 Deaths from Cholera

The size of each territory shows the numbers of deaths from cholera.

MOST CHOLERA DEATHS

Rank	Territory	Deaths[a]
1	Tuvalu	80.0
2	Maldives	73.3
3	Marshall Islands	11.5
4	Papua New Guinea	8.3
5	Djibouti	4.6
6	Zambia	3.5
7	Chad	3.3
8	Tonga	2.0
9	Mali	1.6
10	Swaziland	1.5
11	Guinea	1.4
12	Central African Republic	1.3
13	Sierra Leone	0.9
14	Cameroon	0.9
15	Albania	0.8
16	Tanzania	0.7
17	Togo	0.7
18	Lebanon	0.6
19	Laos	0.6
20	Mozambique	0.6

[a] deaths caused by cholera per 100,000 people, 2004 or most recent data available

Cholera deaths result from severe dehydration caused by diarrhoea. Cholera is treatable: in 2004 the number of cholera deaths was only 2.5% of the number of cholera cases that year. Distributions of cholera cases and deaths differ due to differing availability of treatment.

In 2004 in the Central African Republic 36% of people contracting cholera died. By contrast, there were 73 territories where nobody died from cholera, because of good sanitation, clean water and prompt treatment.

'The cholera outbreak has continued...water provided by the tankers is not enough and they try to boost their supply from the wells, which are not covered. The rain washes faeces and other pollutants into the wells...' Pierre Kahozi, WHO official working in Mozambique, 2004

CHOLERA DEATHS
deaths from cholera, 2004

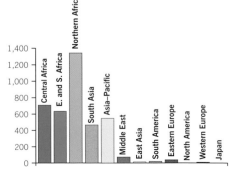

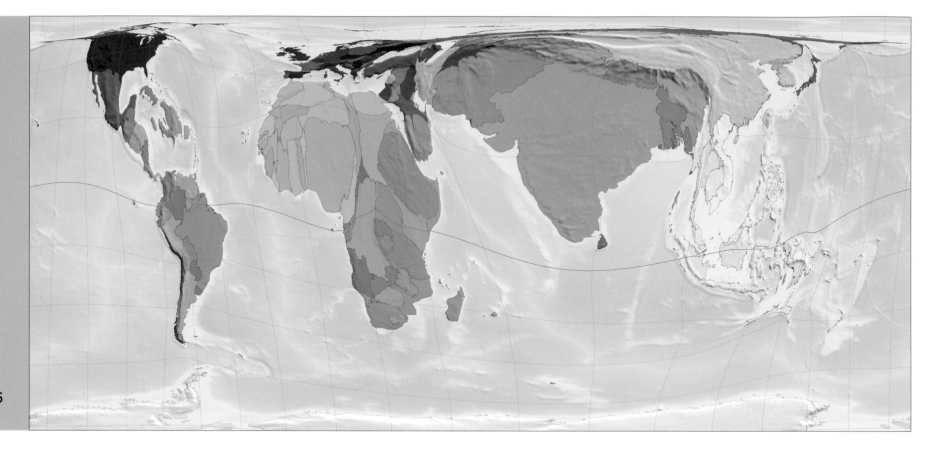

275 Childhood Diarrhoea

The size of each territory shows the number of cases of diarrhoea in children below 4 years of age.

Diarrhoea is common among children everywhere. In an average two-week period, an estimated 82 million children worldwide under the age of 4 years have diarrhoea. Diarrhoea varies in its severity: some children recover quickly, while a small proportion die. Access to clean water and salts for rehydration can reduce the prevalence of diarrhoea and minimize its impact.

The highest prevalence of diarrhoea among children is in Niger, where 4 in every 10 children were affected in a typical two-week period. Most children in Niger will have many episodes of diarrhoea each year, which can ultimately cause chronic debility.

CHILDHOOD DIARRHOEA
% prevalence

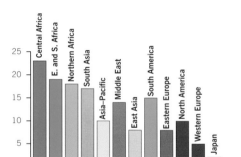

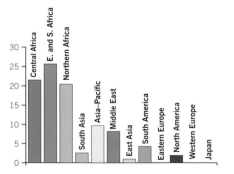

ᵃ % of children with evidence of trachoma, 2003

276 Blinding Disease

The size of each territory shows the number of children with evidence of trachoma, a bacterial disease of the eye that can cause blindness.

Trachoma affects 128 million children worldwide and more than twice as many adults. The bacterium that causes the disease is transmitted from one person's eyes to another's by flies or through contact with fluids.

Trachoma is cheap to treat in its early stages, but in later stages surgery may be necessary. The WHO notes that trachoma is found principally in areas with poor socioeconomic conditions, such as overcrowding, limited access to water and poor sanitation.

TRACHOMA IN CHILDREN
cases per 100 children, 2003

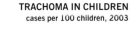

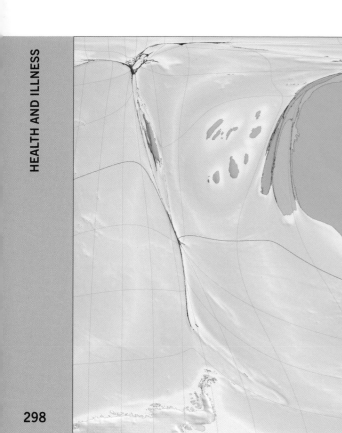

HIGHEST RATES OF POLIO

Rank	Territory	Cases[a]
1	Cape Verde	1.87
2	Nigeria	0.29
3	Chad	0.22
4	Congo	0.22
5	DR Congo	0.20
6	Somalia	0.19
7	Niger	0.16
8	Central African Republic	0.15
9	Angola	0.15
10	Sierra Leone	0.12

[a] cases per 100,000 people, 2000–5

HIGHEST NUMBER OF POLIO CASES

Rank	Territory	Cases[a]
1	Nigeria	77
2	Sudan	24
3	Yemen	22
4	India	14
5	Pakistan	7
6	Ethiopia	5
7	Indonesia	4
8	Cameroon	1
8	Niger	1

[a] cases reported in 2005 (no cases were reported in 2005 in any country other than these 9)

277 Polio Cases

The size of each territory indicates the number of cases of polio, a viral infection that can permanently paralyse groups of muscles.

Poliomyelitis – usually called just polio – cannot be cured, although it is possible to vaccinate against it. Polio usually affects children aged 5 and under.

Since the 1980s the number of polio cases has decreased significantly as a result of widespread vaccination in the latter half of the 20th century. Between 2000 and 2005 the annual number of new polio cases worldwide fell from 2,971 to 155, and in 2005 polio was reported in just 9 territories.

India and Nigeria recorded the largest numbers of polio cases between 2000 and 2005 – 2,508 cases in India and 2,120 in Nigeria. Most territories had no cases at all, and so are not visible on the map.

'[The] Democratic Republic of Congo...could compromise the certification of global eradication of poliomyelitis, because of the economic and social crises that have prevailed...since the early 1990s.' Matthieu Kamwa, *African Health Monitor*, 2002

POLIO CASES
new cases per year, 2000–5

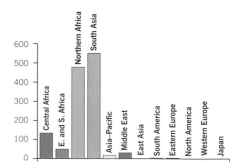

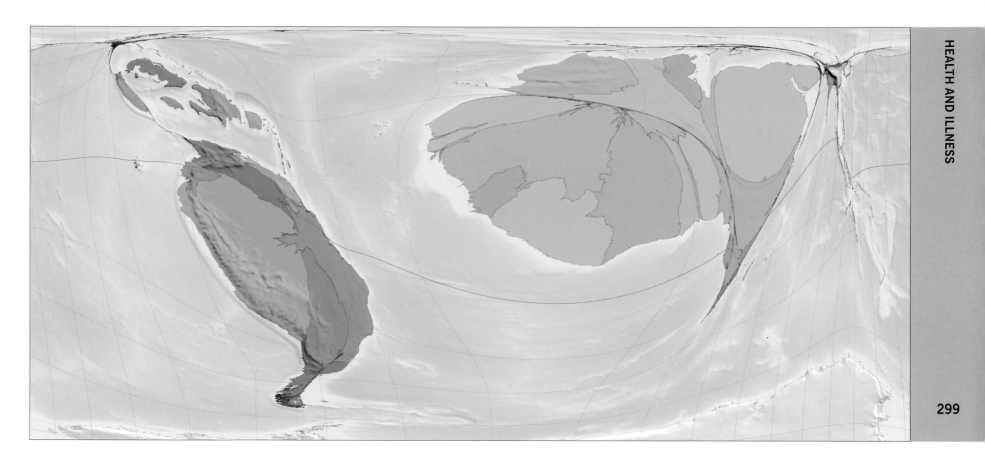

278 Yellow Fever

The size of each territory indicates the number of cases of yellow fever, a disease carried – like malaria – by mosquitoes.

Unlike malaria, which is also carried by mosquitoes, yellow fever is not found in South Asia. It occurs almost exclusively in Northern Africa (which has 68% of cases) and South America (31%).

Yellow fever is also known as black vomit. Both names refer to the disease's more severe symptoms: 'yellow fever' refers to the fever and jaundice that can occur, 'black vomit' to the congealed blood in the vomit of its victims. A vaccine against yellow fever exists, though not everyone has access to it. 1 in 10 yellow fever cases lead to death.

YELLOW FEVER
cases per year, 1995–2004

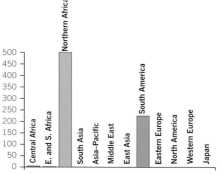

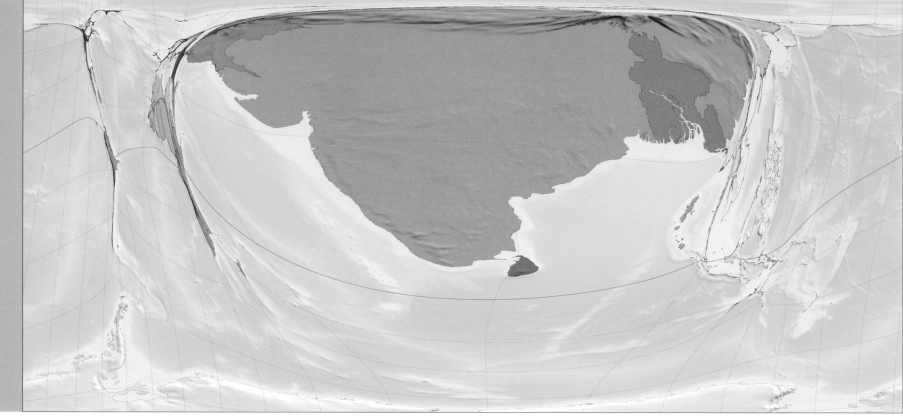

HIGHEST RATE OF RABIES DEATHS

Rank	Territory	Deaths[a]
1	India	2.0
2	Maldives	1.7
3	Bangladesh	1.0
4	Sri Lanka	0.5
5	Namibia	0.4
6	Philippines	0.4
7	Viet Nam	0.3
8	Myanmar	0.3
9	Pakistan	0.2
10	Eritrea	0.1

[a] deaths per 100,000 people, 1994–2004

HIGHEST NUMBER OF RABIES DEATHS

Rank	Territory	Cases[a]
1	India	17,000
2	China	2,009
3	Bangladesh	1,550
4	Myanmar	1,100
5	Philippines	284
6	Uganda	105
7	Sri Lanka	76
8	Nepal	44
9	Indonesia	40
10	Eritrea	34

[a] cases reported in 2003

279 Deaths from Rabies

The size of each territory indicates the numbers of human deaths from rabies, usually caused by a dog bite.

Rabies occurs primarily in animals but can be transmitted to humans, for example when someone is bitten by a rabid animal. Thorough cleaning and vaccinations after being bitten can prevent the development of rabies; however, once symptoms appear (sometimes as long as 90 days after the bite), rabies is almost always fatal.

Most human cases of rabies are caused by dog bites. Where rabies is common, dogs are often greatly feared.
85% of all human rabies deaths between 1995 and 2004 occurred in India, which saw an average of 21,404 deaths per year from rabies during that decade.

'About 3.5 million dog bites are registered every year in India. The Government cannot give vaccine free of cost to all people. By 2006, the price of vaccine is expected to increase…' K. Sandeep, *The Hindu*, 2002

RABIES DEATHS
deaths per year, 1995–2004, thousands

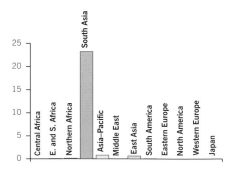

301

280 Outbreaks of Influenza

The size of each territory shows how many people were exposed to outbreaks of influenza, for how long, over a six-year period.

Influenza, commonly known as the flu, greatly fluctuates in prevalence over time. This map shows the length of time, measured in weeks, that people were exposed to influenza outbreaks between 2000 and 2005, multiplied by the number of people exposed. The populations of Russia and Colombia experienced influenza outbreaks for 20% of this period, so these territories appear particularly large on the map. Influenza outbreaks affect whole communities, not just those who catch the disease, because many people who are ill cannot work or care for children, and others are needed to care for those who are ill.

PEOPLE AFFECTED BY INFLUENZA OUTBREAKS
millions living with influenza outbreaks, weeks per year, 2000–5

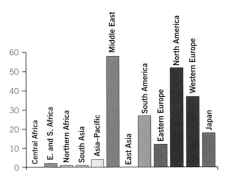

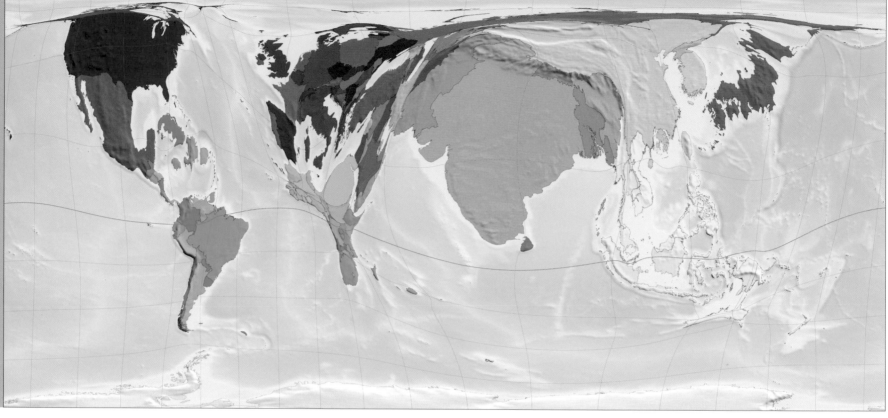

281 **Prevalence of Diabetes**

The size of each territory shows the number of people over 15 years of age with diabetes, a disease characterized by inability either to produce or to use the hormone insulin.

Insulin is a hormone that regulates the body's ability to absorb sugar. A person with Type 1 diabetes is unable to produce enough insulin, and so must inject insulin to survive. Type 2 diabetes, which is more common, occurs when insulin is present but the body lacks the ability to use it properly; this type of diabetes can often be managed through diet and appropriate medication.

As a percentage of total population, the highest prevalence of diabetes cases is in North America. Of these cases, 62% are in the United States, 33% are in Mexico and 4% are in Canada. The territory with the largest total number of diabetes cases in 2001 was India, with 56 million cases.

'Diabetes is responsible for over one million amputations each year. It is a major cause of blindness. It is the largest cause of kidney failure in developed countries and is responsible for huge dialysis costs.' Unite for Diabetes, 2006

DIABETES PREVALENCE
% of population aged 15+ with diabetes

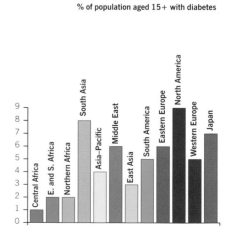

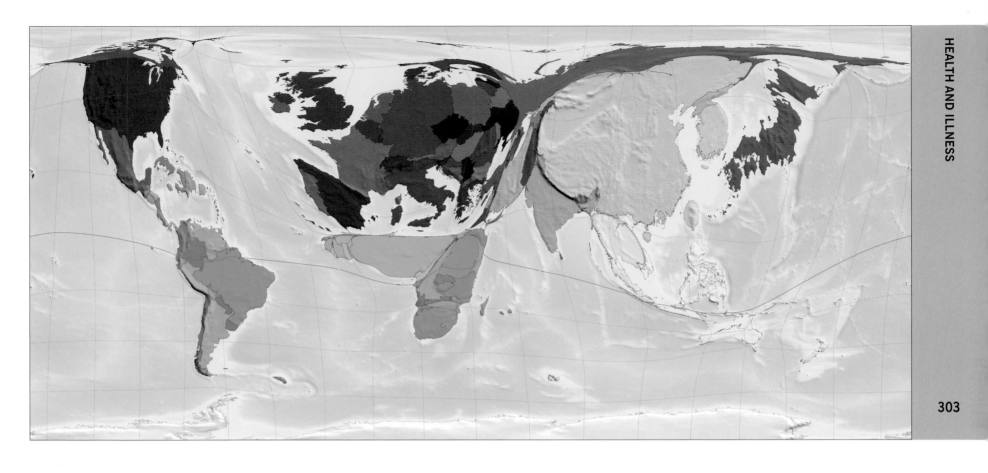

303

282 Alcohol Consumption

The size of each territory shows the total quantity of alcohol consumed. Western Europeans are among the heaviest drinkers.

On average, every Western European drinks one and a third times as much alcohol than a person in any other region. The lowest rate of alcohol consumption is in South Asia, where on average people drink less than a third of the average consumption elsewhere.

In some territories there is practically no alcohol consumption. Many territories in the Middle East and Central Asia and in Northern Africa are not visible on the map for this reason. At the other end of the scale, China, the United States and Russia occupy the largest areas on the map because of the large total amount of alcohol consumed there, a combination of their large populations and generally high consumption rates.

ALCOHOL CONSUMPTION
litres pure alcohol equivalent
per person aged 15+ per year

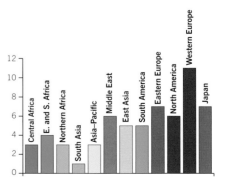

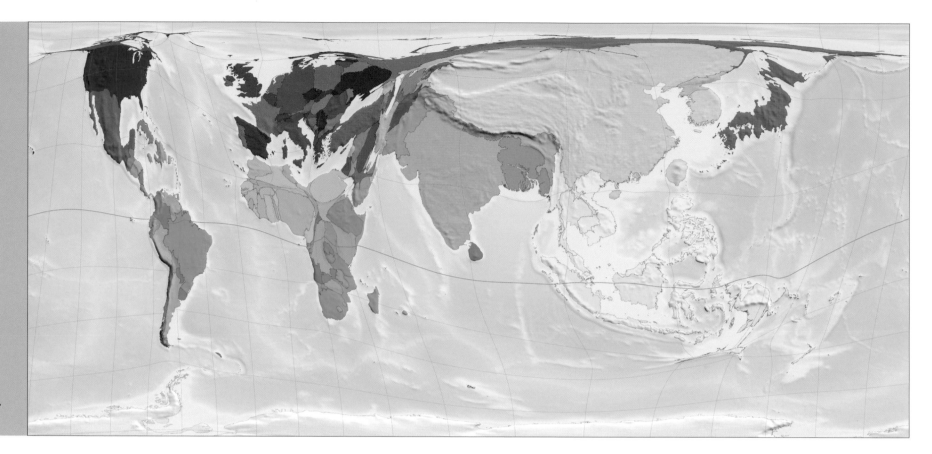

MOST AND FEWEST MALE SMOKERS

Rank	Territory	%ᵃ
1	Yemen	77
2	Indonesia	69
3	Mongolia	68
4	Armenia	68
5	Kenya	67
6	Cambodia	67
7	Russia	64
8	Uruguay	62
9	Kyrgyzstan	60
10	Georgia	60
191	Sudan	24
192	Singapore	24
193	Canada	24
194	Tanzania	23
195	Iran	22
196	Australia	21
197	Panama	20
198	Saudi Arabia	19
199	Puerto Rico	17
200	Sweden	17

ᵃ % of men who smoked, 2002

283 Men Smoking

The size of each territory indicates the number of male smokers.
Worldwide there are four times as many men who smoke as women.

In 2002 there were 941 million male smokers over 15 years of age, which is 43% of all men in this age bracket. The largest population of male smokers is in China, where men are more likely to smoke than not. Even in Puerto Rico and Sweden, which have the lowest percentages of male smokers, 17% of men smoke.

When smoking is this widespread, smokers do not just damage their own health, but also collectively damage the health of people around them. Passive inhalation of tobacco smoke by children, for example, is believed to increase the risks of asthma, cot deaths and chest infections.

'The prevalence of smoking increased dramatically during the world wars, mainly due to the policy of providing free cigarettes to allied troops as a "morale boosting" exercise.' The Cancer Council, 2006

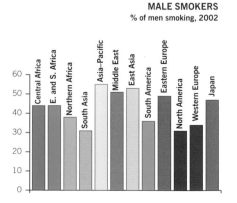

MALE SMOKERS
% of men smoking, 2002

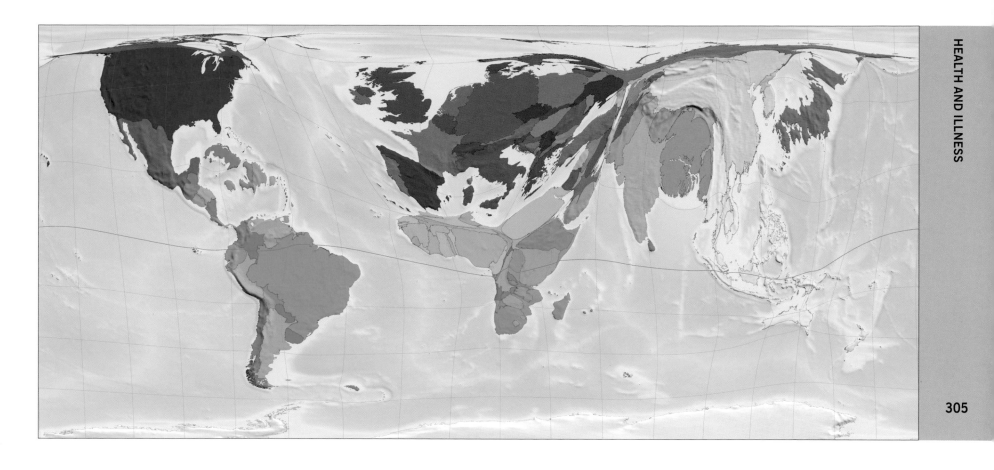

305

MOST AND FEWEST FEMALE SMOKERS

Rank	Territory	%ᵃ
1	Guinea	47
2	Uruguay	39
3	Lebanon	35
4	Chile	34
5	Kenya	32
6	Norway	32
7	Germany	31
8	Ireland	31
9	Hungary	30
10	Yemen	29
191	Indonesia	3
192	Armenia	3
193	Mauritius	3
194	Sudan	2
195	Morocco	2
196	Iran	2
197	Sri Lanka	2
198	Thailand	2
199	Tanzania	1
200	Azerbaijan	1

ᵃ % of women who smoked, 2002

284 Women Smoking

The size of each territory indicates the number of female smokers. There are women who smoke in every territory in the world.

People have smoked or chewed tobacco for at least the last 2,000 years and there has been a well-developed tobacco industry for about the last 300. The industry has grown hugely in the last 100 years with the invention of automated cigarette-rolling machines. With growing recognition of the health risks smokers pose both to themselves and to others, some countries are now starting to ban the smoking of tobacco in certain places.

The numbers of women smokers vary widely across the world, ranging from 1% of the population in Azerbaijan to 47% in Guinea. Of the total world female population over the age of 15 years, 10% are smokers. (Some girls start to smoke before they reach the age of 15, but this map shows only those over 15.) The Americas are home to 29% of the world's 232 million female smokers.

'If the tobacco industry hates a policy then you can guarantee it's successful in reducing the number of cigarettes smoked.'

Simon Chapman, professor of public health, University of Sydney, 2001

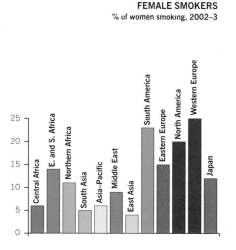

FEMALE SMOKERS
% of women smoking, 2002–3

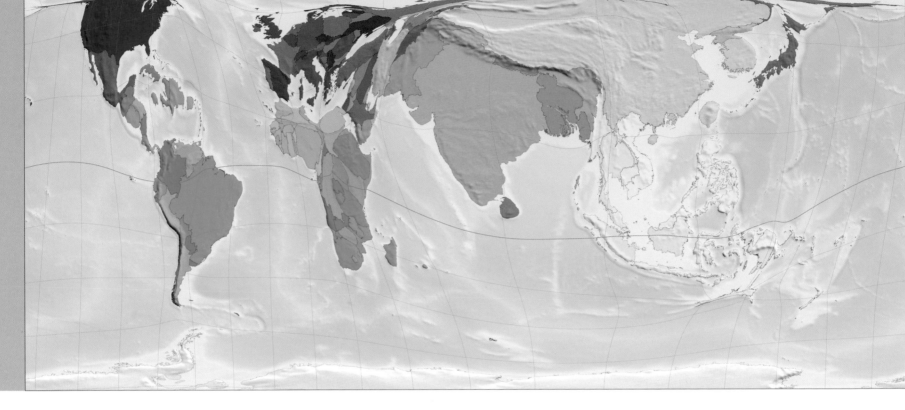

285 Road Deaths

There are about 29 million kilometres of road in the world. The size of each territory in this map indicates the numbers of deaths in road traffic accidents.

In 2001 there were 984,000 deaths caused by road traffic accidents. Half of these occurred in the regions of East and South Asia. The highest rate of road deaths as a percentage of population, however, was in South America. The lowest rates were in Northern Africa and Japan.

On average each year, one person dies in a traffic accident on every 30-kilometre stretch of road. This is only an average, however; in practice some stretches of road experience many more accidents than others. The world's most dangerous road is in Bolivia. It is 70 kilometres long and approximately 100 people die there every year.

ROAD TRAFFIC DEATHS
deaths per 100,000 people, 2001

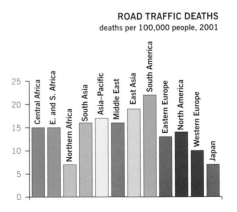

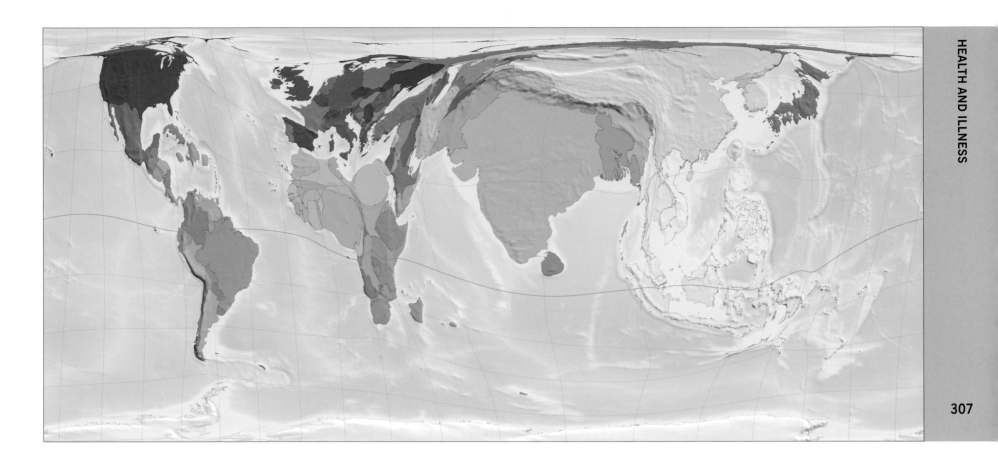

**MOST AND FEWEST YEARS
OF UNHEALTHY LIFE**

Rank	Territory	Yearsª
1	Azerbaijan	20.4
2	Tajikistan	19.2
3	Sao Tome and Principe	18.5
4	Kyrgyzstan	17.0
5	Turkmenistan	16.7
6	Uzbekistan	16.1
7	Maldives	16.1
8	Eritrea	15.2
9	Albania	15.0
10	Mauritius	14.7
191	Liberia	4.4
192	Haiti	4.3
193	Kenya	3.8
194	Namibia	3.5
195	Côte d'Ivoire	3.2
196	Mozambique	2.2
197	Zimbabwe	1.9
198	Zambia	1.6
199	Lesotho	1.4
200	Swaziland	0.3

ª years of unhealthy life per person, 2000–5

286 Years of Poor Health

The size of each territory indicates the total number of years
the current population is expected to live in poor health.

The WHO's Health-Adjusted Life Expectancy
Estimate measures the number of years a
person can expect to live in good health. By
comparing this figure to total life expectancy
we can calculate the average number of
years a person lives in poor health. The map
shows this figure multiplied by the size of
the population, giving the total number of
years of poor health that will be experienced
by the current population of each territory.

Worldwide, the average member of the
population can expect to spend about 10
years in poor health. Figures for individual
territories vary, but not always for the
reasons one might imagine. In Azerbaijan
the average person lives over 20 years in
poor health. In Swaziland, by contrast, the
average person lives for less than 1 year in
poor health – primarily because total life
expectancy there is only 36 years.

*'Children surviving multiple bouts of diarrheal disease,
respiratory infection, helminthic infections, and malaria, may
well suffer lifetime impairments in physical and cognitive
capacities.'* Jeffrey D. Sachs, economist, 2000

UNHEALTHY LIFE EXPECTANCY
years per person

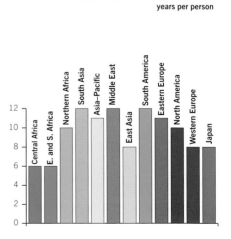

The Perilous World

Death and Disaster

War and Crime

Pollution and Depletion

Extinction and Endangerment

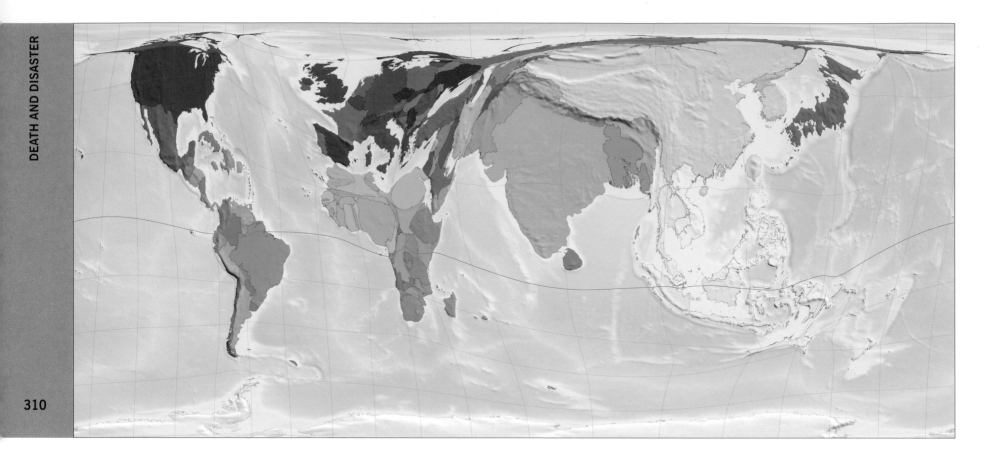

287 Life Expectancy

The size of each territory indicates the total expected years of life of the current population, based on available facts and assuming the continuance of prevailing trends.

This map shows the aggregate number of years the current population of the world is expected to live, assuming all those now living have the same life expectancy as a child born in the same place in 2002. Life expectancy figures are inevitably only an estimate, and there are many complicating factors that can make such predictions difficult. Men and women live for different lengths of time on average. And life expectancy depends on how old you are: a newborn infant may expect on average to live to 50 in a particular territory, but clearly a 60-year-old in the same territory can safely say their life expectancy is longer. Many other factors also matter, including personal health, income and environment.

The longest life expectancy at birth is 81 years 6 months, in Japan. The shortest is 32 years 8 months, in Zambia. The world average life expectancy is 67 years.

'The adult mortality rate [in Zambia] has increased in the last decade... an adult has lost about 11 years of survival due to the AIDS problem...'

Buleti Nsemukila, director of Zambian Central Statistical Office, 2003

LIFE EXPECTANCY
years, for those born in 2002

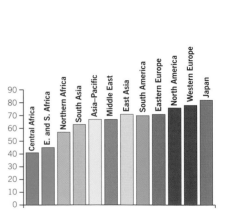

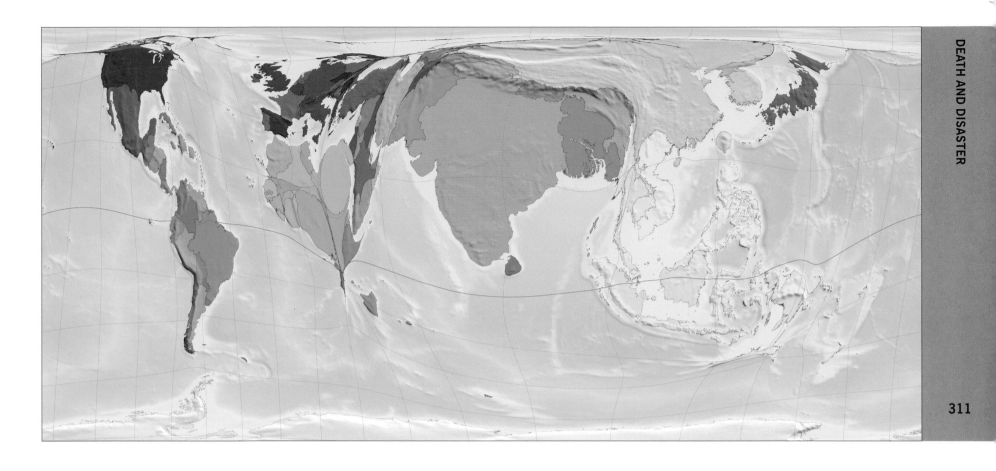

288 Increase in Life Expectancy

The size of each territory indicates the total number of extra years of life that the current population can expect as a result of the increase in life expectancy from 1972 to 2002.

Between 1972 and 2002 life expectancy increased in 88% of territories. Large increases occurred in Northern Africa (an extra 10 years), South Asia (an extra 13 years) and the Asia–Pacific and Australasia region (an extra 14 years). Increases were smaller where life expectancy was already relatively high, for example in Western Europe and North America. Worldwide the average increase was 9 years.

Territories in which life expectancy has stayed the same or decreased have size zero on this map and so do not appear. This applies to 24 territories, most of them in sub-Saharan Africa, where many people live with HIV and die young.

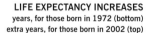
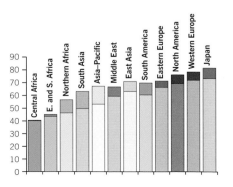

LIFE EXPECTANCY INCREASES
years, for those born in 1972 (bottom)
extra years, for those born in 2002 (top)

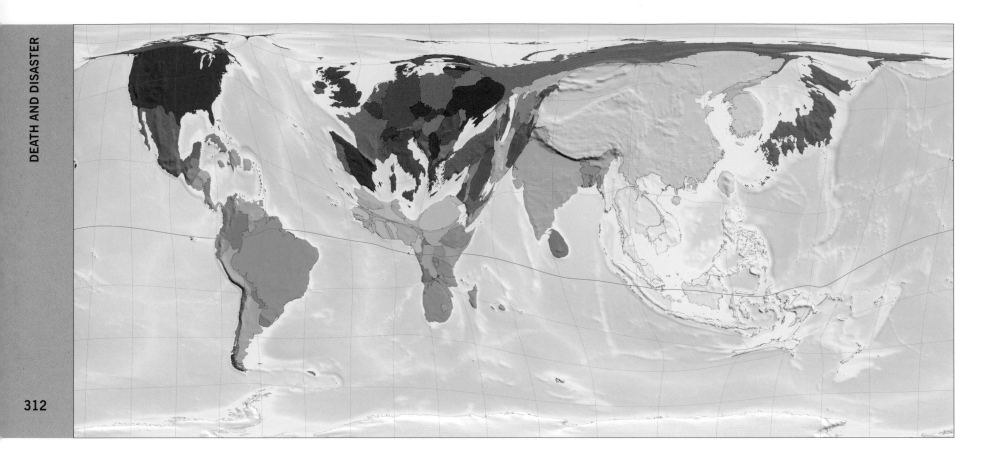

NUMBER OF YEARS BY WHICH WOMEN OUTLIVE MEN

Rank	Territory	Years[a]
1	Russia	12.3
2	Kazakhstan	11.1
3	Latvia	10.7
4	Belarus	10.5
5	Estonia	10.4
6	Ukraine	10.1
7	Lithuania	10.1
8	Brazil	8.6
9	Hungary	8.3
10	Thailand	8.2
191	Nigeria	0.8
192	Malawi	0.7
193	Guinea	0.7
194	Côte d'Ivoire	0.6
195	Niger	0.6
196	Pakistan	-0.3
197	Zambia	-0.4
198	Nepal	-0.5
199	Zimbabwe	-0.8
200	Maldives	-0.9

[a] **average time by which females born in 2002 can expect to outlive males**

289 Women Outliving Men

The size of each territory indicates the total number of extra years the female population can expect to live, compared to their male compatriots.

In 2002, life expectancy at birth was greater for women than for men in almost all territories. The biggest differences were in Russia (12 years) and Kazakhstan (11 years). In only five territories can men expect to live longer than women: Pakistan, Zambia, Nepal, Zimbabwe and the

Maldives. Among these the biggest difference in life expectancy is in the Maldives, where male life expectancy is 11 months longer than female.

Territories in which women have lower life expectancy than men have size zero on this map and so do not appear.

'Up until very recently in human terms, life expectancy for men was greater than for women...'

Carol J. Hogue, professor of maternal and child health, and of epidemiology, Rollins School of Public Health, Emory University, 2006

YEARS WOMEN LIVE LONGER THAN MEN
additional years of life expectancy per female born in 2002

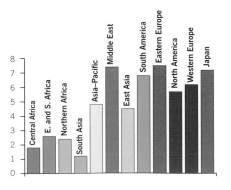

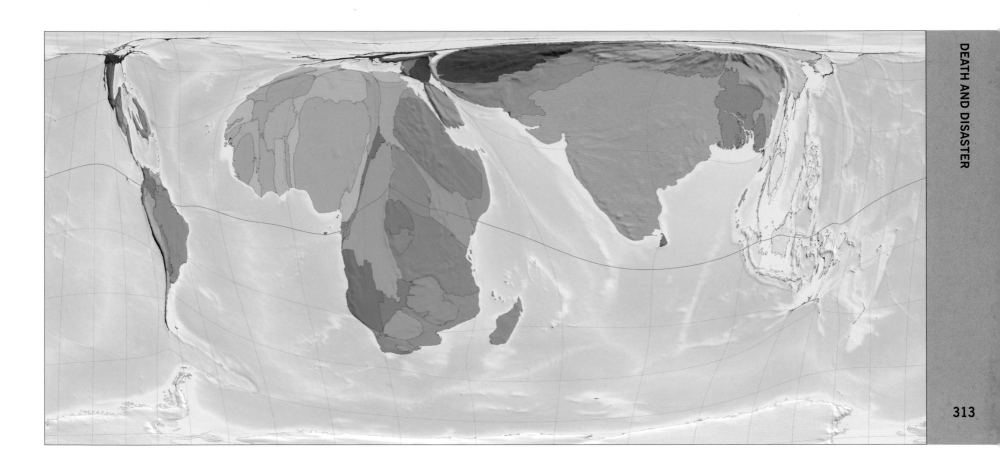

290 # Maternal Mortality

The size of each territory indicates the number of pregnancy-related deaths among women, defined as deaths while pregnant or within 6 weeks of the end of a pregnancy.

In 2000 more than 513,000 women died of pregnancy-related causes. As this map shows, most of these deaths occurred in South Asia and Africa. The smallest numbers of maternal deaths were in territories in Western Europe and Japan.

The territory with the highest rate of maternal deaths as a percentage of the total number of births is Sierra Leone, where 2 mothers die per 100 births. At the other extreme, both Malta and Iceland reported no maternal deaths at all in 2000. The world average rate is 386 deaths for every 100,000 births.

HIGHEST AND LOWEST RATES OF MATERNAL MORTALITY

Rank	Territory	Deaths[a]
1	Sierra Leone	2,000
2	Afghanistan	1,900
3	Malawi	1,800
4	Angola	1,700
5	Niger	1,600
6	Tanzania	1,500
7	Rwanda	1,400
8	Mali	1,200
9	Somalia	1,100
10	Chad	1,100
191	Kuwait	5
192	Portugal	5
193	Denmark	5
194	Ireland	5
195	Spain	4
196	Austria	4
197	Slovakia	3
198	Sweden	2
199	Malta	0
200	Iceland	0

[a] pregnancy-related maternal deaths per 100,000 births, 2000

MATERNAL MORTALITY RATES
maternal deaths per 100,000 births, 2000

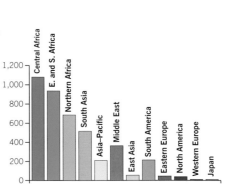

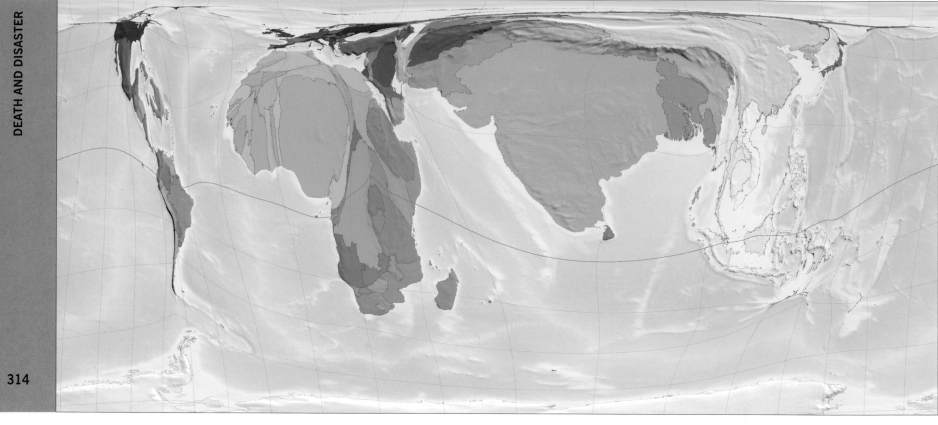

291 Stillbirths

The size of each territory indicates the number of children born dead, past 28 weeks of gestation.

In 2000 there were 3.3 million stillbirths worldwide. More than a third of these occurred in South Asia.

The rate at which stillbirths occur varies greatly between regions. In Central Africa there are 42 stillbirths for every 1,000 births. In Western Europe the rate is a tenth of this. As a territory, Mauritania has the highest rate of stillbirths at 63 per 1,000.

STILLBIRTHS
stillbirths per 1,000 births, 2000

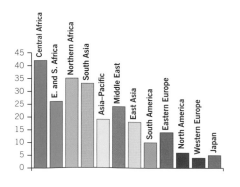

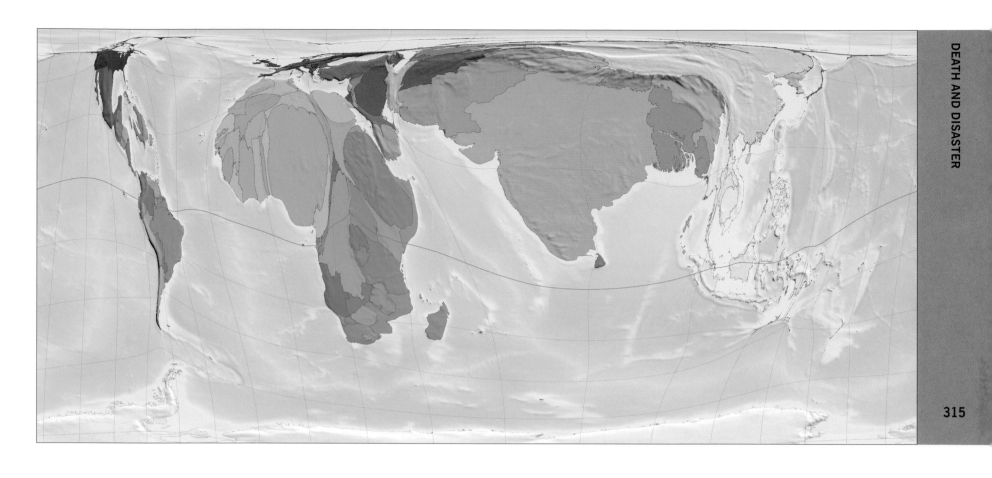

292 Early Neonatal Mortality

The size of each territory indicates the number of deaths occurring within the first week after birth.

In 2000, 3 million babies died during their first week of life. Put another way, 2.3% of all babies born lived less than a week. The rate of early neonatal death ranges from 1 in 1,000 in Japan to 1 in 20 in Mauritania.

According to the WHO the three main causes of neonatal deaths are asphyxia at birth, low birthweight (including prematurity) and infections, all of which can be reduced substantially by access to effective healthcare.

'We must count newborn deaths and make them count, instead of accepting these deaths as inevitable.'
Francisco Songane, director of WHO Partnership for Maternal, Newborn and Child Health, 2006

EARLY NEONATAL MORTALITY
deaths in first week per 1,000 live births, 2000

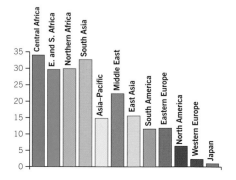

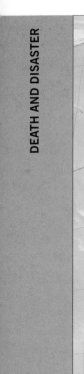

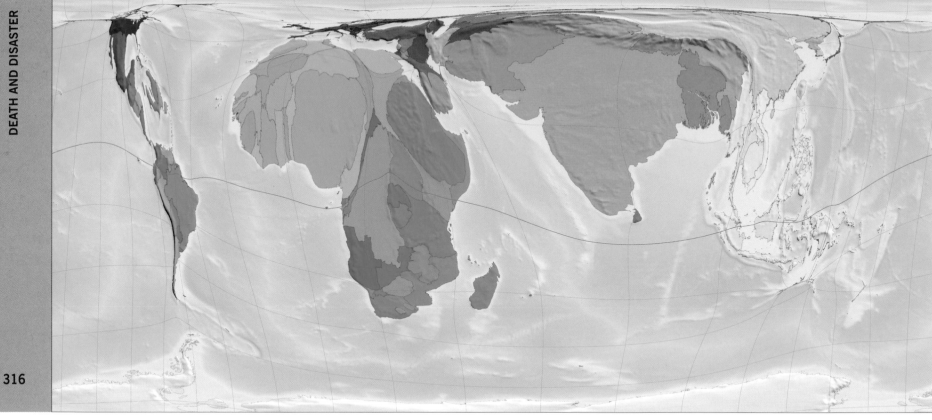

HIGHEST INFANT MORTALITY

Rank	Territory	Deaths[a]
1	Sierra Leone	165
2	Niger	156
3	Angola	154
4	Guinea-Bissau	130
5	DR Congo	129
6	Mozambique	125
7	Mali	122
8	Mauritania	120
9	Chad	117
10	Central African Republic	115
11	Ethiopia	114
12	Malawi	114
13	Burundi	114
14	Nigeria	110
15	Guinea	109
16	Zambia	108
17	Burkina Faso	107
18	Swaziland	106
19	Tanzania	104
20	Côte d'Ivoire	102

[a] **deaths within 12 months of birth per 1,000 live births, 2002**

293 Infant Mortality

The size of each territory indicates the number of deaths occurring during the first year after birth.

In 2002, 7.2 million infants – 5.4% of all babies born – died within a year of birth. 2.3% of all those born died in their first week.

The territory with the largest number of infant deaths was India, where there were 1.7 million in 2002, accounting for 24% of the world total. Of every 100 babies born in India, 7 die in their first 12 months.

In 22 territories the infant mortality rate is over 1 in 10. All of these 22 territories are in Africa. The highest infant mortality rate of any territory is in Sierra Leone, where 16 babies die out of every 100 born.

INFANT MORTALITY
deaths in first year, per 1,000 live births, 2002

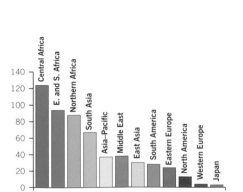

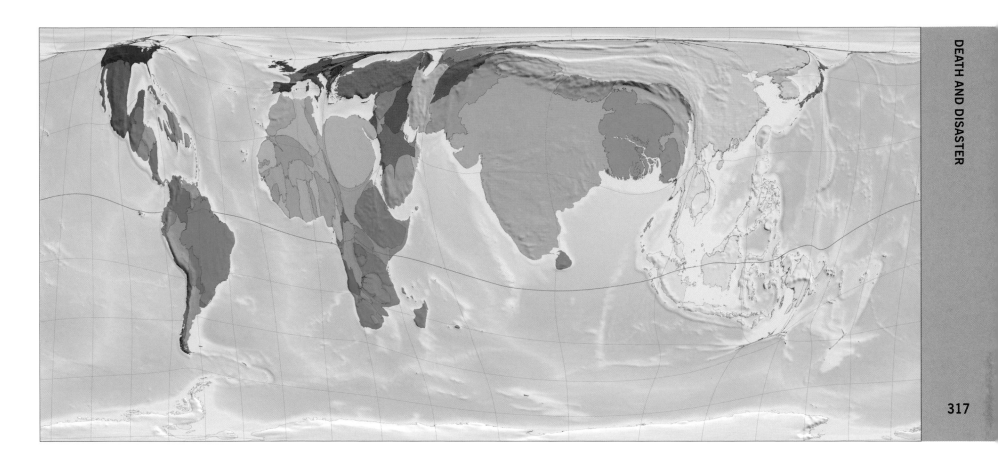

LARGEST AND SMALLEST IMPROVEMENTS IN INFANT MORTALITY RATES

Rank	Territory	Survivals[a]
1	Egypt	122
2	Yemen	115
3	Oman	115
4	Turkey	114
5	Tunisia	114
6	Algeria	104
7	Mali	103
8	Comoros	100
9	Nepal	99
10	Maldives	99
191	Switzerland	10
192	Iceland	10
193	Finland	9
194	Norway	9
195	Netherlands	8
196	Sweden	8
197	Ukraine	6
198	Belarus	5
199	Latvia	4
200	Zambia	1

[a] reduction in infant mortality per 1,000 live births, 1970–2002

294 Decrease in Infant Mortality

The size of each territory indicates how far infant mortality declined between 1970 and 2002.

Since 1970 the world average number of infant deaths per 1,000 live births has fallen by 51. Were infant mortality rates still at their 1970 level, an additional 6.8 million of the children born in 2002 would not have survived to see their first birthdays.

Infant mortality has fallen in every territory where data are available. The biggest improvements have been in the Middle East and Central Asia and in Eastern Europe.

*'We thought it was evil spirits that made our babies sick...
we have been taught it is the lack of clean water
and the absence of cleanliness.'*

Zeytu, an Ethiopian, quoted in *The Rough Guide to a Better World*, 2004

IMPROVEMENT IN INFANT MORTALITY RATES
reduction in infant mortality per
1,000 live births, 1970–2002

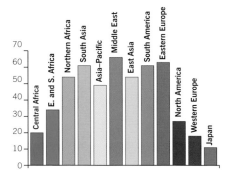

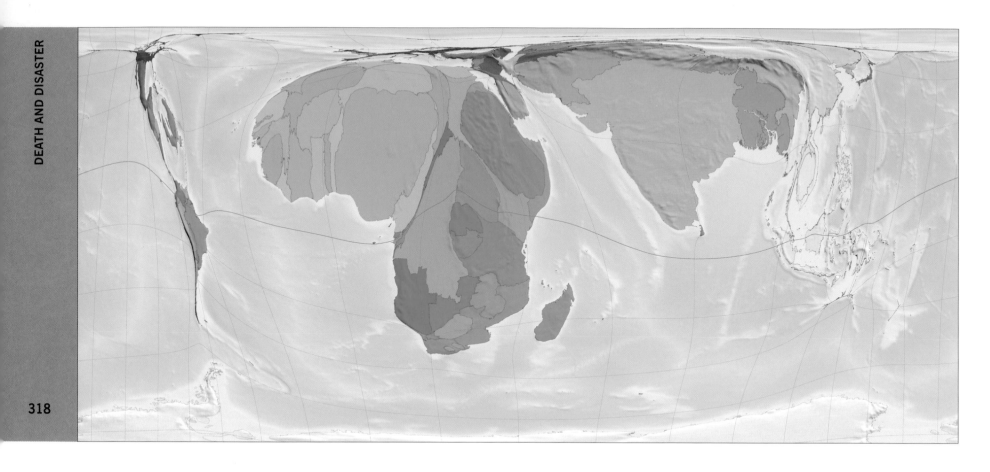

295 Child Mortality

The size of each territory indicates the number of deaths among children under 5 years
of age. Almost 50 times as many die at this age in Central Africa as in Western Europe.

In 2002, 3.2 million children between the ages of 1 and 5 died. Of these, 29% were in South Asia, 26% in Northern Africa and 18% in East and Southern Africa. The territory with the largest number of deaths was India, with 670,000.

The region with the highest death rate, measured as a percentage of the number of children in this age range, was Central Africa, where 79 out of every 1,000 children born die between the ages of 1 and 5. This is 3 times the world average. The region with the lowest death rate is Western Europe, where there are 1.4 deaths per 1,000 children.

CHILD MORTALITY
deaths age 1–4, per 1,000 live births, 2002

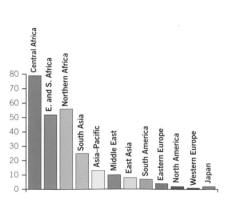

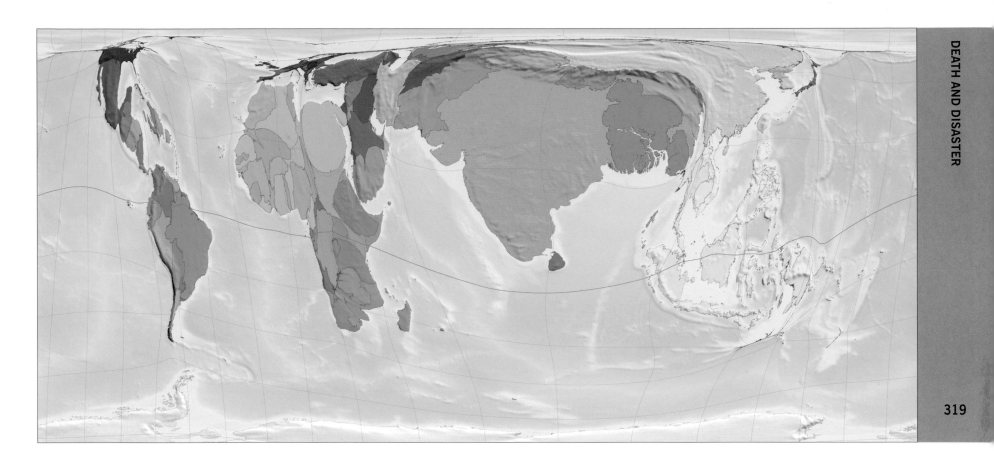

LARGEST AND SMALLEST IMPROVEMENTS IN CHILDREN'S MORTALITY RATES

Rank	Territory	Survivals[a]
1	Gambia	101
2	Bhutan	91
3	Guinea	88
4	Yemen	81
5	Algeria	81
6	Bolivia	81
7	Maldives	79
8	Mali	75
9	Egypt	72
10	Malawi	72
191	Netherlands	2
192	Finland	2
193	Switzerland	2
194	Norway	2
195	Ukraine	1
196	Italy	1
197	Latvia	1
198	Zambia	0
199	Rwanda	0
200	Iceland	0

[a] reduction in mortality rate of children aged 1–4 per 1,000 live births, 1970–2002

296 Decrease in Child Mortality

The size of each territory indicates how far child mortality fell between 1970 and 2002. Almost all territories experienced a drop in child mortality during these years.

Only Rwanda, Zambia and Iceland recorded no decrease in child mortality over this period, and no territories recorded an increase.

Worldwide, there were 34 fewer deaths per 1,000 children born in 2002 than in 1970. The biggest improvements were in the Gambia, Bhutan and Guinea. In the Gambia, for example, there were 101 fewer deaths per 1,000 live births in 2002 than in 1970.

Among the regions of the map the biggest improvement was in South Asia, where an additional 50 children survived per 1,000 in 2002 compared to the figure for 1970. The smallest improvement was in Western Europe, with just 4 extra children surviving per 1,000, primarily because the child mortality rate was already relatively low there.

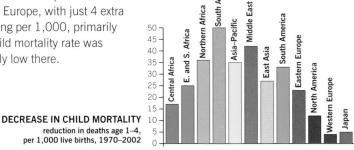

DECREASE IN CHILD MORTALITY
reduction in deaths age 1–4,
per 1,000 live births, 1970–2002

'...millions of young children can be saved by basic, cost-effective measures such as vaccines, antibiotics, micronutrient supplementation, insecticide-treated mosquito nets and improved breastfeeding practices.' Carol Bellamy, executive director of UNICEF, 2004

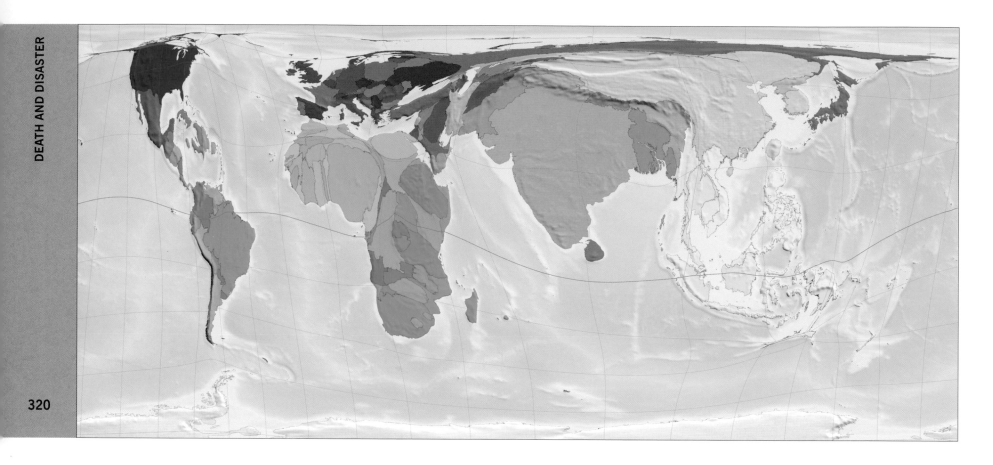

HIGHEST AND LOWEST MORTALITY RATES, MALES AGED 15–60

Rank	Territory	%ª
1	Lesotho	91.2
2	Swaziland	89.4
3	Botswana	85.0
4	Zimbabwe	83.0
5	Zambia	71.9
6	Burundi	65.4
7	Malawi	65.2
8	South Africa	64.2
9	Central African Republic	64.1
10	Mozambique	62.1
191	Netherlands	9.3
192	Israel	9.2
193	Switzerland	9.0
194	Australia	8.9
195	Singapore	8.7
196	Malta	8.4
197	Iceland	8.1
198	Sweden	7.9
199	Kuwait	7.3
200	San Marino	7.3

ª % not expected to reach the age of 60, 2003

297 Mortality of Males Aged 15–60 Years

The size of each territory indicates the number of males who die between the ages of 15 and 60. The ravages of AIDS have hit this age group very hard.

In 2003 the largest numbers of deaths of men and boys between the ages of 15 and 60 years occurred in India and China, followed by Russia, Indonesia and Nigeria. Part of the reason for these high numbers is the large population of these territories. As a percentage, the highest death rates are found in Africa. Lesotho has the highest rate at 91 deaths per 100, meaning that hardly any men will live to the age of 60. At the other extreme, the lowest death rates in this age range are found in Sweden, Kuwait and San Marino.

'There are no grown-ups living with us. I need a bathroom tap and clothes and shoes...But especially, somebody to tuck me and my sister in at night-time.'

Apiwe, aged 13, speaking at the National Children's Forum on HIV/AIDS in South Africa, 2001

PREMATURE MALE MORTALITY
male mortality aged 15–60, per 1,000

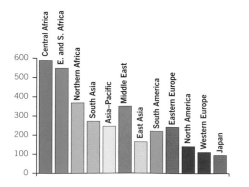

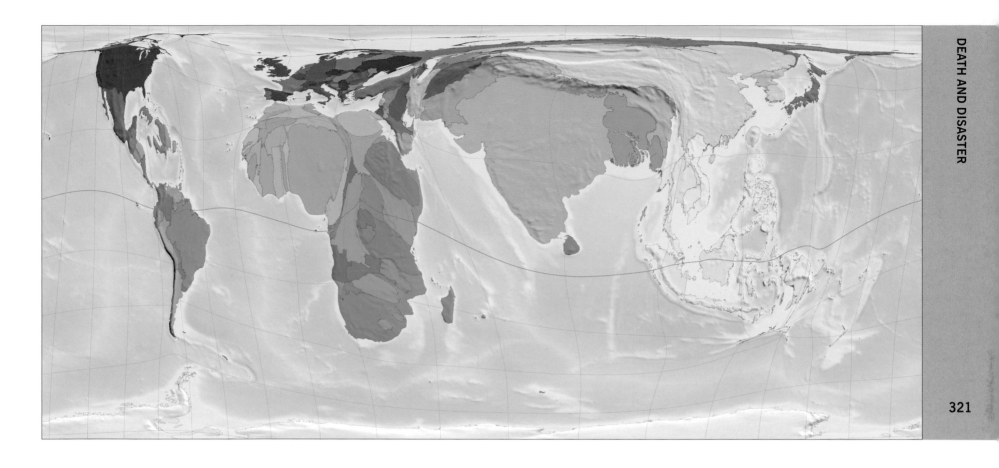

HIGHEST AND LOWEST MORTALITY
RATES, FEMALES AGED 15–60

Rank	Territory	%ᵃ
1	Botswana	83.9
2	Zimbabwe	81.9
3	Swaziland	79.0
4	Lesotho	78.1
5	Zambia	68.5
6	Malawi	61.5
7	Central African Republic	59.0
8	South Africa	57.9
9	Tanzania	55.0
10	Mozambique	54.3
191	Sweden	5.0
192	Malta	4.9
193	Greece	4.8
194	Monaco	4.7
195	Italy	4.7
196	Cyprus	4.7
197	Spain	4.6
198	Japan	4.5
199	Andorra	4.1
200	San Marino	3.2

ᵃ % not expected to reach the age of 60, 2003

298 Mortality of Females Aged 15–60 Years

The size of each territory indicates the number of females who die between the ages of 15 and 60. Again, the effect of AIDS looms large.

In 2002, 125 of the 200 territories recorded female life expectancies in excess of 60 years; it is thus generally expected that women will live beyond their sixtieth birthdays. Nonetheless, as this map shows, significant numbers do not: of all women and girls alive in 2003, over 300 million are expected to die before reaching age 60.

The highest rates of female mortality in this age range are in East and Southern Africa, particularly in Botswana, Zimbabwe and Swaziland. These territories also have some of the highest rates of HIV prevalence in the world.

PREMATURE FEMALE MORTALITY
female mortality aged 15–60, per 1,000

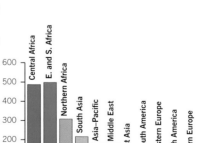

'One death is a tragedy. A million deaths is a statistic.'
Joseph Stalin, general secretary of the Communist Party of the Soviet Union, undated

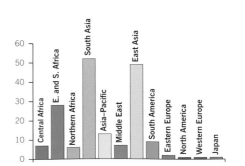

Rank	Territory	Affected[a]
1	Botswana	79
2	Bangladesh	71
3	Zimbabwe	71
4	Mauritania	68
5	Malawi	65
6	Djibouti	65
7	India	58
8	Swaziland	54
9	Mozambique	53
10	China	53
11	Hong Kong (China)	49
12	Antigua and Barbuda	48
13	Fiji	47
14	Laos	46
15	Tonga	46
16	Samoa	43
17	Vanuatu	42
18	Ethiopia	41
19	Albania	37
20	Eritrea	36

[a] number affected by disasters per 1,000 people, 1975–2004

299 People Affected by Disasters

The size of each territory indicates the numbers of people affected by disasters between 1975 and 2004. Almost 90% of them were in Africa and Asia.

Disasters are defined by international agencies as events that overwhelm the local capacity to deal with them, necessitating a request for outside help. They include droughts, epidemics, volcanoes, storms, fires, accidents and events caused indirectly by wars (but not direct effects of war such as battlefield death or injury). Being affected by a disaster is here defined as requiring assistance to survive, such as shelter, water, sanitation, medication or food.

Of all the people affected by disasters between 1975 and 2004, 43% were in South Asia, 41% in East Asia and 5% in East and Southern Africa.

'We cannot prevent disasters...What we hope to do is to be more proactive, to be better prepared so that we can react better, faster.'
Winston Choo, chair of Singapore Red Cross, 2006

PEOPLE AFFECTED BY DISASTERS
per 1,000 people per year, 1975–2004

Central Africa, E. and S. Africa, Northern Africa, South Asia, Asia–Pacific, Middle East, East Asia, South America, Eastern Europe, North America, Western Europe, Japan

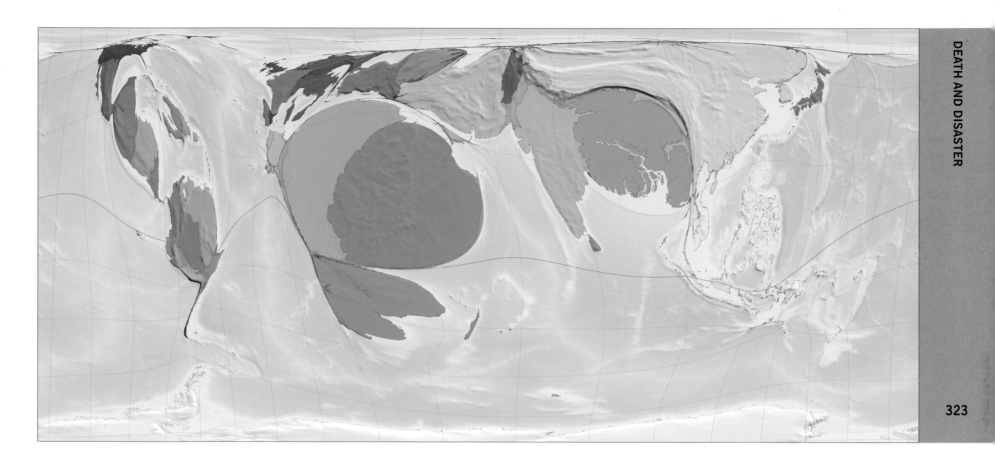

300 People Killed by Disasters

The size of each territory indicates the number of deaths caused by disasters
between 1975 and 2004, a period that saw some cataclysmic events.

This map shows deaths that are known
or presumed to be caused by disasters.
The period covered includes the 1976
earthquake in China, the 1984 drought in
Sudan and Ethiopia, and the 1991 cyclone
in Bangladesh.

Deaths caused by disasters can be
exacerbated by human factors. For example,
the direct cause of the 1984 Ethiopian
famine was a lack of rain, but the problems
were aggravated by civil war and an initial
shortage of international assistance.

DEATHS FROM DISASTERS
deaths per million people per year, 1975–2004

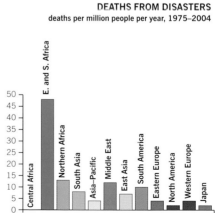

*'We are preparing ourselves for up to 1,000 dead bodies from this
flood alone...'* Inspector Daniel Gezahenge on floods in Ethiopia in 2006, quoted on BBC website

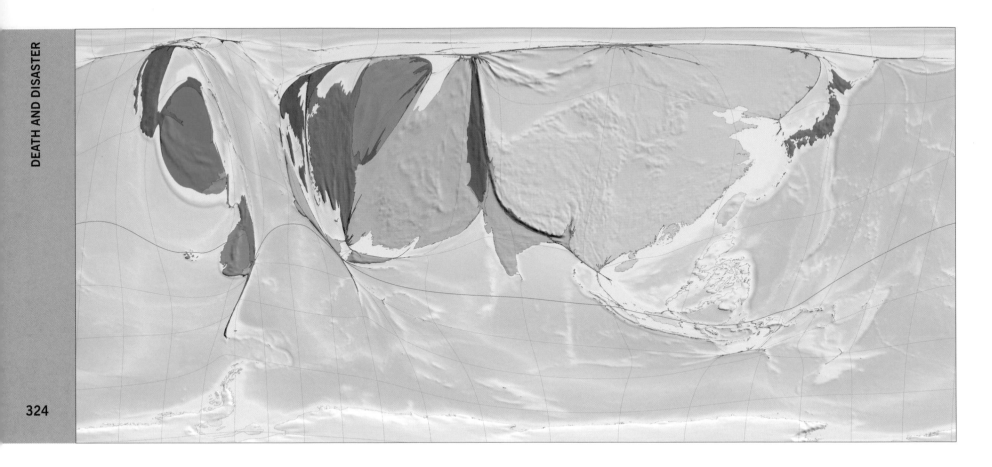

MOST PEOPLE KILLED IN EARTHQUAKES

Rank	Territory	Value[a]
1	Armenia	310
2	Guatemala	74
3	Iran	39
4	Afghanistan	14
5	Turkey	13
6	China	7
7	Hong Kong (China)	7
8	El Salvador	7
9	Philippines	4
10	Taiwan	4
11	Italy	3.8
12	Algeria	3.5
13	Mexico	3.4
14	Yemen	3.0
15	Romania	2.8
16	Solomon Islands	2.6
17	Vanuatu	2.3
18	Georgia	2.1
19	Tajikistan	1.9
20	Colombia	1.9

[a] people killed in earthquakes per million people per year, 1975–2000

301 People Killed in Earthquakes

The size of each territory indicates the number of people killed in earthquakes between 1975 and 2000. Certain areas are particularly prone to these sudden movements of the Earth's crust.

Earthquakes are most likely to occur in zones close to active boundaries of the planet's tectonic plates. Damaging earthquakes occur commonly in, among other places, Colombia, China, Iran, Indonesia, India, Japan, the Philippines and Peru.

Between 1975 and 2000 there were an estimated 471,000 earthquake-related deaths worldwide. Of these, 52% occurred in China (most of them during the 1976 Tangshan earthquake) and 16% in Iran. The number of deaths that result from an earthquake depends not only on the magnitude and duration of the quake itself but also on how well prepared the infrastructure is to cope with the effects. The high death toll of almost 18,000 in the 1999 Izmit earthquake in Turkey, for example, was attributed in part to poor-quality housing that could not withstand the tremors.

DEATHS CAUSED BY EARTHQUAKES
deaths per million people per year, 1975–2000

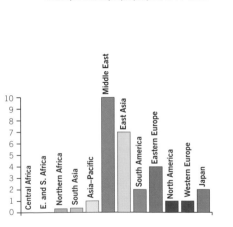

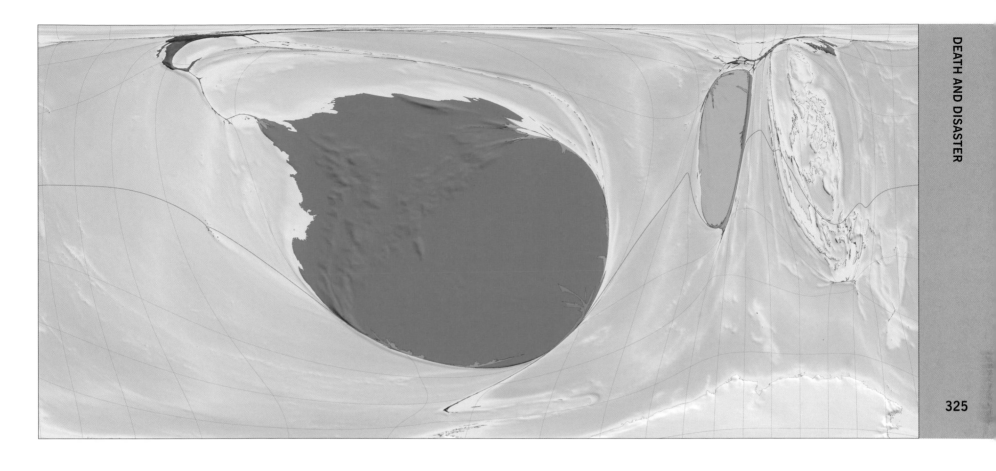

TOTAL VOLCANO DEATHS

Rank	Territory	Deaths[a]
1	Colombia	877.31
2	Cameroon	68.58
3	Philippines	27.65
4	Indonesia	25.35
5	Mexico	4.62
6	United States	3.46
7	Ethiopia	2.46
8	DR Congo	2.35
9	Japan	1.81
10	Papua New Guinea	0.35

[a] volcano deaths per year, 1975–2000

HIGHEST RATES OF VOLCANO DEATHS

Rank	Territory	Deaths[a]
1	Colombia	20.17
2	Cameroon	4.37
3	St Vincent and the Grenadines	0.77
4	Philippines	0.35
5	Indonesia	0.12
6	Timor-Leste	0.09
7	Papua New Guinea	0.06
8	Comoros	0.05
9	DR Congo	0.05
10	Mexico	0.05

[a] volcano deaths per million people per year, 1975–2000

302 People Killed by Volcanoes

The size of each territory indicates the number of people killed by volcanoes between 1975 and 2000, excluding deaths from earthquakes and tsunamis caused by volcanic activity.

Volcanic activity can endanger human life in various ways: lava flows, mud flows, pyroclastic flows (flows of hot ash, rocks and gases), landslides and the settling of ash can all pose serious risk to life.

There were volcano-related deaths in only 17 territories between 1975 and 2000. 86% of those occurred in Colombia, primarily in the town of Armero, which was inundated by mud flows following the eruption of the Nevado del Ruiz volcano on 13 November 1985. Almost 22,000 deaths were recorded there.

DEATHS CAUSED BY VOLCANOES
deaths per million
people per year, 1975–2000

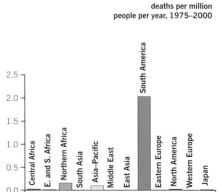

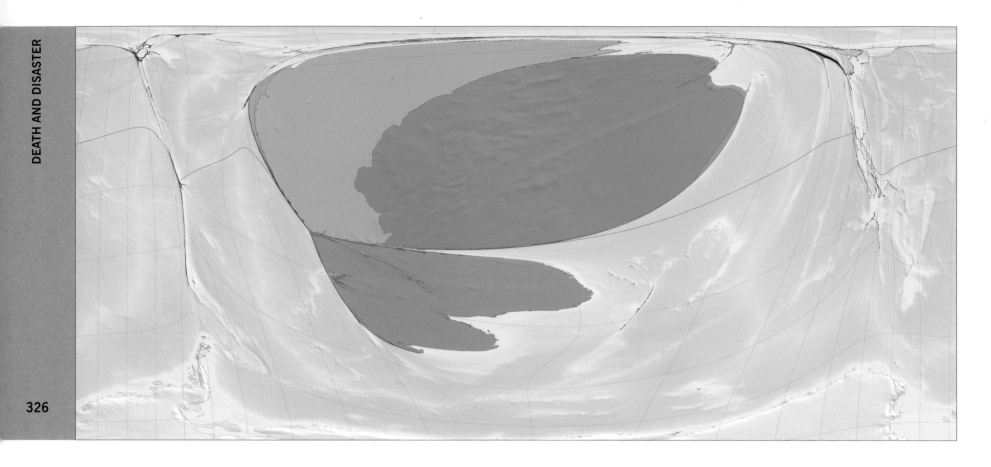

MOST PEOPLE KILLED BY DROUGHT

Rank	Territory	Value[a]
1	Mozambique	208.00
2	Sudan	175.36
3	Ethiopia	167.43
4	Swaziland	17.48
5	Chad	13.90
6	Somalia	2.52
7	Papua New Guinea	0.67
8	Indonesia	0.24
9	Uganda	0.18
10	Kenya	0.10
11	China	0.10
12	Timor-Leste	0.10
13	Hong Kong (China)	0.09
14	Guinea	0.06
15	Pakistan	0.04
16	Burundi	0.04
17	India	0.02
18	Bangladesh	0.01
19	Brazil	<0.01
20	Philippines	<0.01

[a] **deaths related to drought per million people per year, 1975–2000**

303 People Killed by Drought

Sustained drought can cause crop failure, the death of livestock and ultimately deaths of people.
The size of each territory indicates the number of people killed by drought between 1975 and 2000.

Unlike other disasters, droughts are
sometimes slow to develop and can
continue for years.

Worldwide there were an estimated
560,000 deaths from drought between
1975 and 2000. Although these deaths
were spread over 20 of the 200 territories
on the map, most of them – more than
98% – occurred in just three territories:
Ethiopia, Sudan and Mozambique. (Note
that Eritrea, where many died during the
drought of 1984, was at that time a part
of Ethiopia.)

DEATHS CAUSED BY DROUGHT
deaths per million people per year, 1975–2000

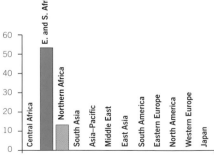

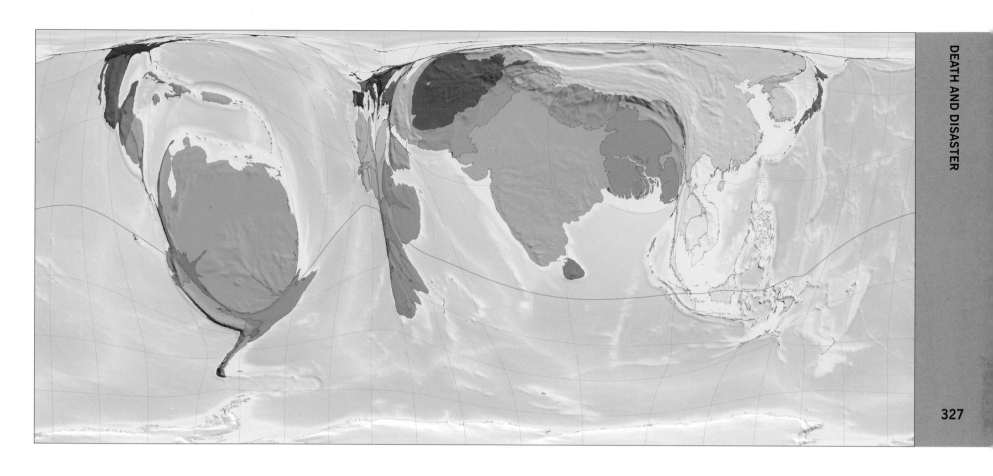

MOST PEOPLE KILLED IN FLOODS

Rank	Territory	Deaths[a]
1	Venezuela	46.14
2	Afghanistan	15.10
3	Somalia	10.02
4	Djibouti	9.89
5	Tajikistan	8.90
6	Nepal	6.75
7	Puerto Rico	5.11
8	Bhutan	3.88
9	Honduras	3.76
10	El Salvador	3.38
11	Peru	2.98
12	Cambodia	2.84
13	Bangladesh	2.65
14	Guatemala	2.57
15	Mozambique	2.51
16	Yemen	1.95
17	Ecuador	1.93
18	Seychelles	1.92
19	Jamaica	1.72
20	Fiji	1.59

[a] deaths related to floods, per million people per year, 1975–2000

304 People Killed in Floods

Flooding kills people in every part of the world. The size of each territory indicates the number of people killed in floods between 1975 and 2000.

Over 170,000 people were killed in floods between 1975 and 2000, with the highest death tolls occurring in South America, South Asia and East Asia. At the other end of the scale, Central Africa, Japan and Western Europe each experienced less than 0.6% of worldwide flood deaths during this interval.

The territory with the largest number of flood deaths as a percentage of population was Venezuela. 99% of these occurred in 1999 following several days of unusually heavy rains. 1999 was also the year with the largest total number of flood deaths worldwide during this period.

DEATHS CAUSED IN FLOODS
deaths per million
people per year, 1975–2000

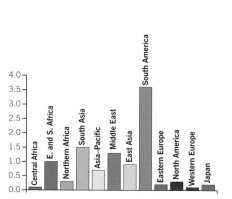

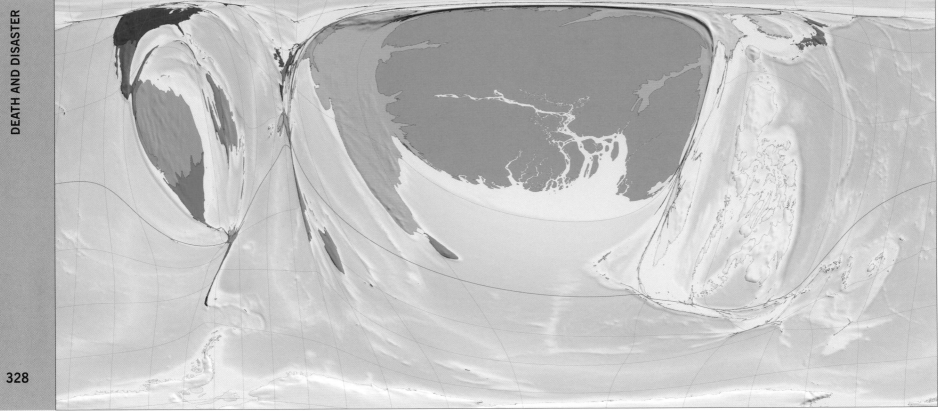

MOST PEOPLE KILLED BY STORMS

Rank	Territory	Deaths[a]
1	Honduras	83
2	Cook Islands	53
3	Bangladesh	42
4	Nicaragua	25
5	St Lucia	22
6	Vanuatu	18
7	Dominica	17
8	Philippines	9
9	Fiji	8
10	Solomon Islands	8
11	Dominican Republic	8
12	Haiti	8
13	St Kitts–Nevis	5
14	Viet Nam	4
15	Samoa	4
16	Comoros	3
17	Tonga	3
18	El Salvador	3
19	Antigua and Barbuda	3
20	Madagascar	3

[a] storm deaths per million people per year, 1975–2000

305 People Killed by Storms

The size of each territory indicates the numbers of people killed in storm disasters between 1975 and 2000. South Asia, especially Bangladesh, has suffered particularly badly.

Storms include cyclones, hurricanes, tornadoes, tropical storms, typhoons and winter storms. The majority of storm deaths between 1975 and 2000 occurred in Bangladesh, where 138,987 people died in a cyclone in 1991. Of all storm deaths during this period, 72% were in South Asia.

High death rates from storms are often experienced in island territories or territories with significant lengths of coastline, and tend to occur between the tropics of Cancer and Capricorn: the 20 territories with the highest death rates are all located wholly or partly in the tropics.

DEATHS CAUSED BY STORMS
deaths per million people per year, 1975–2000

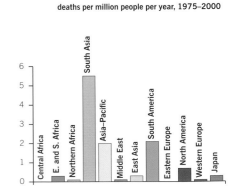

'The weather that accompanies a cyclone is torrential rain along with high winds; inundation by surge is, however, the most terrifying event.'
Sazedur Rahman, former director of SAARC Meteorological Research Centre, Dhaka, 2001

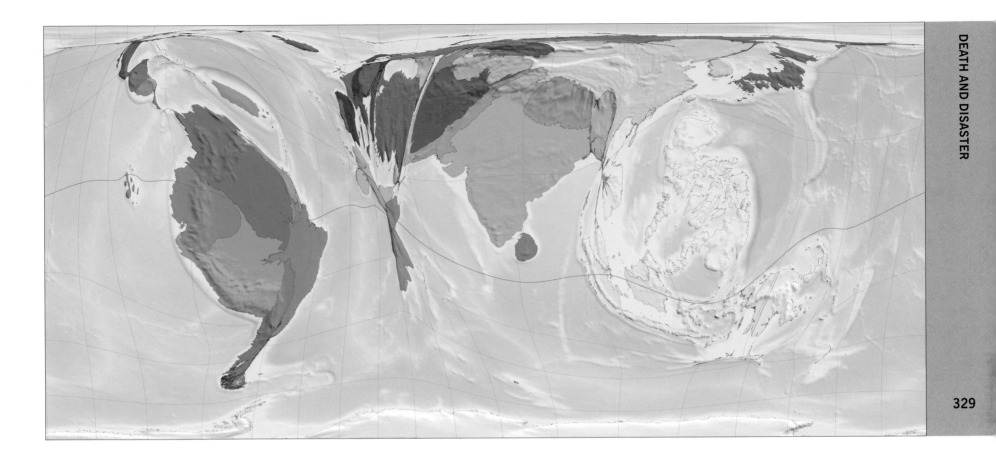

MOST PEOPLE KILLED IN SLIDE DISASTERS

Rank	Territory	Deaths[a]
1	Iceland	4.9
2	Ecuador	2.4
3	Papua New Guinea	2.2
4	Tajikistan	2.2
5	Peru	1.6
6	Colombia	1.3
7	Puerto Rico	1.2
8	Kyrgyzstan	1.2
9	Afghanistan	1.2
10	Nepal	0.9
11	Albania	0.7
12	Bolivia	0.7
13	Philippines	0.6
14	Liberia	0.5
15	Georgia	0.5
16	Chile	0.5
17	Guyana	0.5
18	Austria	0.4
19	Turkey	0.3
20	El Salvador	0.3

[a] landslide and avalanche deaths per million people per year, 1975–2000

306 People Killed by Avalanches and Landslides

The size of each territory indicates the numbers of deaths in avalanches and landslides between 1975 and 2000. These events can begin with a small movement and rapidly grow to devastating power.

Avalanches and landslides are the downhill movement of snow, ice, soil or rock. Often the avalanche is triggered by a particular event such as a passing skier or an earth tremor, and is then amplified by the effects of gravity and building momentum.

Between 1975 and 2000 almost 17,000 people were killed in avalanches and landslides, which is 1 death every year per 10 million people on the planet. The largest numbers of deaths occurred in India, China, Colombia and the Philippines. 69% of territories experienced no avalanche or landslide deaths at all.

'On 9 May 1993, at 1:30 pm... "Las Brisas" — comprising around 150 shanty hous[es], an estimated 300 people — were reported to have been buried by a gigantic landslide from "El Tierrero" mountain.'

UNDHA, on torrential rains, tremor and landslides in the Nambija area of Ecuador, 1993

LANDSLIDE AND AVALANCHE DEATHS
deaths per million people per year, 1975–2000

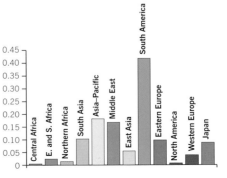

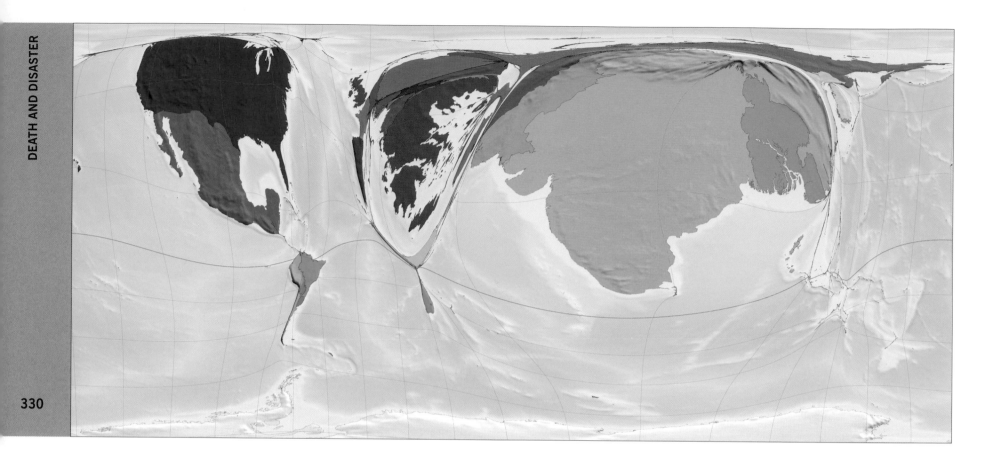

MOST PEOPLE KILLED BY EXTREME TEMPERATURES

Rank	Territory	Deaths[a]
1	Greece	3.8
2	Cyprus	2.7
3	Albania	0.8
4	Mexico	0.4
5	Afghanistan	0.4
6	Lithuania	0.4
7	Poland	0.4
8	Croatia	0.3
9	United States	0.3
10	Bangladesh	0.3
11	India	0.3
12	Pakistan	0.2
13	Russia	0.2
14	Romania	0.2
15	Serbia and Montenegro	0.1
16	Jordan	0.1
17	Kyrgyzstan	0.1
18	Uruguay	0.1
19	Spain	0.1
20	Peru	0.1

[a] deaths caused by extreme temperatures per million people per year, 1975–2000

307 People Killed by Extreme Temperatures

The size of each territory indicates the number of people killed by extreme temperatures, of either heat or cold, between 1975 and 2000.

Over 15,000 people died in severe heatwaves or extremes of cold between 1975 and 2000. The territories where the most deaths occurred were India, the United States, Greece and Mexico.

Very hot and very cold temperatures can occur in the same place at different times: while South Asian territories have hot summers, for instance, winters here can also be very cold.

DEATHS CAUSED BY EXTREME TEMPERATURES
deaths per million people per year, 1975–2000

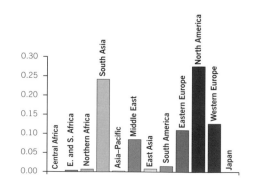

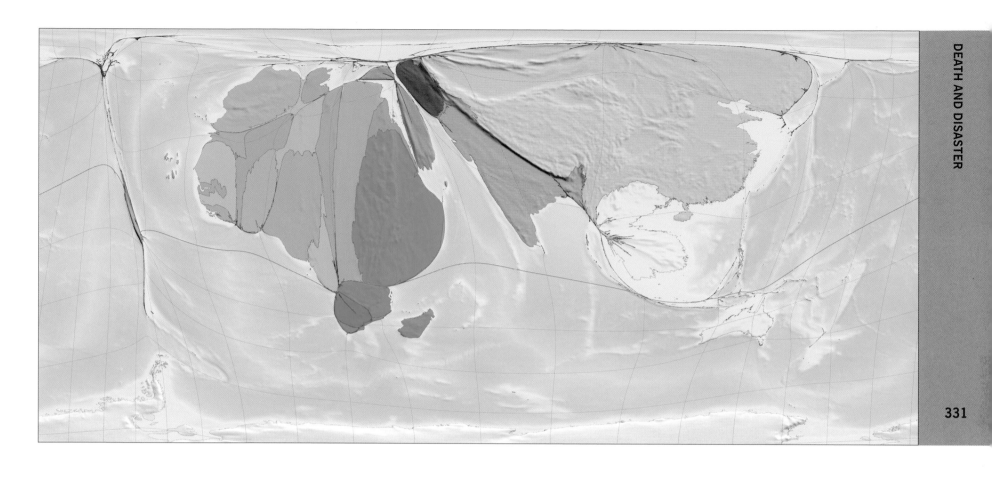

MOST PEOPLE SERIOUSLY AFFECTED BY PEST INFESTATIONS

Rank	Territory	Affected[a]
1	Chad	129
2	Gambia	129
3	Senegal	98
4	Eritrea	90
5	Mauritania	89
6	Burkina Faso	81
7	Guinea-Bissau	81
8	Sudan	80
9	Ethiopia	74
10	Mali	74
11	Niger	67
12	Cape Verde	65
13	Morocco	58
14	Cameroon	40
15	Zambia	39
16	Yemen	39
17	Tunisia	36
18	Viet Nam	32
19	Afghanistan	32
20	Botswana	32

[a] people affected by infestations per 1,000 people per year, 1975–2000

308 People Affected by Infestations

The size of each territory indicates the number of people affected by disasters caused by infestations, usually of insects, between 1974 and 2004.

The map shows only those insect and vermin infestations that were so severe that the local population required external assistance to survive. The main insects that cause substantial damage are locusts, grasshoppers and army worms (insect larvae). One case, in Chad in 1987, involved an infestation of rats. Disasters occur when the number of insects is so large as to cause widespread devastation. The three territories with the largest populations affected by insect infestations are China, Ethiopia and India.

'And the locusts went up over all the land of Egypt...and there remained not any green thing.' The Bible, Exodus 10: 14–15

PEST INFESTATIONS
severely affected per 1,000 people per year, 1974–2004

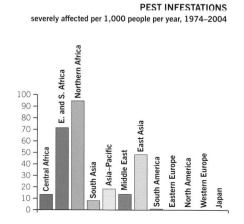

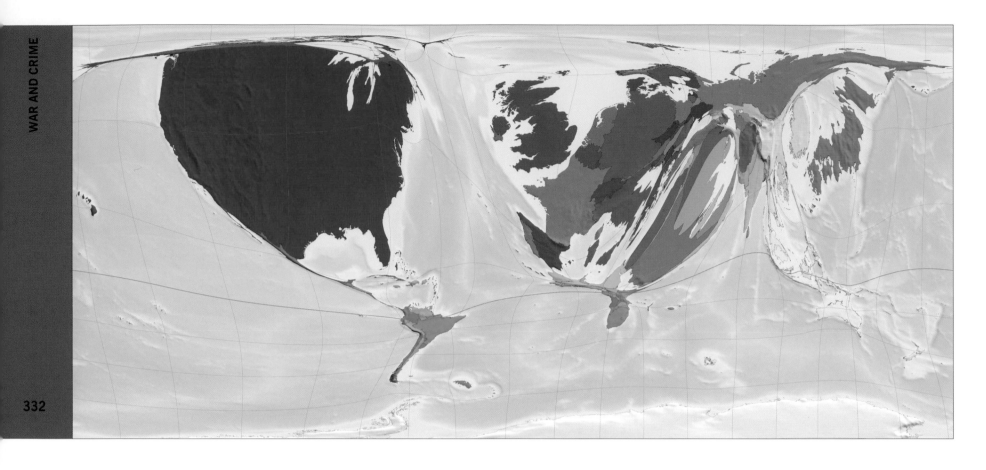

HIGHEST AND LOWEST MILITARY SPENDING IN 1990

Rank	Territory	US$ᵃ
1	Kuwait	12,968
2	Qatar	6,763
3	United Arab Emirates	3,768
4	Israel	2,132
5	Saudi Arabia	1,856
6	Oman	1,779
7	United States	1,688
8	Norway	907
9	Singapore	848
10	United Kingdom	821
191	Madagascar	4.9
192	Nigeria	4.8
193	Myanmar	4.0
194	Gambia	4.0
195	Bangladesh	3.1
196	Malawi	2.6
197	Nepal	2.0
198	Ghana	1.3
199	Iceland	0
200	Costa Rica	0

ᵃ spending per person, 1990

309 Military Spending 1990

The size of each territory shows how much its government spent on its military forces in 1990, in US dollars. The top three spenders outside the United States spent more than the rest of the world put together.

Military spending by governments covers the costs of military personnel, including recruitment, training and salaries, and the costs of equipment, supplies, weapons and construction. Also shown on this map is spending on military assistance to other territories.

In 1990 military spending by governments was estimated at US$954 billion worldwide. Of this, the largest portion, 45%, was spent by the United States, followed by Russia (then the USSR) at 6%, with Germany (reunified in 1990) at 6%, with Germany (reunified in 1990)

and the United Kingdom at 5% each. These territories each spent more than the total military expenditure in any of the following regions: Central Africa, East and Southern Africa, Northern Africa, South Asia, Asia–Pacific and Australasia, East Asia, South America, Eastern Europe and Japan.

The highest rates of military spending per person were in Middle Eastern territories, such as Kuwait and Qatar. In Kuwait, 10,000 times more was spent per resident than in Ghana.

TOTAL MILITARY SPENDING
US$ billions, 1990

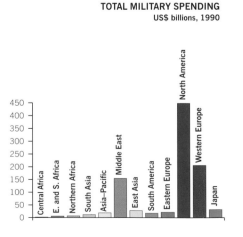

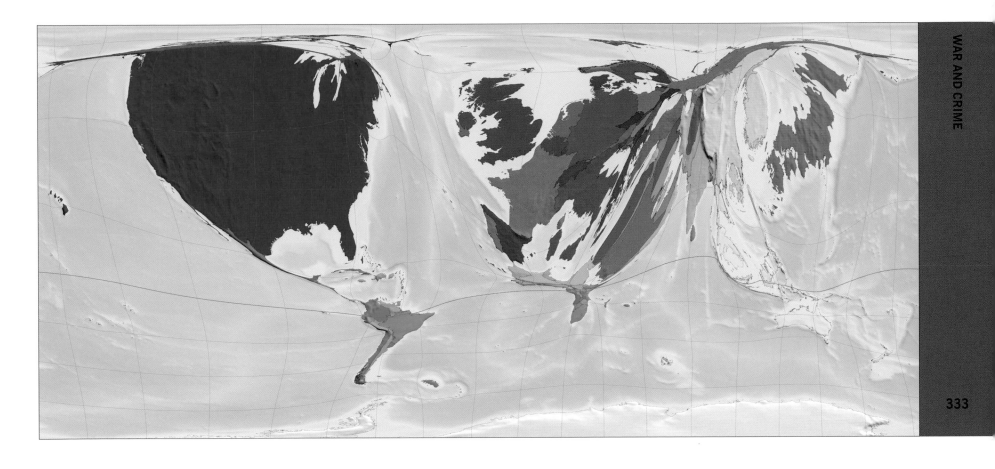

HIGHEST AND LOWEST MILITARY SPENDING IN 2002

Rank	Territory	US$ᵃ
1	Qatar	1,771
2	Kuwait	1,534
3	Israel	1,514
4	United States	1,213
5	Singapore	1,077
6	United Arab Emirates	906
7	Oman	892
8	Norway	889
9	Brunei	820
10	Saudi Arabia	786
191	Nepal	3.1
192	Tajikistan	2.7
193	Central African Republic	2.6
194	Gambia	2.6
195	Niger	2.1
196	Ghana	1.8
197	Moldova	1.5
198	Timor-Leste	0
199	Costa Rica	0
200	Iceland	0

ᵃ **spending per person, 2002**

310 Military Spending 2002

The size of each territory shows how much its government spent on its military forces in 2002, in US dollars. Again the United States is the biggest military spender by far.

In 2002 military spending was estimated at US$789 billion worldwide.

In this year the United States spent US$353 billion, which – as in 1990 – was 45% of all military spending worldwide. This sum amounted to almost 9 times the outlay by the next biggest spender, Japan. The worldwide spending was less than in 1990 as, by 2002, the cold war had ended. However, since 2002 significant new conflicts have begun and have continued to generate more new spending.

'When the rich wage war it is the poor who die.'
Jean-Paul Sartre, *Le Diable et le Bon Dieu (The Devil and the Good Lord)*, 1951

MILITARY SPENDING PER PERSON
average state military spending, US$ per person, 2002

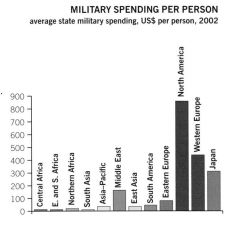

Rank	Territory	US$[a]
1	Russia	39
2	Israel	34
3	Norway	33
4	France	22
5	United States	22
6	Sweden	21
7	Canada	18
8	Netherlands	17
9	United Kingdom	16
10	Czech Republic	5
11	Libya	4
12	Spain	3
12	Poland	2
14	Iceland	2
15	Australia	2
16	Romania	1
16	Turkey	1
18	Qatar	1
19	South Korea	1
20	Bahrain	1

[a] **earnings from arms exports per person, 2003**

311 Arms Exports

The size of each territory shows how much it earns from arms exports, in US dollars.

This map includes major conventional weapons and weapon systems, such as aircraft, ships, armoured vehicles, missiles, artillery, and guidance and radar systems. Small arms and ammunition are not included.

A few territories account for most exports of this weaponry. In 2003, five territories were responsible for 80% of arms exports: the United States, Russia, France, Germany and the United Kingdom. The top 22 territories accounted for over 99% of arms exports.

EARNINGS FROM ARMS EXPORTS
US$ million, 2003

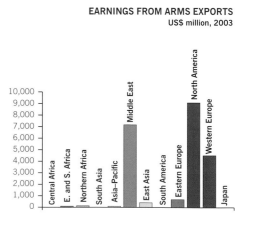

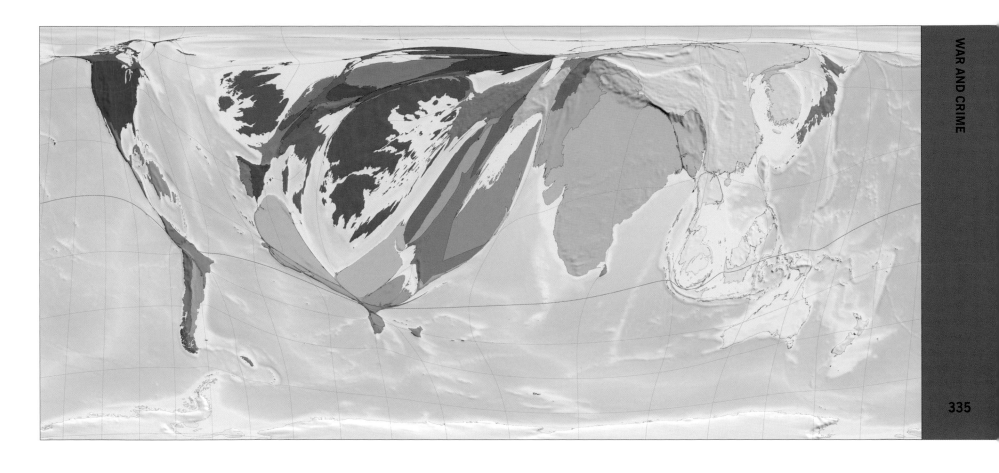

312 Arms Imports

The size of each territory indicates how much
it spends on arms imports, in US dollars.

Included on this map are major
conventional weapons and systems,
but not small arms and ammunition.

India, China and Greece were the three
biggest spenders on imported weapons in
2003, the year depicted in this map.
Together these three territories accounted
for 42% of all arms imports worldwide in

dollar terms. Central African territories
record very little spending on arms imports,
so are barely visible on this map, and
figures for East and Southern Africa and
Japan are also relatively low. To some
extent imports fluctuate from year to year,
so a map for another year might show
a somewhat different distribution.

SPENDING ON ARMS IMPORTS
US$ millions, 2003

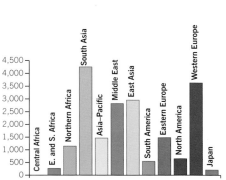

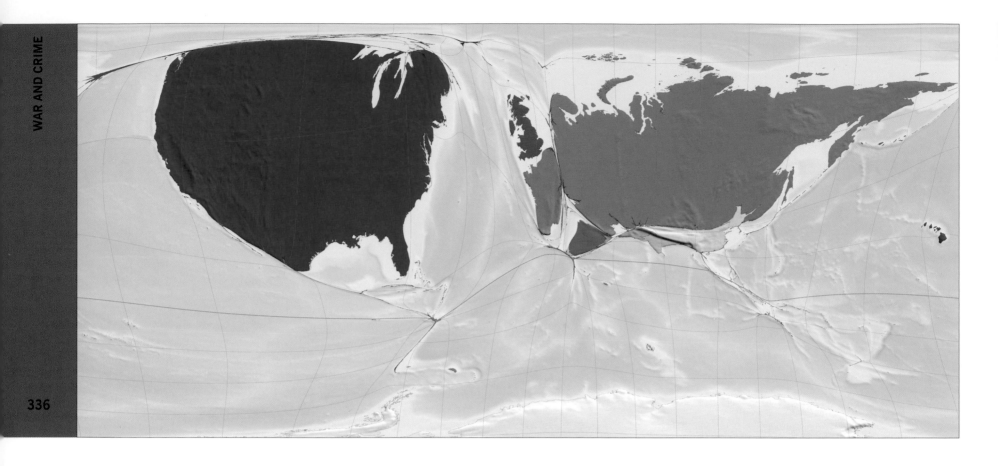

MOST STRATEGIC NUCLEAR WEAPONS BY POPULATION

Rank	Territory	Weapons[a]
1	Russia	41.6
2	United States	29.7
3	Israel	23.8
4	France	5.9
5	United Kingdom	3.0
6	Pakistan	0.2
7	China	0.2
8	India	0.1

[a] weapons per million people, 2002

MOST STRATEGIC NUCLEAR WEAPONS BY NUMBER

Rank	Territory	Weapons[a]
1	United States	8,646
2	Russia	6,000
3	France	350
4	China	250
5	United Kingdom	180
6	Israel	150
7	India	60
8	Pakistan	36

[a] weapon count, 2002

313 Nuclear Weapons

The size of each territory indicates the number of known or suspected strategic nuclear weapons. Despite efforts at non-proliferation, the number of nuclear-armed countries has risen.

In 2002 eight territories were known or suspected to have strategic nuclear weapons: the United States, Russia, France, China, the United Kingdom, Israel, India and Pakistan. If other territories do have strategic nuclear weapons, they probably have considerably fewer than the territories listed here, so their holdings would not much alter the appearance of this map.

The United States, which has the largest number of nuclear weapons, has 240 times more than Pakistan, which has the fewest. The international Treaty on the Non-Proliferation of Nuclear Weapons, adopted in 1968, was designed to stop the spread of nuclear weapons. By March 2002, 187 parties had signed this treaty.

'Nuclear weapons are clearly inhumane...it is inevitable that the horror of Hiroshima and Nagasaki will be repeated – somewhere, sometime – in an unforgivable affront to humanity itself.' Takashi Hiraoka, mayor of Hiroshima, Hiroshima Peace Declaration, 1995

STRATEGIC NUCLEAR WEAPONS
estimated total, 2002

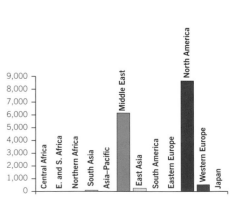

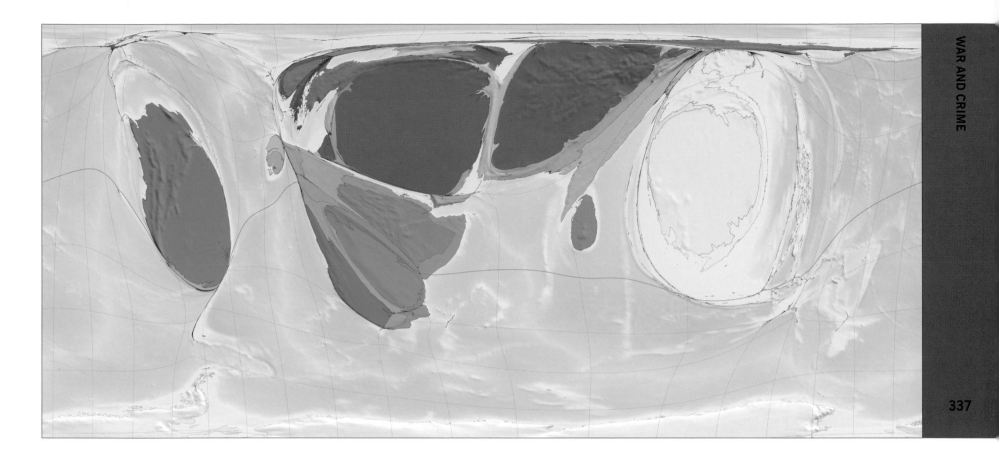

314 Landmine Deaths and Injuries

The size of each territory indicates the number of deaths and injuries caused by landmines between 2003 and 2005. Landmines laid in times of war can wreak havoc long after hostilities are over.

Landmines are explosive devices buried (not very deeply) in the ground and rigged to explode when stepped upon or driven over. The possible effects include death or the loss of one or more limbs. Although used primarily as weapons of war, landmines can remain in place for years after war has ended. Ongoing landmine casualties are thus a common and tragic legacy of past wars. Between 2003 and 2005, there were almost 7,000 landmine deaths and injuries a year worldwide. The most were in Iraq, Afghanistan, Cambodia and Colombia. Together these four territories were the site of almost 4,000 casualties during this period.

LANDMINE CASUALTIES
deaths and injuries, annual average, 2003–5

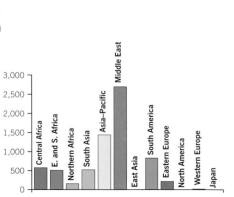

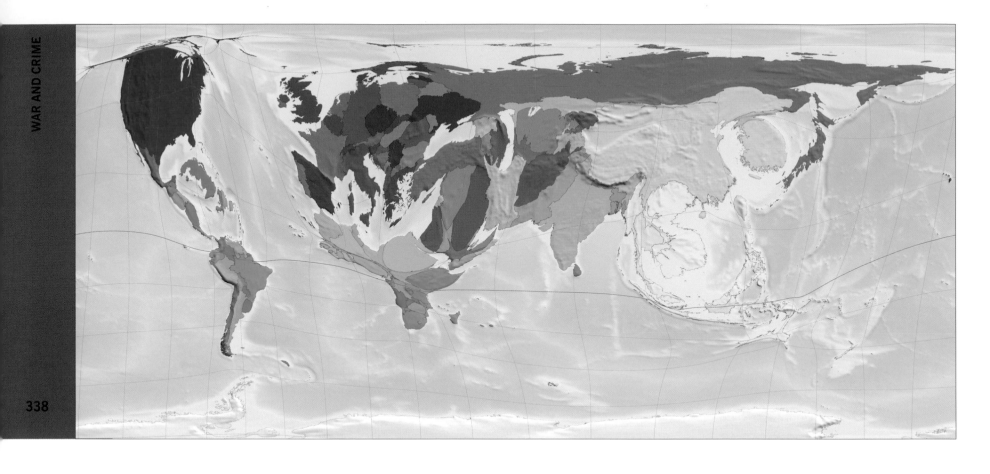

LARGEST AND SMALLEST ARMED FORCES DURING THE COLD WAR

Rank	Territory	Armed forces[a]
1	Russia	37
2	Seychelles	24
3	Syria	23
4	Israel	23
5	Czech Republic	20
6	Bulgaria	18
7	Greece	18
8	United Arab Emirates	15
9	Cuba	15
10	Brunei	14
191	Bangladesh	0.64
192	Rwanda	0.63
193	Papua New Guinea	0.55
194	Cameroon	0.46
195	Gambia	0.45
196	Kenya	0.43
197	Malawi	0.42
198	Mali	0.37
199	Burkina Faso	0.31
200	Niger	0.18

[a] armed forces in 1985 per 1,000 in 2002 population

315 Armed Forces in 1985

In 1985 there were over 29 million people serving in the armed forces worldwide. The sizes of territories here show how these were distributed across the globe.

This map includes army, navy and air force personnel as well as military administration. 33% of these military personnel were in Middle Eastern and Central Asian territories, including Russia (then the USSR).

In 1985 the three territories with the smallest numbers of military personnel per resident were Niger, Burkina Faso and Mali, all in Northern Africa. Among these territories there was less than 1 person in the armed forces for every 1,000 people living there. At the other extreme there were 37 armed forces personnel for every 1,000 people in Russia.

ARMED FORCES IN 1985
millions of people

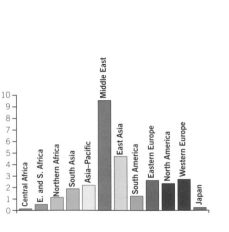

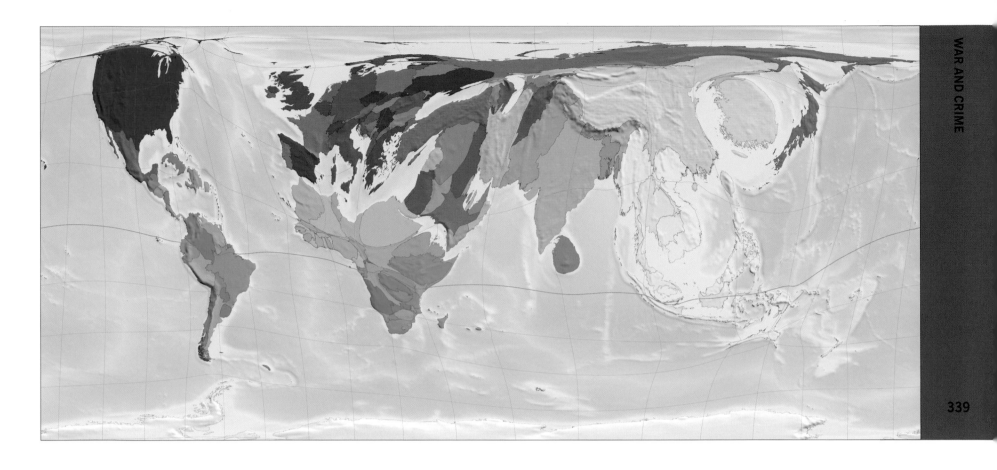

LARGEST AND SMALLEST ARMED FORCES IN 2002

Rank	Territory	Armed forces[a]
1	Eritrea	43
2	Israel	26
3	Brunei	23
4	Lebanon	20
5	Qatar	20
6	Jordan	19
7	Syria	18
8	Greece	16
9	Bahrain	16
10	Oman	15
191	Benin	0.76
192	Tanzania	0.74
193	Gambia	0.71
194	Nigeria	0.65
195	Mozambique	0.59
196	Mali	0.56
197	Papua New Guinea	0.54
198	Niger	0.43
199	Malawi	0.42
200	Ghana	0.34

[a] armed forces in 2002 per 1,000 in 2002 population

316 Armed Forces in 2002

Between 1985 and 2002 the total number of people serving in the armed forces worldwide fell by a third. The size of each territory here shows the number of armed forces personnel serving in 2002.

The geographical balance also shifted: the territory with the largest number of armed forces personnel in 2002 was China, followed by the United States and then India. Two-fifths of the worldwide reduction in armed forces between 1985 and 2002 occurred in Russia alone.

Worldwide there were a total of 19 million people in the armed forces in 2002, an average of 3 for every 1,000 people alive. The territory with the highest proportion of armed forces relative to its population was Eritrea, at 43 in 1,000. The lowest was Ghana, with 0.34 in 1,000. Despite the worldwide reduction in armed forces, however, the proportion of both the Eritrean and Ghanaian populations in the armed forces has increased since 1985, as have those of many other countries.

'Older men declare war. But it is the youth who must fight and die.'
Herbert Hoover, former US president, 1944

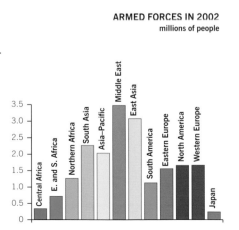

ARMED FORCES IN 2002
millions of people

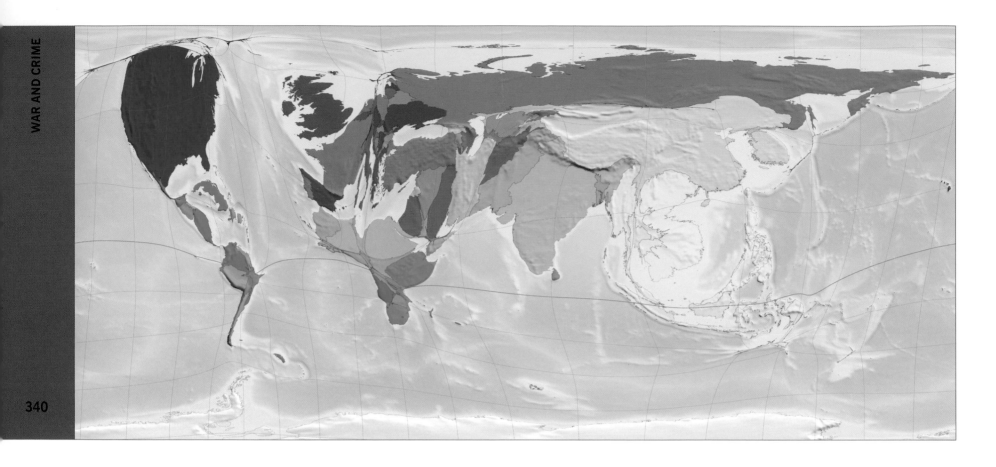

LARGEST NUMBER OF ARMED FORCES AT WAR SINCE SECOND WORLD WAR

Rank	Territory	Armed forces[a]
1	Russia	18.3
2	Israel	13.5
3	Nicaragua	8.2
4	Viet Nam	7.3
5	Syria	6.2
6	Laos	6.2
7	France	5.4
8	United Kingdom	5.3
9	Iraq	5.3
10	Afghanistan	5.1
11	Turkey	4.5
12	Azerbaijan	4.3
13	Armenia	4.3
14	United States	4.3
15	Tajikistan	4.1
16	Cuba	4.1
17	Uzbekistan	4.0
18	Iran	3.9
19	Myanmar	3.8
20	Georgia	3.5

[a] armed forces fighting per
1,000 people per year, 1945–2004

317 Armed Forces at War 1945–2004

The size of each territory indicates the time spent by
its armed forces fighting wars between 1945 and 2004.

For the purposes of this map, time is
measured in person-years. A person-year
means one person fighting for one year.
Thus large size on this map could indicate
a small force fighting for many years or a
large force fighting for a few years, or
anything in between. Years not at war,
when armed forces were on standby, are
not included. Note that the forces in
question may not have been in their home
territory when they were fighting.

On average, between 1945 and 2004,
12 million person-years were spent in wars
every year. Over half of the person-years in
this period were accounted for by armed
forces from just four territories: 22% from
Russia (which constituted part of the USSR
for some of this time), 14% from China,
11% from the United States, and 8%
from India. The largest armed forces
that avoided war throughout the period
were those of Japan, followed by those
of Sweden, Austria and Singapore.

ARMED FORCES IN STATES AT WAR
average number of military personnel in territories
at war, thousands per year, 1945–2004

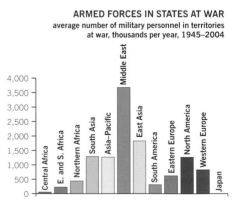

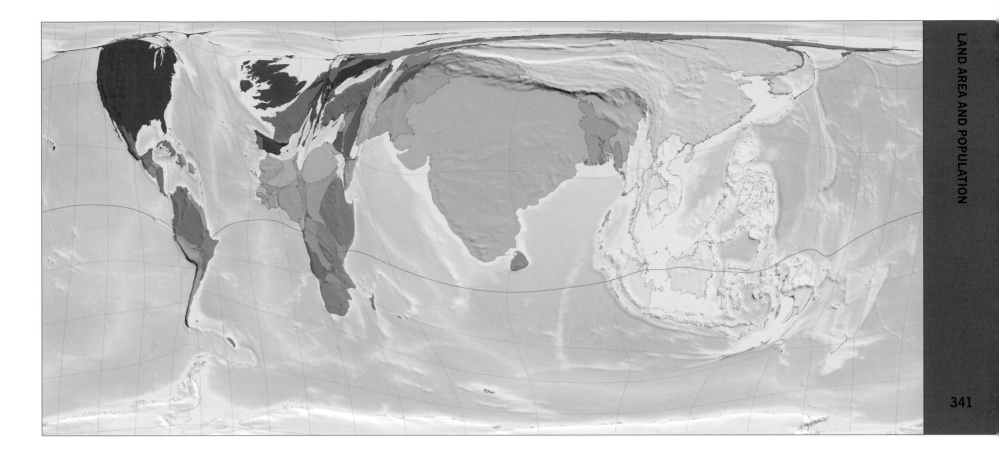

MOST TIME AT WAR SINCE SECOND WORLD WAR

Rank	Territory	%[a]
1	Myanmar	100
2	United Kingdom	93
3	Indonesia	87
4	Colombia	85
5	Philippines	80
6	India	77
7	Cambodia	75
8	Angola	70
9	France	70
10	Sudan	68
11	Nicaragua	68
12	Chad	67
13	Laos	63
14	Ethiopia	62
15	Israel	60
16	Pakistan	58
17	United States	58
18	Viet Nam	57
19	Iraq	55
20	Afghanistan	53

[a] **% of years 1945–2004 where the country was at war for some time**

318 Population at War 1945–2004

The size of each territory indicates the time it spent involved in wars between 1945 and 2004.
But wars are fought by people, not by land, so this map focuses on populations of warring territories.

Every year, on average, between 1945 and 2004, 2.8 billion people were living in territories at war. India accounts for 29% of these, China for 20%.

Between 1945 and 2004, the average territory had 48 peaceful years out of 60. Myanmar (known as Burma for most of this interval) had none. The United Kingdom had only 4. But 22 territories avoided war for all 60 years.

PROPORTION OF TIME AT WAR
% of years 1945–2002 in which wars were fought, weighted by population

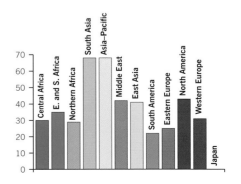

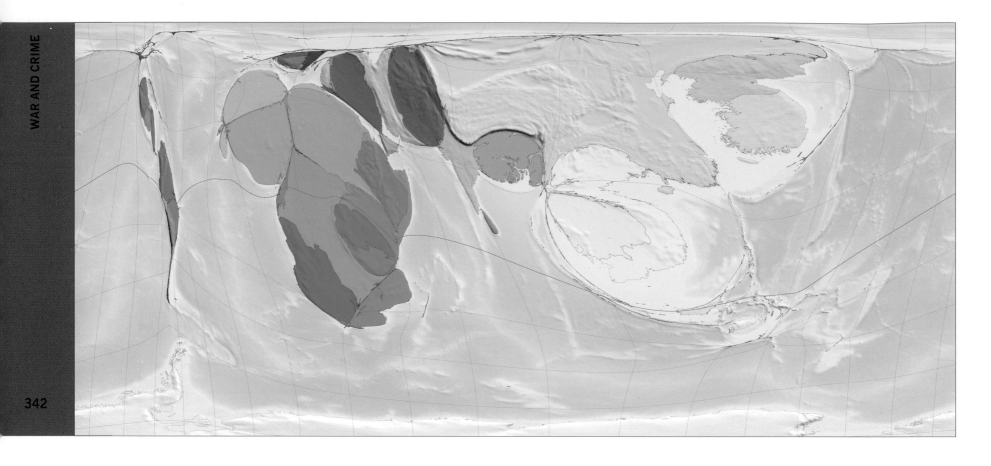

MOST WAR DEATHS SINCE SECOND WORLD WAR

Rank	Territory	Deaths[a]
1	Cambodia	16.1
2	Timor-Leste	14.4
3	Angola	12.9
4	Rwanda	10.3
5	North Korea	10.0
6	Afghanistan	8.7
7	Sudan	8.4
8	Burundi	7.6
9	DR Congo	5.9
10	Mozambique	5.0
11	South Korea	4.7
12	Viet Nam	4.7
13	Iraq	4.6
14	Western Sahara	4.4
15	Lebanon	3.7
16	Algeria	3.5
17	Liberia	3.1
18	Uganda	2.8
19	Congo	2.8
20	Somalia	2.7

[a] war deaths 1945–2000 as % of 2002 population

319 War Deaths 1945–2000

The size of each territory indicates the number of deaths directly attributable to armed conflicts between 1945 and 2000.

An estimated 51 million people died in wars between 1945 and 2000, almost a third of them in China. After China, the highest numbers of war deaths occurred in Viet Nam, the Democratic Republic of Congo and Sudan. Very few occurred in Japan, Western Europe or North America, and relatively few in Eastern Europe and South America, although some territories in these regions did have high death counts, such as Croatia, Bosnia Herzegovina, Serbia and Montenegro, Colombia, Bolivia and Guatemala.

WAR DEATHS 1945–2000
millions of people killed in wars

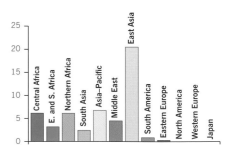

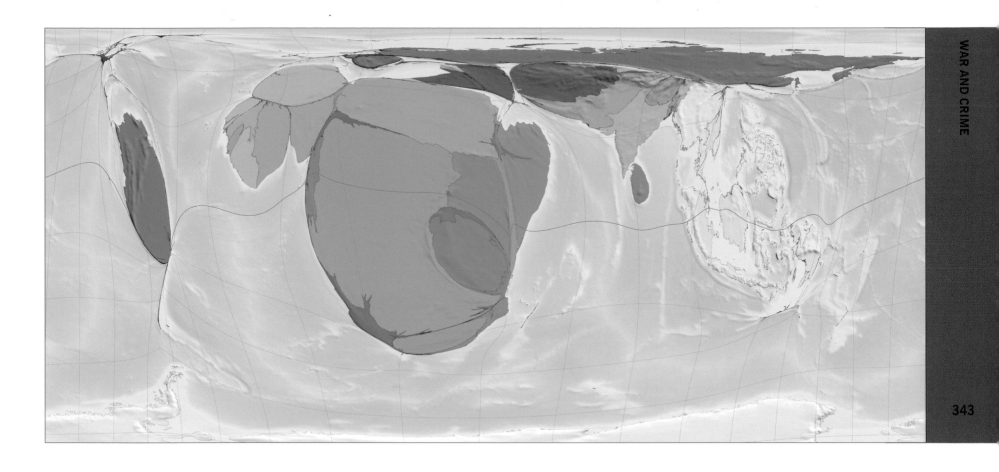

MOST WAR DEATHS IN 2002

Rank	Territory	Deaths[a]
1	Burundi	1,246
2	DR Congo	860
3	Somalia	722
4	Liberia	596
5	Sudan	464
6	Macedonia, FYR of	401
7	Congo	359
8	Gaza and West Bank	314
9	Uganda	265
10	Afghanistan	257
11	Côte d'Ivoire	235
12	Zimbabwe	232
13	Colombia	190
14	Algeria	161
15	Angola	125
16	Central African Republic	122
17	Russia	119
18	Guinea	117
19	Tajikistan	94
20	Myanmar	88

[a] war deaths per million people, 2002

320 War Deaths in 2002

The size of each territory indicates the number of deaths directly attributable to armed conflict in 2002.

In 2002 there were an estimated 172,000 war deaths worldwide. The majority of territories recorded none; the deaths shown here occurred in only 80 of the 200 territories, and 70% of them occurred in just 9 territories. The Democratic Republic of Congo bore the largest share of war deaths in 2002, at 26%. As a fraction of total population Burundi was hardest hit, with 1.2 people in every 1,000 dying.

The following territories saw the largest numbers of war deaths in their regions: Somalia in East and Southern Africa, Indonesia in the Asia–Pacific and Australasia, Colombia in South America, Sudan in Northern Africa, India in South Asia, and Russia in the Middle East and Central Asia.

'Because they tried to change this state of things...men and women have died throughout the continent...'

Gabriel García Márquez, receiving the 1982 Nobel Prize for Literature, 1982

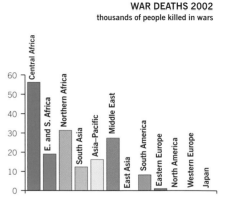

WAR DEATHS 2002
thousands of people killed in wars

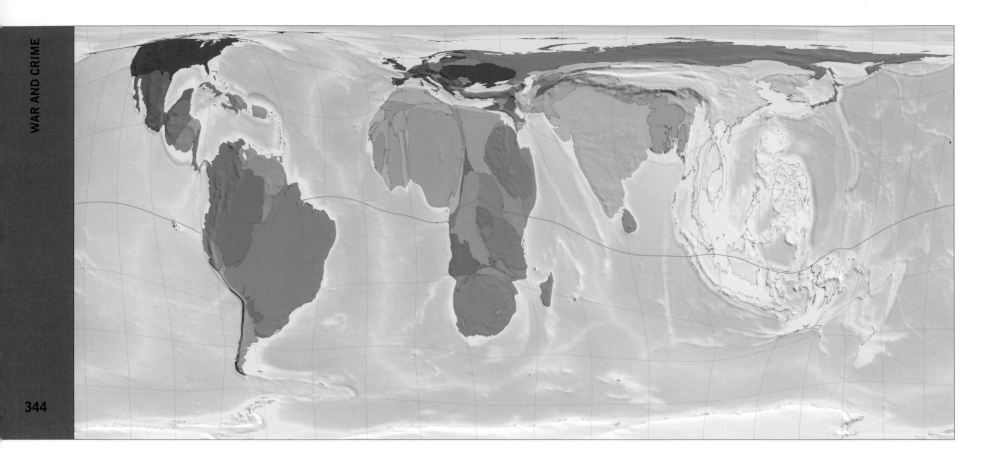

MOST AND FEWEST HOMICIDES

Rank	Territory	Deaths[a]
1	Colombia	724
2	Sierra Leone	500
3	South Africa	431
4	Angola	395
5	El Salvador	385
6	Guatemala	372
7	Venezuela	353
8	Somalia	331
9	Russia	329
10	Liberia	329
191	Singapore	8.3
192	Slovenia	7.7
193	Tonga	7.6
194	Germany	7.2
195	Israel	7.1
196	France	6.8
197	Japan	6.2
198	Iceland	6.0
199	Cyprus	2.3
200	San Marino	0.0

[a] violent deaths per million people, 2002

321 Violent Deaths

The size of each territory indicates the number of deaths occurring as a result of violence, excluding war deaths.

This map shows numbers of violent deaths, including homicide (murder, infanticide and manslaughter) but not deaths resulting directly from war, which are shown separately on Map 320.

In 2002 over half a million people died violent deaths. The largest numbers were in Brazil and India, each with over 57,000 that year. The territories where the highest proportion of people died violently were Colombia, Sierra Leone and South Africa.

The region with the most violent deaths was South America, followed by South Asia and Northern Africa. There were relatively few violent deaths in Japan and Western Europe.

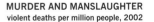

MURDER AND MANSLAUGHTER
violent deaths per million people, 2002

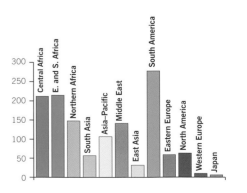

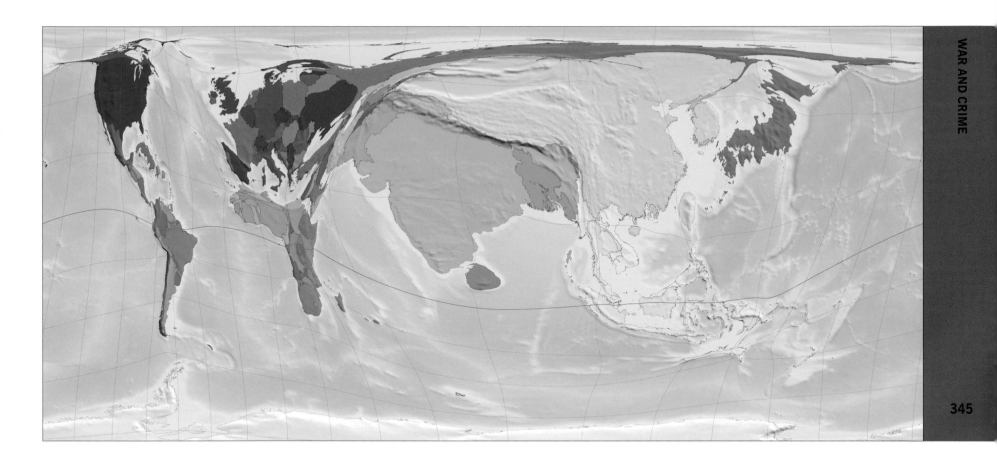

MOST AND FEWEST SUICIDES

Rank	Territory	Deaths[a]
1	Lithuania	451
2	Russia	410
3	Belarus	383
4	Kazakhstan	371
5	Ukraine	358
6	Sri Lanka	319
7	Latvia	308
8	Estonia	295
9	Slovenia	293
10	Hungary	282
191	Philippines	17
192	Egypt	15
193	Haiti	7
194	Cyprus	7
195	Syria	6
196	Antigua and Barbuda	4
197	Jamaica	1
198	Kiribati	0
199	Dominica	0
200	St Kitts–Nevis	0

[a] **self-inflicted deaths per million people, 2002**

322 Suicides

The size of each territory indicates the number of suicides.

In 2002 there were almost 900,000 suicides – self-inflicted deaths – worldwide. Over half of these were in China and India. A further 7% were in Russia and 4% in Japan.

On average, 141 per million people across the world cause their own deaths every year. Among the regions of the map, Northern Africa has the lowest suicide rate at 44 per million. The highest regional rate is in Japan, which has 245 suicides per million people.

SUICIDES
self-inflicted deaths per million people, 2002

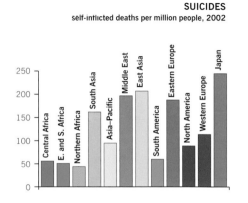

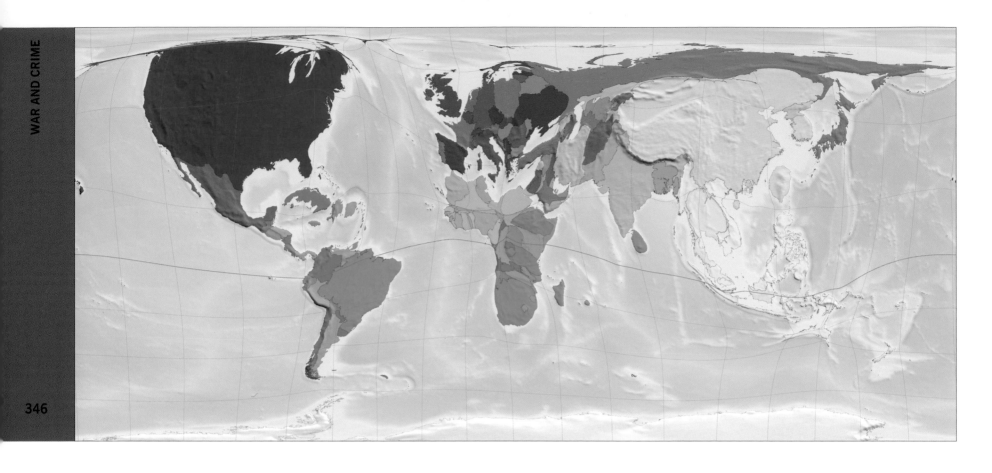

MOST AND FEWEST PRISONERS

Rank	Territory	Prisoners[a]
1	United States	7,513
2	Russia	5,985
3	St Kitts–Nevis	5,095
4	St Lucia	5,030
5	Bahamas	5,000
6	Cuba	4,867
7	Palau	4,850
8	Suriname	4,833
9	Turkmenistan	4,583
10	Belize	4,530
191	Micronesia	361
192	Nigeria	335
193	Gambia	321
194	Mali	321
195	India	320
196	Nepal	290
197	Comoros	286
198	Congo	255
199	Nauru	231
200	Burkina Faso	222

[a] prisoners per million people, 2006

323 Prisoners

The size of each territory indicates the number of people in prison there.
A quarter of all people imprisoned across the world are in the United States.

During 2006 there were an estimated 9.3 million people in prison around the world at any given time. Half of these were in just three territories, the United States (24%), China (17%) and Russia (9%). Worldwide, 0.15% of all people are in prison, or 1 person in every 670.

The highest rates of incarceration as a proportion of population are also in the United States (0.75%) and Russia (0.60%). The lowest are in Burkina Faso and Nauru. In both the latter two territories little more than 0.02% of the population is imprisoned.

'The thoughts of a prisoner – they're not free either. They keep returning to the same things.'
Alexander Solzhenitsyn, *One Day in the Life of Ivan Denisovich*, 1962

PEOPLE IN PRISON
prisoners per 1,000 people, 2006

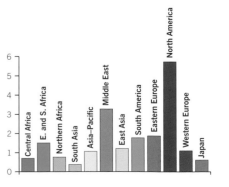

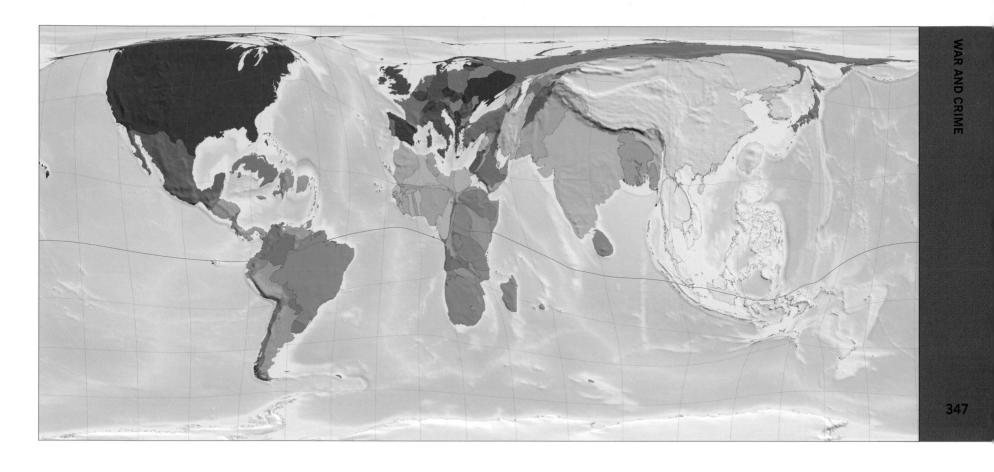

Rank	Territory	Unconvicted prisoners[a]
1	Panama	2,243
2	Bahamas	2,150
3	Suriname	2,001
4	United States	1,593
5	Cuba	1,561
6	Swaziland	1,463
7	St Lucia	1,429
8	St Kitts–Nevis	1,406
9	Puerto Rico	1,401
10	Libya	1,240
11	Uruguay	1,217
12	Georgia	1,142
12	Honduras	1,082
14	Belize	1,065
15	Trinidad and Tobago	1,055
16	South Africa	1,033
16	Lebanon	1,028
18	Russia	1,012
19	Mexico	905
20	Argentina	873

[a] prisoners awaiting trial per million people, 2006

324 Prisoners Awaiting Trial

The size of each territory indicates the number of people who are imprisoned but have not yet been tried for the crimes of which they are accused.

In 2006 there were an estimated 2.8 million prisoners awaiting trial: 30% of all prisoners worldwide. The territories with the highest proportions of prisoners who have not been tried are Haiti (84% of all prisoners), Andorra (77%) and Bolivia (75%). The largest numbers of prisoners awaiting trial are in China and the United States, with China having slightly more (although it also has a far larger total population). The third largest number of prisoners awaiting trial is in India. In the United States, 16 people are in prison awaiting trial for every 10,000 members of the population. The corresponding figures for China and India are 4 and 2 respectively.

PRE- AND POST-TRIAL PRISONERS
top: prisoners awaiting trial per 1,000 people
bottom: prisoners after trial per 1,000 people

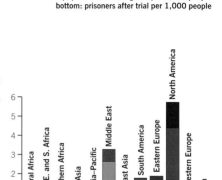

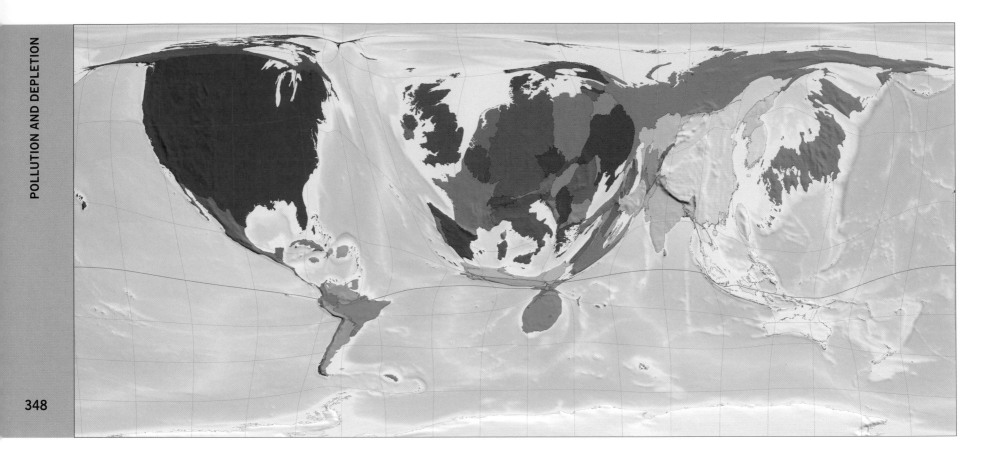

HIGHEST AND LOWEST CARBON DIOXIDE EMISSIONS IN 1980

Rank	Territory	Tonnes[a]
1	Luxembourg	28.9
2	Bahamas	27.3
3	Brunei	25.5
4	Qatar	23.0
5	United States	16.3
6	Canada	13.4
7	Germany	13.1
8	Belgium	12.8
9	Trinidad and Tobago	12.4
10	Denmark	11.7
191	Comoros	0.050
191	Uganda	0.050
193	Niger	0.049
194	Bhutan	0.031
194	Burundi	0.031
194	Central African Republic	0.031
197	Nepal	0.030
198	Cambodia	0.029
199	Chad	0.028
200	Ethiopia	0.027

[a] **CO2 emissions per person**

325 Carbon Dioxide Emissions in 1980

The size of each territory indicates the amount of carbon dioxide, the main gas implicated in global warming, emitted in 1980.

Roughly 60% of the global warming resulting from human activities is thought to be caused by the emission of carbon dioxide, a colourless, odourless gas produced primarily by the burning of organic matter such as fossil fuels.

16 billion tonnes of carbon dioxide were emitted worldwide in 1980, 70% of the amount emitted in 2000. Although quantities emitted have increased in recent years, the worldwide distribution of emissions has remained roughly the same. The territories emitting the largest amounts in 1980 were the United States and China; they are the largest emitters today. Even among the biggest emitters there is huge variation, however. For example, the United States emitted 14 times as much carbon dioxide per member of the population as China.

'...the world need[s] to differentiate between the survival emissions of the poor and luxury emissions of [the] rich.'
Sunita Narain, director of the Centre for Science and Environment, New Delhi, 2002

CARBON DIOXIDE EMISSIONS IN 1980
billion tonnes

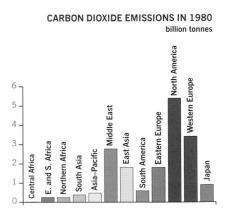

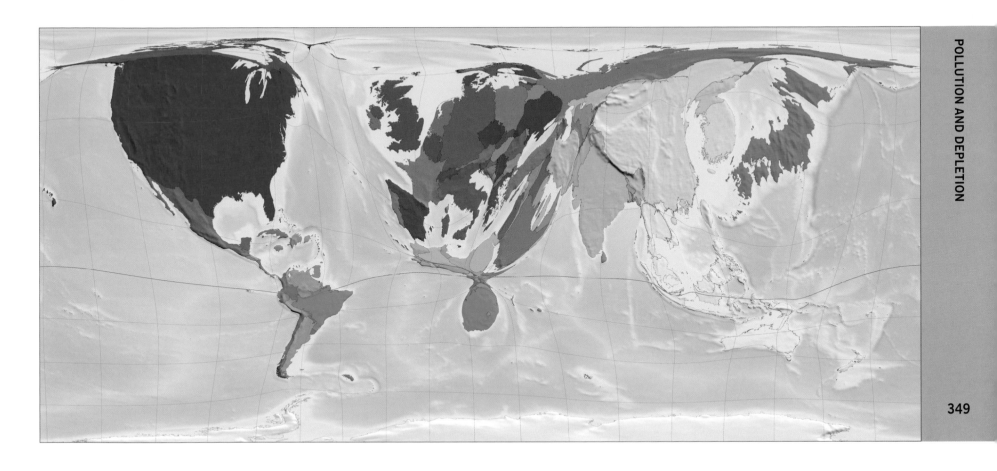

HIGHEST AND LOWEST CARBON DIOXIDE EMISSIONS IN 2000

Rank	Territory	Tonnes[a]
1	Qatar	64
2	Bahrain	27
3	Brunei	21
4	Kuwait	21
5	Trinidad and Tobago	20
6	Luxembourg	19
7	United States	19
8	Australia	18
9	United Arab Emirates	18
10	Saudi Arabia	17
191	DR Congo	0.095
192	Comoros	0.094
192	Malawi	0.094
192	Niger	0.094
192	Tanzania	0.094
192	Uganda	0.094
197	Burundi	0.048
197	Cambodia	0.048
199	Chad	0.047
200	Afghanistan	0.040

[a] CO2 emissions per person

326 Carbon Dioxide Emissions in 2000

The size of each territory indicates the amount of carbon dioxide emitted in 2000.
The United States and China remain at the head of the absolute rankings.

Emissions of carbon dioxide vary widely from place to place because of differences in lifestyles and means of producing energy, as well as the type and level of industrialization, types of transport in use and fossil fuel consumption. The most polluting territories emit 1,000 times more carbon dioxide per person than the least polluting.

In the year 2000 almost 23 billion tonnes of carbon dioxide were emitted into the atmosphere worldwide. Among the regions on the map the largest emitter was North America, which was responsible for 28% of total carbon dioxide emissions. The smallest was Central Africa at 0.09%. 21 territories emitted over 10 tonnes of carbon dioxide per member of their population. 66 territories emitted less than 1 tonne per person.

CARBON DIOXIDE EMISSIONS IN 2000
billion tonnes

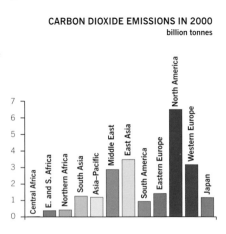

GREATEST INCREASES IN CARBON DIOXIDE EMISSIONS, 1980–2000

Rank	Territory	Tonnes*
1	Qatar	41.1
2	Bahrain	15.6
3	Kuwait	11.7
4	Saudi Arabia	11.2
5	United Arab Emirates	10.1
6	Trinidad and Tobago	7.7
7	Australia	7.1
8	Singapore	6.4
9	Israel	6.3
10	South Korea	6.3
11	Libya	5.7
12	Oman	5.4
13	Malta	4.9
14	Malaysia	4.7
15	Ireland	4.6
16	Cyprus	4.2
17	Greece	3.8
18	New Zealand	3.4
19	Turkmenistan	3.4
20	Portugal	3.3

*increase in CO2 emissions per person

327 Increase in Emissions of Carbon Dioxide

The size of each territory indicates the increase in carbon dioxide emissions between 1980 and 2000. Nearly three-quarters of all territories saw an increase over this period.

Between 1980 and 2000, 72% of territories increased their emissions of carbon dioxide, resulting in a combined increase of 6.6 billion tonnes a year. The remaining territories reduced their emissions by a combined 1.9 billion tonnes a year. Territories whose emissions stayed the same or decreased have size zero on this map and so do not appear.

The greatest increases in carbon dioxide emissions over this period were in China, the United States and India. 42% of the world population lives in these 3 territories, which collectively are responsible for 45% of the global increase in emissions. Per member of the population, however, the increase in emissions from the United States was over 3 times as large as China's and over 4 times as large as India's.

The greatest increases in carbon dioxide emissions per person were in Qatar, followed by Bahrain.

'...emissions of carbon dioxide – the most important cause of climate change – continue to rise in many parts of the world...'

Michel Jarraud, secretary general of the WMO, 2005

CHANGE IN CARBON EMISSIONS 1980–2000
millions of tonnes per year

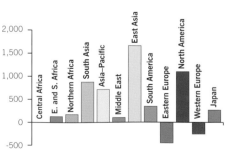

only increases are included in totals

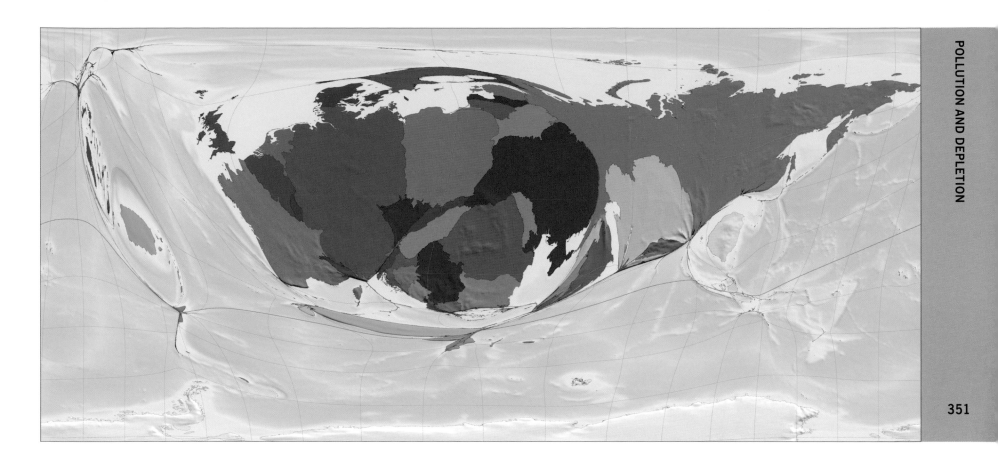

Rank	Territory	Tonnes[a]
1	Bahamas	21.6
2	Luxembourg	9.5
3	Kazakhstan	7.7
4	Estonia	6.7
5	Puerto Rico	4.6
6	Brunei	4.5
7	Romania	4.4
9	Serbia and Montenegro	4.3
9	Bulgaria	3.9
10	Ukraine	3.9
11	Poland	3.8
12	Germany	3.5
13	Denmark	3.4
14	Azerbaijan	3.3
14	Russia	3.3
16	Belgium	2.8
16	Moldova	2.8
16	Sweden	2.8
19	Belarus	2.7
19	Lithuania	2.7

[a] reduction in CO2 emissions per person

328 Decrease in Emissions of Carbon Dioxide

The size of each territory indicates the decrease in carbon dioxide emissions between 1980 and 2000. Little more than a quarter of territories managed a reduction during this interval.

28% of territories reduced their annual carbon dioxide emissions between 1980 and 2000. Together these territories brought their annual emissions down by 1.9 billion tonnes a year; however, this decrease is more than offset by increases in other territories, which come to a combined total over 3 times as large. Territories whose emissions stayed the same or increased have size zero on this map and so do not appear.

Almost half of the reductions in emissions came about in territories of the former Soviet Union, although Russia was still the third largest emitter of carbon dioxide in 2000. Substantial cuts in emissions were also made by Germany (15%), Poland (8%) and France (6%). Declines in industrial production and closure of factories contributed to some of these decreases.

CHANGE IN CARBON EMISSIONS, 1980–2000
tonnes per person per year

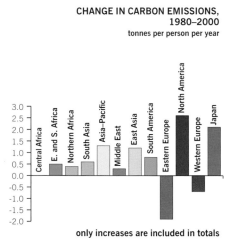

only increases are included in totals

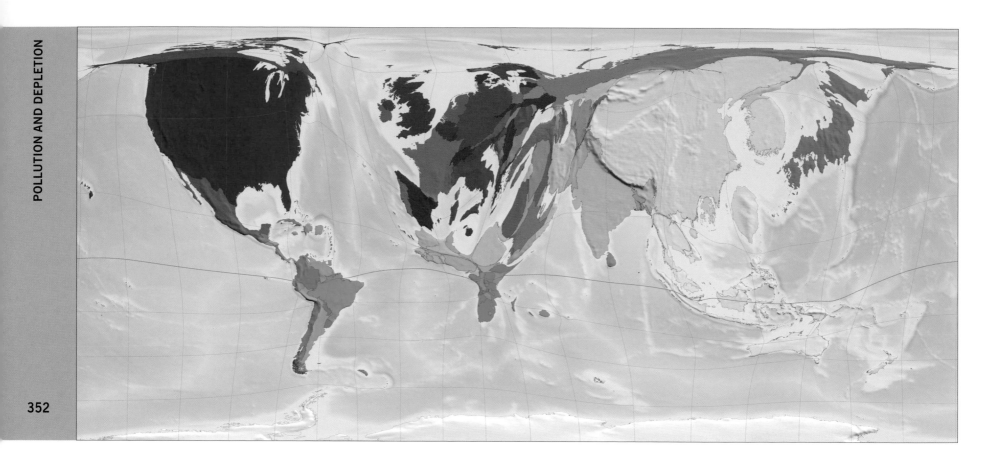

329 Projected Carbon Dioxide Emissions in 2015

The size of each territory indicates the amount of carbon dioxide it is predicted to emit in 2015.

It is predicted that by 2015 an additional 10 billion tonnes of carbon dioxide will be emitted into the atmosphere annually compared to emissions in the year 2000. This prediction comes from the United States Energy Information Administration. The amount actually emitted and its precise distribution may differ from the prediction depending on future events and human action.

China is expected to register the largest total increase in carbon emissions by 2015 – an additional 4 billion tonnes – followed by the United States (an additional 1 billion tonnes) and then Russia, France and India (half a billion tonnes each). If the predictions turn out to be correct, then by 2015 North America and East Asia will each contribute 24% of worldwide emissions. At the other end of the scale, African territories will contribute just 4%.

Emissions may also fall in some areas, either because of a decline in economic activity or because of deliberate actions taken to reduce carbon output.

COST OF CARBON DAMAGE
estimated at US$20 per tonne carbon: US$ billion, 2003

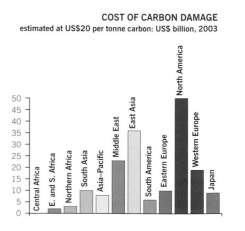

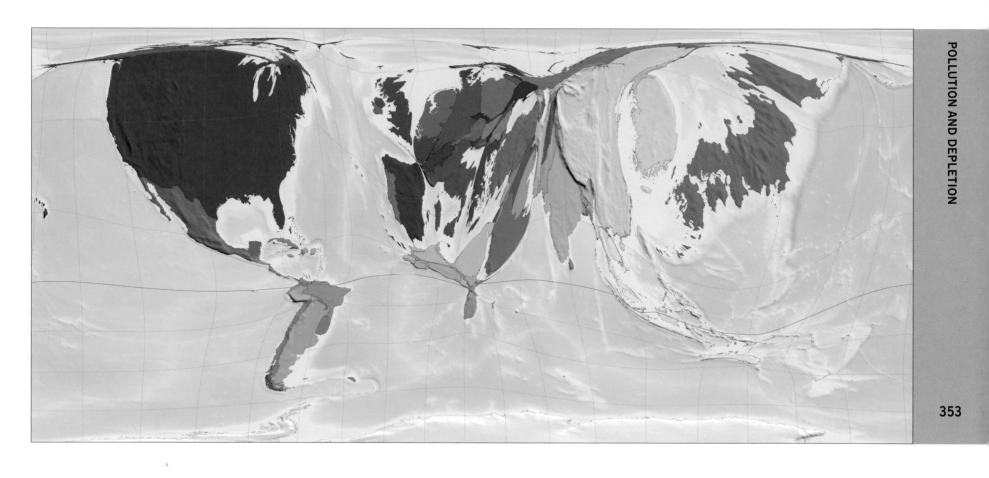

330 Particulate Damage

The size of each territory indicates the human cost of airborne pollution in the form of particulates, measured as the amount (in US dollars) that would have to be spent cleaning up the air to prevent human deaths resulting from particulates.

Particulate pollution is pollution in the form of dust particles under 10 microns in diameter. The main sources of particulates are fossil-fuel power plants, vehicles, heating systems and industrial processes. Particulates have been linked to human health problems including cardiopulmonary disease, lung cancer and acute respiratory infections. Damage of this kind is harder to quantify financially than damage to infrastructure or environment. This map shows one estimate, based on World Bank figures, of how much money would have to be spent to improve air quality enough to prevent deaths from particulate damage.

Few territories are willing or able to expend the resources needed to prevent every single death, and they differ in exactly how far they are willing to go towards this goal. The World Bank estimates include a factor that allows for this variation, effectively placing a higher cost on particulate pollution in territories that are willing to spend more to clean it up. Thus in this map a country may show a high cost of particulate damage because it has a substantial problem with particulates or because it places a high value on the human lives put at risk.

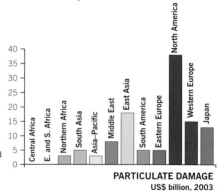

PARTICULATE DAMAGE
US$ billion, 2003

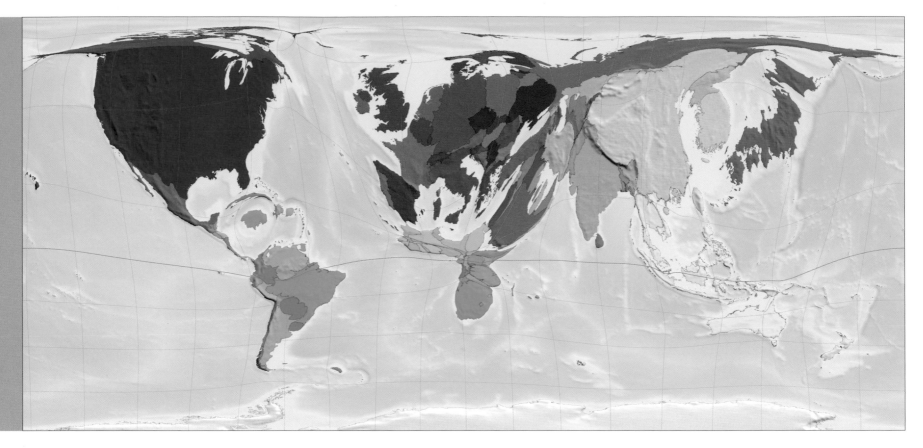

331 Greenhouse Gases

Greenhouse gases trap heat in the earth's atmosphere, causing it to warm up. The size of each territory indicates its greenhouse gas emissions.

This map shows emissions of the three leading greenhouse gases – carbon dioxide, methane and nitrous oxide – which between them account for 98% of the greenhouse effect. Other greenhouse gases, not shown here, include various fluorocarbons and sulphur hexafluoride. Quantities of gases are weighted according to their global warming potential because some have a stronger effect than others.

The territories emitting the largest amounts of greenhouse gases are the United States, China, Russia and Japan. The highest emissions per person are from Qatar, which emits the equivalent of 86 tonnes of carbon dioxide per person per year. Qatar has significant oil and gas reserves but a population of less than a million people.

GREENHOUSE GAS EMISSIONS
equivalent tonnes of carbon dioxide, per person per year

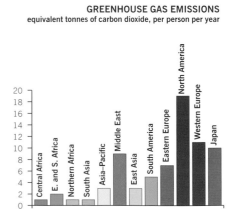

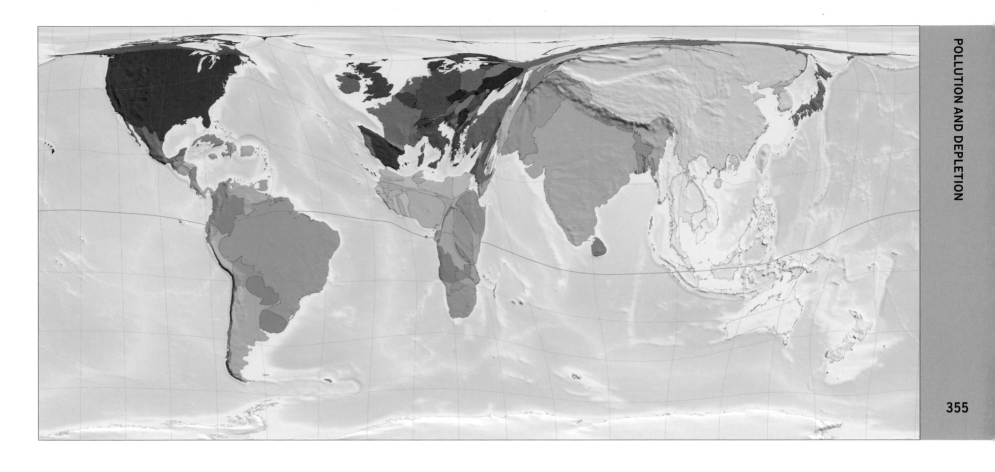

HIGHEST AND LOWEST EMISSIONS OF AGRICULTURAL GASES

Rank	Territory	Tonnes[a]
1	New Zealand	9.7
2	Brunei	8.6
3	Uruguay	7.7
4	Australia	5.4
5	Paraguay	5.1
6	Ireland	4.8
7	Argentina	3.0
8	Botswana	2.8
9	Bolivia	2.7
10	St Vincent and the Grenadines	2.6
191	Sierra Leone	0.044
192	Mozambique	0.031
193	Madagascar	0.031
194	Tanzania	0.031
195	Kiribati	0.006
196	Afghanistan	0.003
197	Nauru	<0.001
198	Grenada	<0.001
199	United Arab Emirates	<0.001
200	Singapore	<0.001

[a] emissions per person, carbon dioxide equivalents, 2002

332 Agricultural Methane and Nitrous Oxide

The size of each territory indicates emissions of methane and nitrous oxide – the main agricultural contributions to greenhouse gas emissions.

Methane is produced by animals and their manure. Nitrous oxide is emitted from soil, particularly when nitrogen fertilizers are used.

Nitrous oxide is a more powerful greenhouse gas, weight for weight, than methane. Both are also more powerful than carbon dioxide: methane has 21 times the greenhouse effect of carbon dioxide and nitrous oxide 310 times the effect. For this reason the emissions on this map are measured not simply in weight but in terms of the equivalent amount of carbon dioxide emissions that would produce the same effect.

South America is the region with the highest (carbon dioxide equivalent) emissions of agricultural methane and nitrous oxide per person living there. The territory with the highest emissions per person is New Zealand, with the equivalent of 9.7 tonnes of carbon dioxide being emitted each year by 2002.

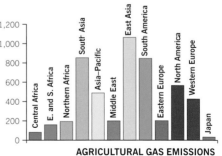

AGRICULTURAL GAS EMISSIONS
methane and nitrous oxide emissions, million tonnes of carbon dioxide equivalent per year

'The emission of 1 tonne of methane into the atmosphere has the same effect on the climate system over 100 years as the emission of about 23 tonnes of carbon dioxide.'

Dominic Ferretti, National Institute of Water and Atmospheric Research, New Zealand, 2005

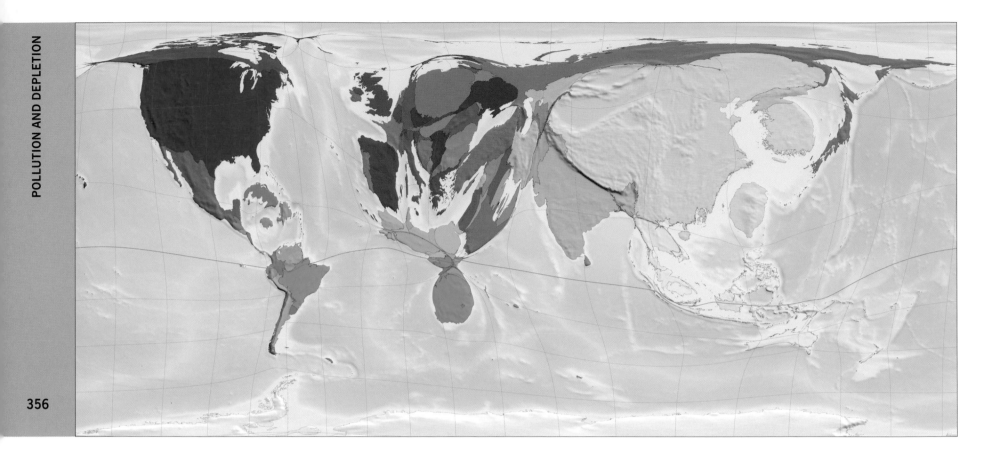

HIGHEST AND LOWEST EMISSIONS
OF SULPHUR DIOXIDE

**HIGHEST AND LOWEST EMISSIONS
OF SULPHUR DIOXIDE**

Rank	Territory	kg[a]
1	Bulgaria	121
2	Estonia	72
3	Malta	65
4	Israel	49
5	Spain	47
6	Antigua and Barbuda	45
7	Greece	45
8	United States	43
9	Jamaica	40
10	Cuba	39
191	Monaco	1.3
192	Bolivia	1.2
193	Georgia	1.0
194	Comoros	0.6
195	Costa Rica	0.5
196	Tajikistan	0.5
197	Honduras	0.4
198	Yemen	0.2
199	Ethiopia	0.2
200	DR Congo	0.0

[a] sulphur dioxide emitted from burning fuel,
per person, 2002

333 Sulphur Dioxide

The size of each territory indicates the sulphur dioxide emissions
from the burning of coal, lignite and petroleum products.

Sulphur dioxide is released when fossil
fuels containing sulphur are burned. It is
also produced naturally by volcanoes and
forest fires. Sulphur dioxide can be harmful
to plants, animals and buildings, and
sulphurous smog from coal fires can be
fatal to people. However, the principal
damage from sulphur dioxide arises when it
dissolves in atmospheric water to become
sulphuric acid, which then falls as acid
rain. Acid rain can severely damage forests
and water-based ecosystems as well
as buildings.

In 2002, 97 million tonnes of sulphur
dioxide from the burning of fuel were
emitted worldwide. The highest emissions
per person were in Bulgaria. The lowest
were in the Democratic Republic of Congo.

*'A bluish haze hangs over the Indian capital...vehicles crawl bumper
to bumper in rush hour, emitting diesel fumes, lead, sulphur dioxide
and carbon monoxide.'* Sanjoy Hazarika, *New York Times*, 1994

SULPHUR DIOXIDE EMISSIONS
kg per person per year, 2002

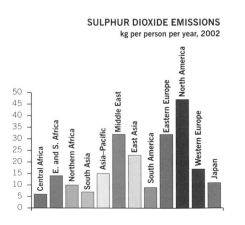

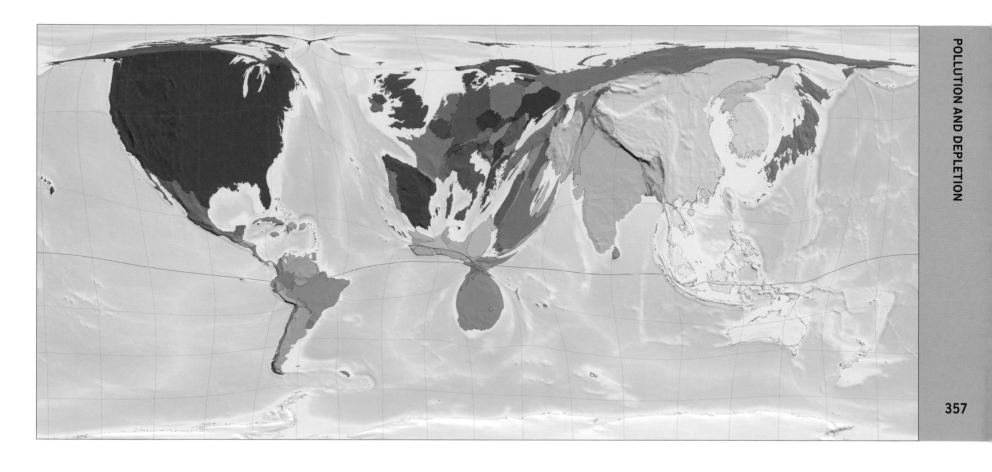

334 Nitrogen Oxides

The size of each territory indicates the quantity of nitrogen oxides
emitted from burning fuels, including those used to power vehicles.

Nitrogen oxides are emitted in the burning
of various types of fuel, and particularly
in traffic fumes. They produce acid rain,
contribute to the greenhouse effect and
ozone depletion, and cause direct harm to
humans when they react with hydrocarbon
vapours and sunlight to form
photochemical smog.

In 2002, 81.6 million tonnes of nitrogen
oxides were emitted worldwide. 27% of
these were from North America; 0.1%
were from Central Africa. North America
also produces the highest nitrogen oxide
emissions per person and Central Africa
the least.

NITROGEN OXIDE EMISSIONS
kg per person per year, 2002

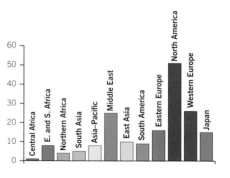

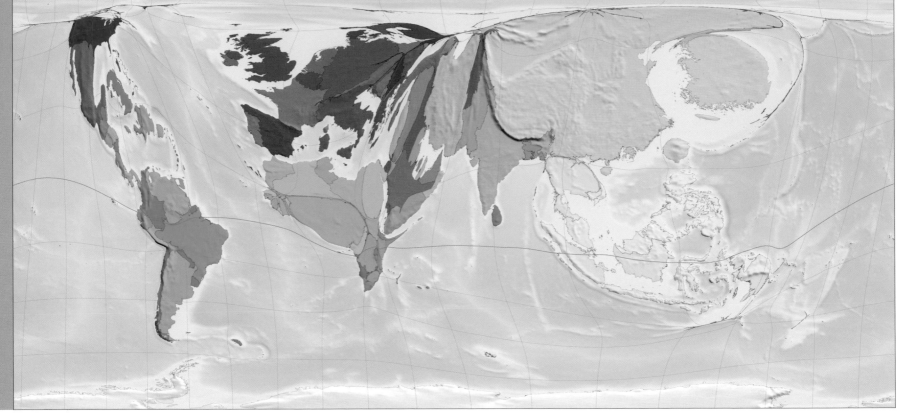

HIGHEST AND LOWEST PRODUCTION OF CHLOROFLUOROCARBONS

Rank	Territory	Grammes[a]
1	Bahamas	183
2	Libya	182
3	Cyprus	165
4	Kuwait	145
5	Brunei	145
6	Qatar	145
7	South Korea	140
8	Lebanon	137
9	Bahrain	135
10	United Arab Emirates	128
191	Switzerland	0.47
192	Madagascar	0.46
193	Ethiopia	0.43
194	Canada	0.40
195	Czech Republic	0.36
196	Singapore	0.20
197	Slovenia	0.20
198	Japan	0.15
199	Slovakia	0.15
200	Hungary	0.03

[a] chlorofluorocarbons produced per person per year, 2002

335 Chlorofluorocarbons

The size of each territory indicates the quantity of chlorofluorocarbons emitted. Despite efforts to phase out the use of these substances, substantial amounts are still reaching the atmosphere.

Chlorofluorocarbons, often abbreviated to CFCs, are a class of chemical compounds that have a broad variety of uses in industry as well as in consumer products and processes such as refrigerators, dry cleaning and aerosols. When released into the atmosphere they have two major adverse environmental effects: they deplete the ozone layer and contribute to global warming. The Montreal Protocol, which came into force in 1989, is an international treaty that lays down a timetable for the phasing out of ozone-depleting substances, including CFCs.

This map shows the worldwide distribution of the estimated 110,000 tonnes of CFCs that were emitted in 2002.

CHLOROFLUOROCARBON EMISSIONS
grammes per person per year

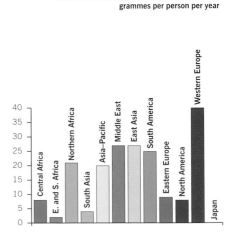

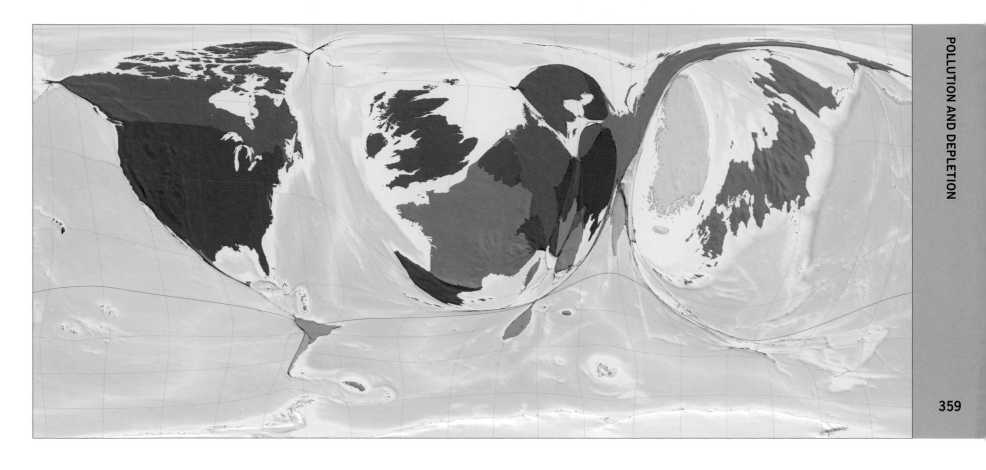

MOST NUCLEAR WASTE PRODUCED PER PERSON

Rank	Territory	Grammes[a]
1	Canada	42
2	Sweden	35
3	France	19
4	Lithuania	14
5	Belgium	14
6	Finland	14
7	South Korea	13
8	United Kingdom	11
9	Slovakia	11
10	Slovenia	10

[a] nuclear waste produced per person, 2001

MOST NUCLEAR WASTE PRODUCED IN TOTAL

Rank	Territory	Tonnes[a]
1	United States	1,630
2	Canada	1,300
3	France	1,146
4	Japan	996
5	United Kingdom	650
6	South Korea	634
7	Germany	420
8	Sweden	310
9	Belgium	144
10	Spain	136

[a] nuclear waste produced, 2001

336 Nuclear Waste

The size of each territory indicates quantities of nuclear waste produced, mainly as a result of power generation in nuclear-fuelled plants.

Around 8,910 tonnes of heavy-metal nuclear waste are generated each year, mainly in nuclear power plants. Three territories produce over 1,000 tonnes a year: the United States, Canada and France. Canada also produces the most waste per person, although Sweden is not far behind.

Of the 200 territories on the map, 179 produce less than the world average amount of nuclear waste per person, which is 1.4 tonnes per year per million people. The remaining 21 territories, most of them located in Europe, account for the rest.

'...we have a 50-year history in this country of not finding any long-term management option for very high-level, relatively dangerous radioactive waste.'
Gordon McKerron, chair of the UK Committee on Radioactive Waste Management, 2006

NUCLEAR WASTE
tonnes of heavy-metal nuclear waste generated per year

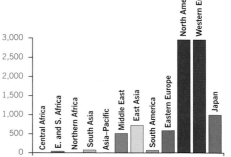

MOST AND LEAST HAZARDOUS WASTE GENERATED

Rank	Territory	kg[a]
1	Estonia	4,774
2	Kyrgyzstan	1,329
3	Uzbekistan	1,108
4	Russia	966
5	Hungary	345
6	Slovakia	307
7	Czech Republic	273
8	Luxembourg	253
9	Finland	231
10	Bahrain	200
191	Gaza and West Bank	4.8
192	Jamaica	3.8
193	St Lucia	3.4
194	Moldova	2.8
195	Belize	2.7
196	Iran	2.5
197	Sri Lanka	2.1
198	Niger	2.1
199	Azerbaijan	2.0
200	Turkey	1.0

[a] hazardous waste generated per person, 2001

337 Hazardous Waste

The size of each territory indicates the amount of hazardous waste generated. Much of it is dealt with in other territories.

This map shows where flammable, oxidizing, infectious, radioactive, poisonous or corrosive wastes are produced. The location at which waste is generated, however, is not necessarily where it ends up: some territories send their waste to be managed elsewhere.

The 3 biggest producers of hazardous wastes are Russia, the United States and Uzbekistan. Among the regions, the biggest producer is the Middle East and Central Asia. In contrast, South Asia, the Asia–Pacific and Australasia, and Central Africa generate very low proportions of hazardous waste.

HAZARDOUS WASTE
kg generated per person per year

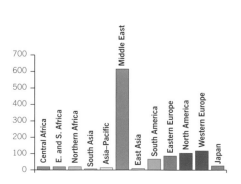

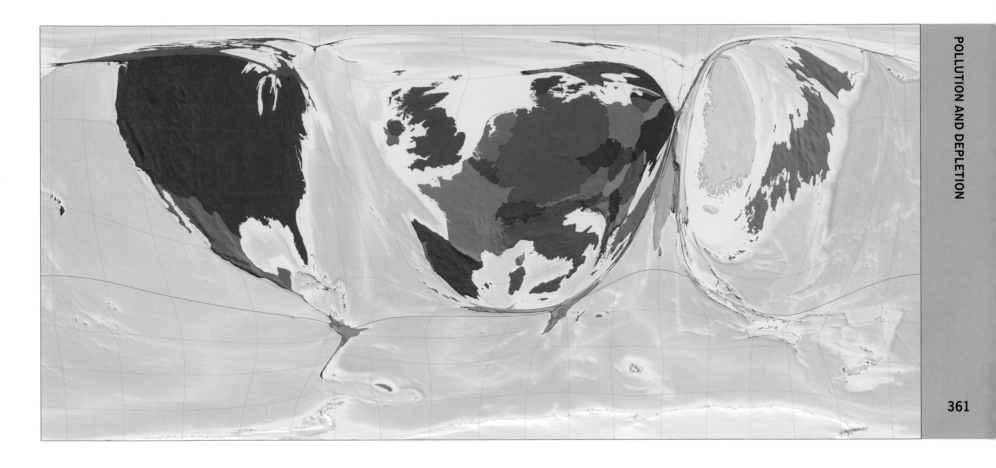

MOST AND LEAST SEWAGE SLUDGE GENERATED IN OECD

Rank	Territory	kg[a]
1	Luxembourg	41
2	South Korea	34
3	Germany	30
4	Denmark	28
5	Switzerland	28
6	Austria	26
7	Sweden	25
8	United States	24
9	Norway	23
10	Netherlands	22
19	Canada	16
20	Japan	13
21	Ireland	10
22	Poland	9
23	Hungary	9
24	Belgium	8
25	Australia	6
26	Greece	3
27	Portugal	3
28	Iceland	1

[a] sewage sludge generated per person per year, 1999 (covers only the then 28 OECD countries)

338 Sewage Sludge

The size of each territory indicates the quantity of sewage sludge that goes into public sewerage systems.

Sewage sludge, as defined here, is domestic or industrial waste that goes into a public sewerage system. In places where there is no public sewerage system, no sewage sludge is recorded. Thus Africa, South America and much of Asia are small on this map.

UNEP notes that while sewage sludge can be a useful fertilizer, it also carries environmental dangers, including the possible presence of heavy metals, organic compounds and disease-causing organisms.

The largest quantities of sewage sludge per person are produced in Luxembourg, partly because this small territory, virtually a city-state, has no room for septic tanks, which are not connected to sewerage systems.

'...a sharp water oxygen decline is caused by sewage, and this sewage contains substances requiring lots of oxygen for the biochemical oxidation process. This...is killing fish en masse.'

Oleksandr Poliakov, chief ichthyologist, Luhansk Regional Fishing Control Authority, Ukraine, 2005

SEWAGE SLUDGE
kg collected per person, 1999

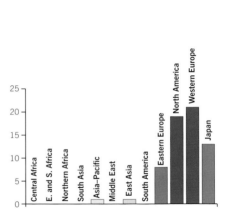

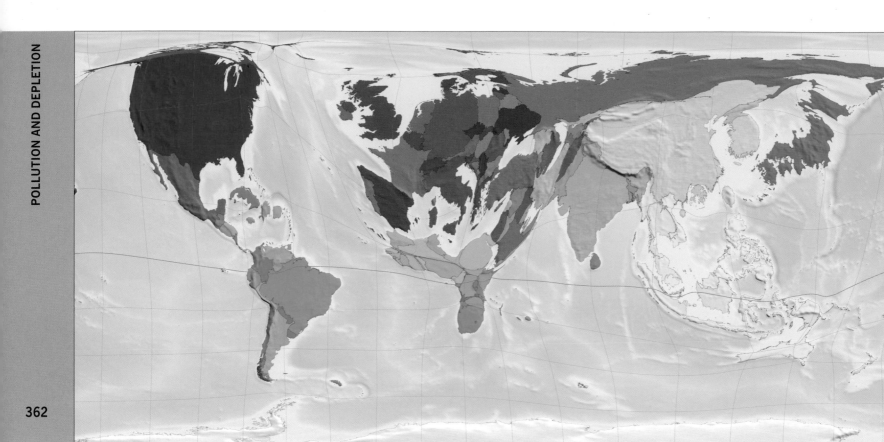

MOST AND LEAST WASTE GENERATED

Rank	Territory	kg[a]
1	Russia	1,439
2	Monaco	1,176
3	Georgia	1,058
4	Singapore	1,048
5	Azerbaijan	928
6	Hong Kong (China)	771
7	Luxembourg	735
8	United States	715
9	Iceland	697
10	Norway	680
191	Colombia	171
192	Morocco	156
193	Benin	149
194	Panama	131
195	Bolivia	77
196	Peru	54
197	Costa Rica	17
198	Nepal	17
199	Burkina Faso	10
200	Madagascar	9

[a] waste produced per person, 2002

339 Waste Generated

The size of each territory indicates the quantity of municipal waste generated by and collected from homes, schools and businesses.

Rubbish includes organic waste, paper, packaging, and bulky items such as fridges or mattresses. The largest total quantity of rubbish is generated in China, where the largest population of any single territory lives. The largest quantity per person is generated in Russia. Very small amounts of rubbish are generated per person in Madagascar, Burkina Faso, Nepal and Costa Rica. In many territories, it is more common to reuse packaging than to discard it, which reduces the amount of rubbish generated.

The regions that produce the greatest volume of waste each year are, in descending order, the Middle East and Central Asia, North America, East Asia and Western Europe.

MUNICIPAL WASTE
kg generated per person, 2002

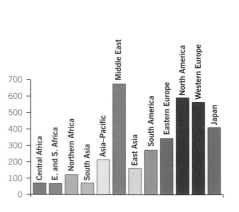

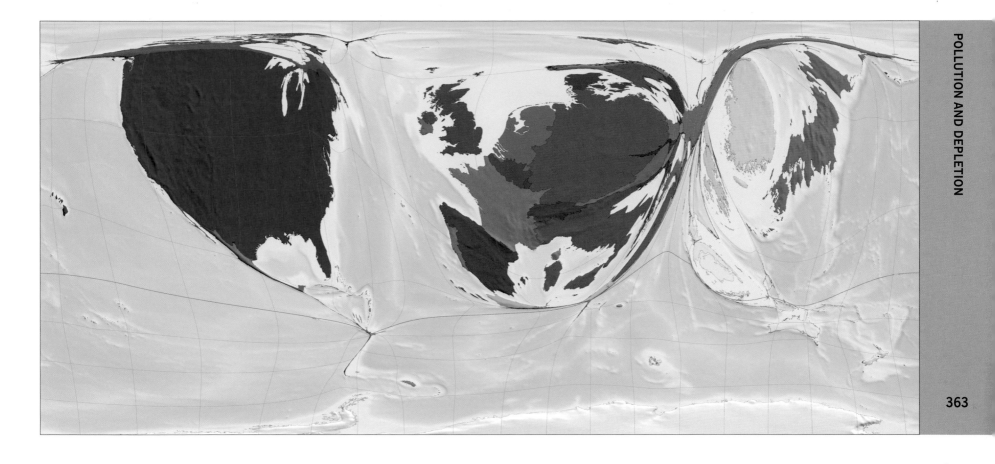

340 Waste Recycled

The size of each territory indicates the quantity of municipal waste that is recycled.
The richer territories tend to recycle most, but also produce far more waste than others.

Worldwide, 6.6% of all municipal waste generated in 2002 was recycled. The territories that recycle the largest quantities of waste are chiefly in North America and Western Europe, but also include Japan and South Korea. Even so, the quantities of non-recycled waste generated in Western Europe and North America, put together, still amount to more than the total generated in most other regions put together.

'One of the great challenges of our time is to collectively agree on what is waste and what are second-hand products – this question extends to end-of-life ships as much as to electronic goods...' Achim Steiner, head of UNEP, 2006

RECYCLED AND NON-RECYCLED WASTE
top: waste recyled; bottom: waste not recycled
million tonnes, 2002

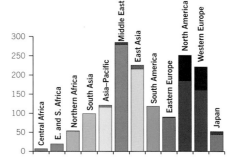

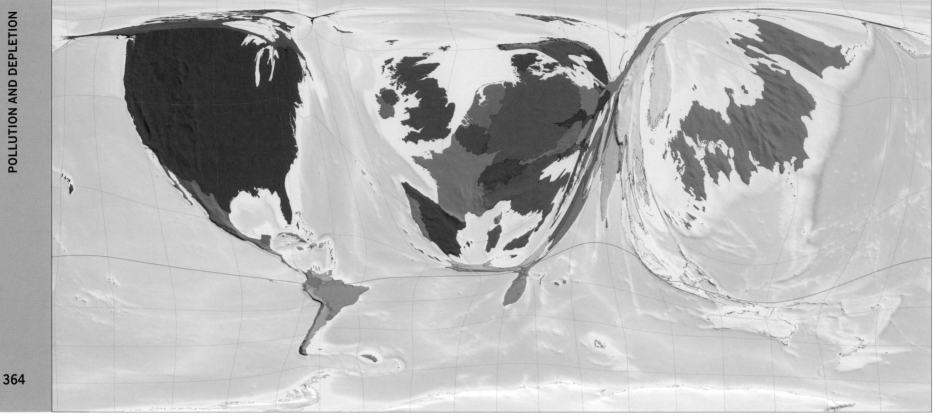

The size of each territory indicates the cost in US dollars of maintaining capital assets such as buildings and equipment.

Capital assets are fixed assets such as buildings and equipment used by businesses or governments. Assets deteriorate over time and have to be maintained or replaced. This map shows the total cost of this maintenance in each territory.

Large, technologically advanced infrastructures are likely to cost more to maintain than those that are small and simple. The United States, Japan, Germany and France have the highest total maintenance costs for fixed assets. Per-person costs are highest in Norway, Switzerland and Denmark, where they are roughly 1,000 times higher than in Nepal, Ethiopia or Burundi.

HIGHEST AND LOWEST COSTS OF MAINTAINING FIXED CAPITAL

Rank	Territory	US$ᵃ
1	Norway	70
2	Switzerland	62
3	Japan	54
4	Denmark	51
5	United States	45
6	Finland	44
7	Sweden	40
8	Netherlands	40
9	Austria	39
10	Germany	38
191	Eritrea	0.13
192	Sierra Leone	0.11
193	Malawi	0.10
194	Guinea-Bissau	0.10
195	Liberia	0.09
196	Haiti	0.07
197	DR Congo	0.07
198	Nepal	0.06
199	Ethiopia	0.05
200	Burundi	0.05

ᵃ capital consumption per person per year, US$ hundreds, 2003

FIXED CAPITAL CONSUMPTION
US$ billion, 2003

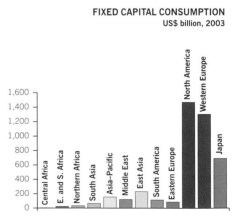

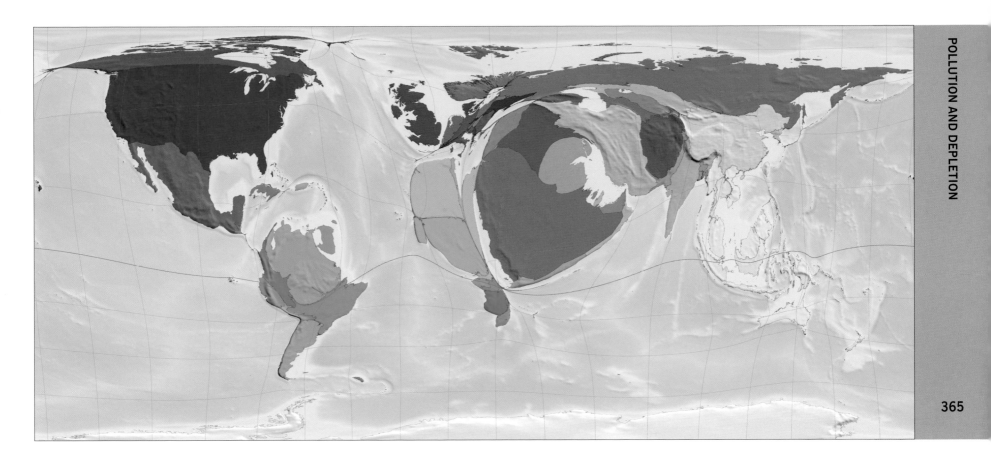

HIGHEST AND LOWEST RATES OF ENERGY DEPLETION

Rank	Territory	US$[a]
1	Kuwait	91
2	Saudi Arabia	44
3	Norway	30
4	Trinidad and Tobago	30
5	Venezuela	13
6	Canada	12
7	Gabon	8
8	Russia	8
9	Algeria	7
10	Kazakhstan	7
191	Comoros	0.05
192	Seychelles	0.05
193	Slovakia	0.05
194	Kyrgyzstan	0.04
195	Georgia	0.03
196	DR Congo	0.02
197	Philippines	0.02
198	Bulgaria	0.02
199	Bosnia and Herzegovina	0.02
200	Tajikistan	0.00

[a] energy depletion due to current extraction of fossil fuels, per person, 2003

342 Fossil Fuel Depletion

The size of each territory indicates the annual rate at which fossil fuel resources are being depleted, measured in terms of US dollar value at current prices.

Fossil fuels – coal, oil and natural gas – are so slow to form that for all practical purposes they can be considered finite resources: once they are extracted they are not replaced. This map shows the rate of depletion of reserves of fossil fuels in each territory.

In terms of total dollar value, the United States has the highest annual rate of depletion of fossil fuels, followed by Russia, Saudi Arabia and Iran. Rates of depletion per person are highest in Kuwait, followed by Saudi Arabia and Norway.

'*Modern energy can greatly increase productivity – by providing lighting that extends the workday and powering machines that increase output and enabling households to engage in activities that generate income.*' World Coal Institute, 2007

FOSSIL FUEL DEPLETION
US$ billion, 2003

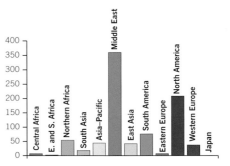

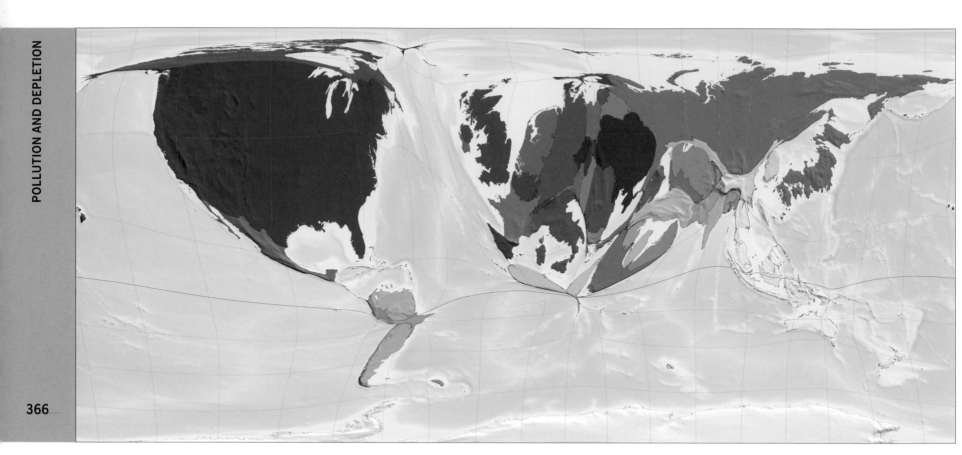

HIGHEST GAS CONSUMPTION

Rank	Territory	Tonnes[a]
1	Qatar	385
2	United Arab Emirates	168
3	Kuwait	81
4	Netherlands	73
5	United States	72
6	Russia	71
7	Canada	68
8	Turkmenistan	60
9	Ukraine	49
10	Romania	43
11	Uzbekistan	41
12	Azerbaijan	37
13	Saudi Arabia	36
14	Belarus	34
15	Luxembourg	33
15	Belgium	33
17	United Kingdom	31
18	Hungary	28
19	Lithuania	28
20	Slovakia	26

[a] gas used per person, in tonnes
of oil equivalent, 1965–2004

343 Gas Depletion

The world's natural gas reserves are finite. The size of each territory indicates the amount of them it used up between 1965 and 2004.

Gas is generated so slowly that for practical purposes there will be no more once existing reserves are exhausted. Over the 40-year period depicted on this map, the equivalent in natural gas of almost 60 billion tonnes of oil was used up. The territories using the most gas were the United States, Russia, Ukraine and Canada.

Almost all the natural gas ever burned as fuel has been burned since 1965. Until around that time, natural gas from oilfields was generally considered useless and was disposed of by flaring. More recently, however, it has found extensive use as a fuel.

40 YEARS OF GAS CONSUMPTION
gas consumed 1965–2004,
billions tonnes oil equivalent

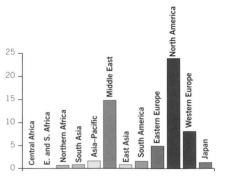

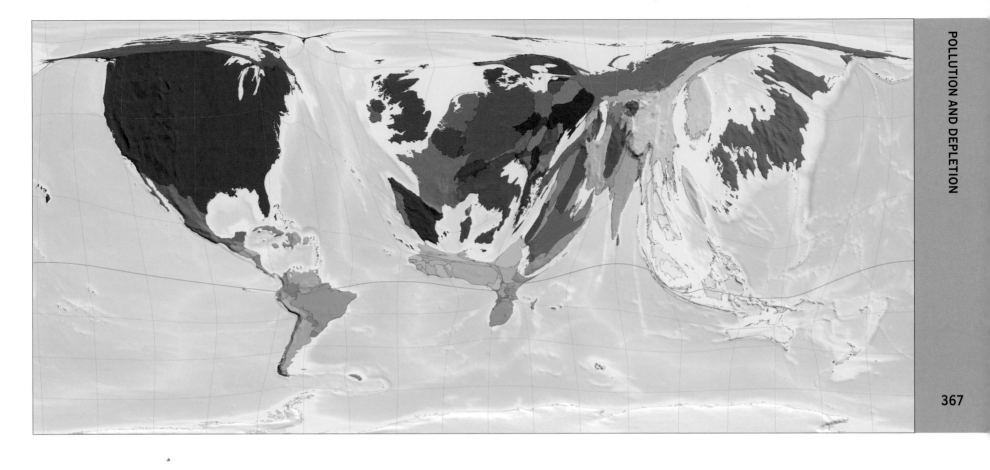

HIGHEST OIL CONSUMPTION

Rank	Territory	Tonnes[a]
1	Singapore	168
2	United Arab Emirates	112
3	Kuwait	108
4	United States	107
5	Canada	101
7	Luxembourg	101
6	Belgium	101
8	Sweden	92
9	Denmark	91
10	Netherlands	90
11	Iceland	87
13	Finland	83
12	Norway	78
14	Saudi Arabia	72
15	Japan	71
16	Belarus	70
17	Qatar	68
18	Switzerland	68
19	Lithuania	64
20	Germany	64

[a] tonnes of oil used per person, 1965–2004

344 Oil Depletion

No one knows how much oil remains in the Earth. The size of each territory indicates the amount burned between 1965 and 2004.

Between 1965 and 2004 an estimated 123 billion tonnes of oil were used worldwide. This is equivalent to an average of 19.7 tonnes per person over this 40-year period (based on the population in 2002). The most profligate users of oil were the United States, Japan, Russia and Germany.

This map shows where oil was used, not where it was extracted. The map also gives no hint of how much oil remains in the ground. We know how much we use, but no one is certain of the exact size of our remaining reserves.

'My father rode a camel. I drive a car. My son flies a jet-plane. His son will ride a camel.' Saudi Arabian saying, undated

40 YEARS OF OIL CONSUMPTION
oil consumed 1965–2004, billions tonnes

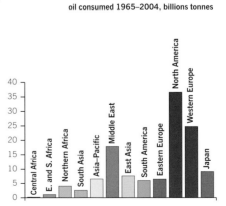

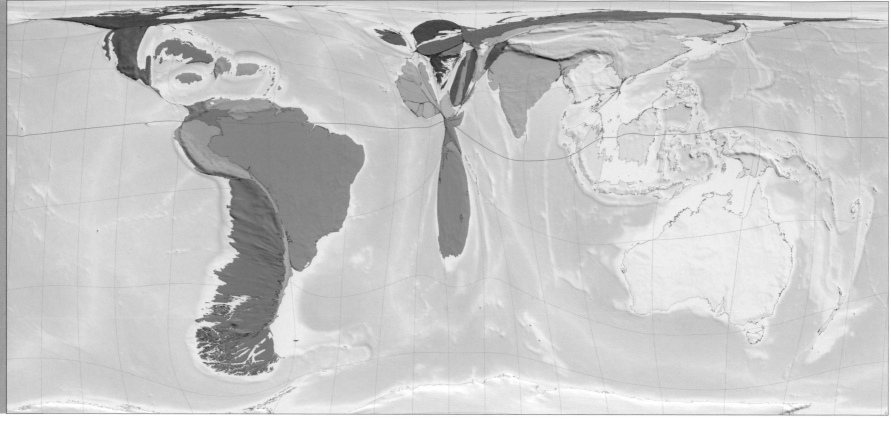

345 Mineral Depletion

The size of each territory indicates the annual rate of depletion of mineral resources, measured in terms of US dollar value at current prices.

Mineral resources include gold, lead, zinc, iron, copper, nickel, silver, bauxite and phosphate. Mineral resources are depleted when minerals are mined either for local use or for export. The territories with the highest total rates of mineral depletion are Australia, Brazil, Chile and China. Australia is the largest producer of bauxite (aluminium ore), Brazil of industrial diamonds, China of tungsten, and South Africa of platinum and gold. The territories that appear small on this map may do so either because they have little in the way of minerals to begin with or because they have already used up what resources they had.

MINERAL DEPLETION
US$ billion, 2003

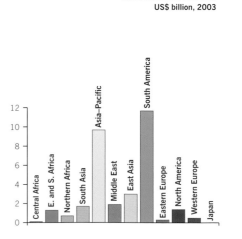

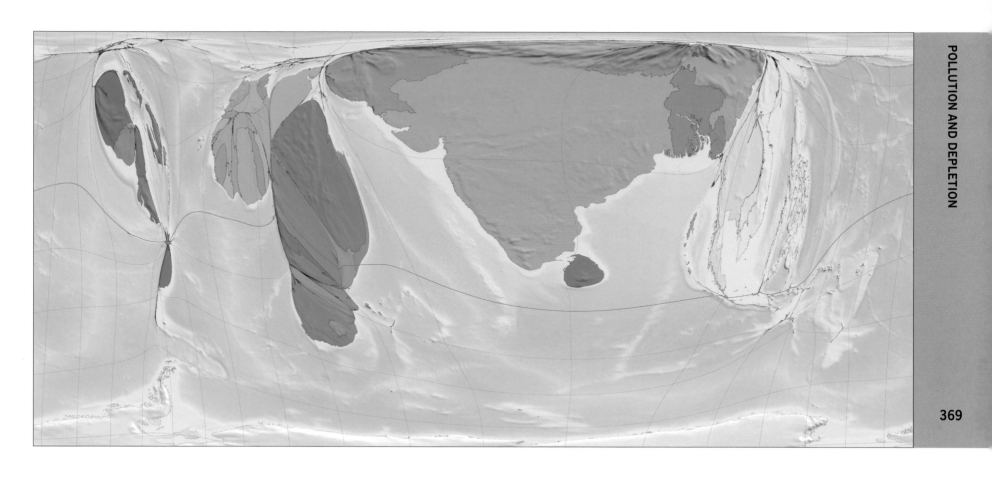

HIGHEST RATES OF FOREST DEPLETION

Rank	Territory	US$[a]
1	Guatemala	18
2	Uganda	15
3	Burundi	13
4	Costa Rica	13
5	El Salvador	12
6	Ethiopia	12
7	Uruguay	11
8	Lesotho	10
9	Togo	10
10	Nepal	10
11	Sierra Leone	9
12	Nicaragua	8
13	Ghana	7
14	Rwanda	7
15	Guinea	7
16	Thailand	7
16	Liberia	7
18	Niger	6
19	India	5
20	Pakistan	5

[a] loss of potential future earnings
due to current sales, per person, 2003

346 Forest Depletion

The size of each territory indicates the annual rate of depletion of forests, measured in terms of US dollar value at current prices.

As quantified here, forest depletion is the net rate of loss of forests as a result of tree felling that is not offset by regrowth. The loss is measured in terms of the market value of the trees in the form of untreated round timber.

In other words this map shows, at the territory level, the dollar value of wood that is not sustainably harvested.

The highest rates of unsustainable harvesting occur in India, Ethiopia, Pakistan and Bangladesh. Almost half of the world total (46%) occurs in India, where the annual timber depletion exceeds that of the next 25 territories combined,

although the population of India is also almost as large as the combined population of those 25 other territories. (Forest depletion per person in India is ranked only 19th among the territories for which data are available.)

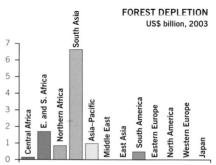

FOREST DEPLETION
US$ billion, 2003

'A lot of the illegal timber is bought by China and Japan, then converted to products that are sold in Europe, North America, Australia and New Zealand. It's...destroying livelihoods — destroying people.' Annie Kajir, CEO of Environmental Law Centre, 2006

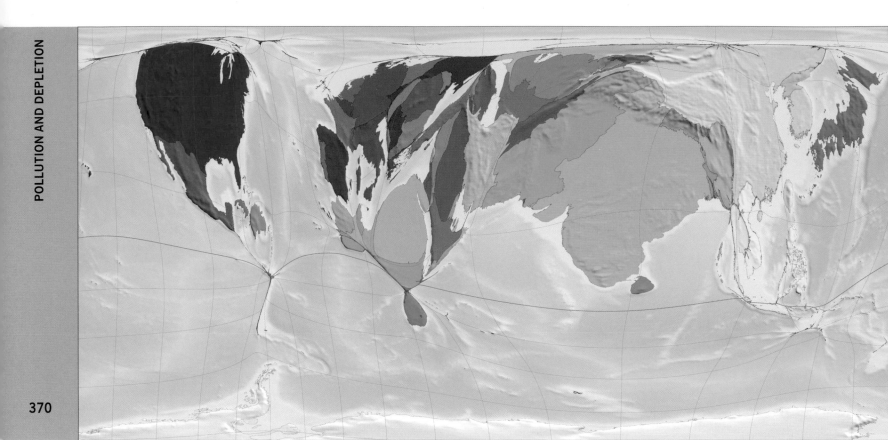

HIGHEST RATES OF WATER DEPLETION

Rank	Territory	cubic metres[a]
1	Turkmenistan	4,938
2	Uzbekistan	2,198
3	Azerbaijan	1,889
4	Kazakhstan	1,694
5	Iraq	1,599
6	Bulgaria	1,470
7	Kyrgyzstan	1,082
8	Pakistan	1,003
9	Romania	975
10	Egypt	933
11	Afghanistan	887
12	Tajikistan	852
13	Serbia and Montenegro	835
14	Iran	833
15	Libya	815
16	Dominican Republic	733
17	Saudi Arabia	715
18	Macedonia, FYR of	700
19	Moldova	674
20	United States	674

[a] amount used above 10% of resource, per person, 2003

347 Excess Water Use

The size of each territory indicates the annual volume of water used in excess of 10% of the renewable internal fresh-water resources.

A territory that uses more than a certain fraction of its internal water resources every year may risk water shortage in the future. The degree of risk varies depending on local conditions, and opinions differ about what fraction of water resources can safely be consumed. This map shows which territories use more than 10% of their water resources annually, and by how much they do so.

Egypt uses 33 times more water annually than its entire internal water resources. This is possible because the river Nile supplies Egypt with rainwater from elsewhere. Other territories with greater internal resources may use more water per person but still consume a smaller fraction of their resources. For example, 4 territories use more per person than Egypt but still use under 5% of their total internal resources annually.

75 territories use less than 10% of their renewable internal fresh-water resources and so do not appear on this map. 51 territories use between 10% and 100% of their internal freshwater annually and 15 territories use 100% or more. Data were missing for 59 territories.

WATER DEPLETION
water use per person above 10% threshold, cubic metres, 2003

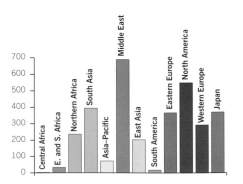

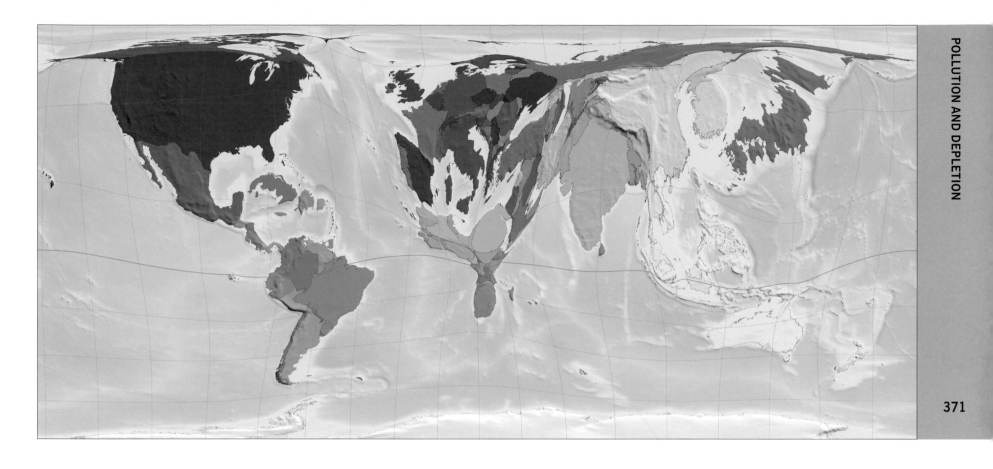

348 Domestic Water Use

The size of each territory indicates the annual domestic water use in cubic metres. While some people do not have clean water even to drink, some others have enough to fill swimming pools.

Domestic water use, as defined here, includes water use in people's homes, for drinking water, for public services and in commercial service establishments such as hotels. 325 billion cubic metres of water are used domestically each year around the world.

The average water use per person is 52 cubic metres per year, but actual usage varies hugely. Between 1987 and 2003 people living in Cambodia, where the majority do not have access to clean water supplies, used an average of 1.8 cubic metres of water each per year. People in Costa Rica used 100 times this amount. And people in Australia used almost 3 times as much as Costa Ricans, much of the extra going to water lawns and fill swimming pools.

'I remember when I was 14, carrying a 20 litre water can on my head, filling it from a river some thirty minutes away. When I came to Canada, I was shocked by the extravagant use of water here.'

Sieru Efrem, originally from Ethiopia but living in Toronto, quoted on the BBC News website, 2003

DOMESTIC WATER USE
cubic metres per person per year, 1987–2003

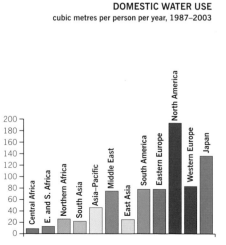

349 Industrial Water Use

The size of each territory indicates the annual industrial water use in cubic metres. Industrial use worldwide far outstrips domestic use.

Between 1987 and 2003 roughly twice as much water was used each year for industry as for domestic purposes: worldwide, industries used 665 billion cubic metres per year on average. Just under a 3rd of this total was used in the United States, and just under a 30th in all 19 territories of East and Southern Africa combined. Central Africa, East and Southern Africa, South Asia, Northern Africa and the Asia–Pacific and Australasia region all have low industrial water use per person.

'More than one-half of the world's major rivers are being seriously depleted and polluted, degrading and poisoning the surrounding ecosystems, thus threatening the health and livelihood of people...'
Ismail Serageldin, chairman of the World Commission on Water for the 21st Century, 1999

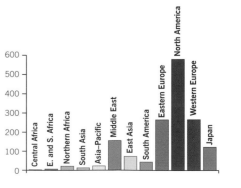

INDUSTRIAL WATER USE
cubic metres per person per year, 1987–2003

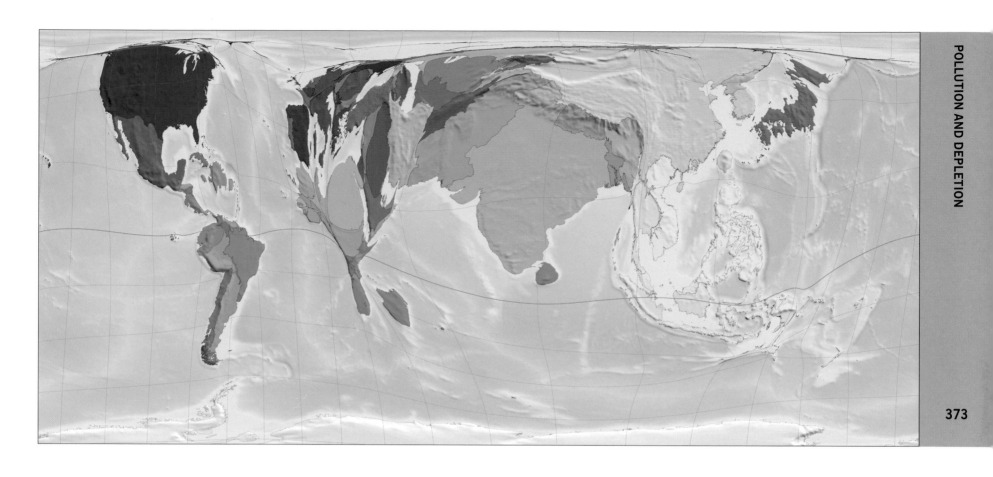

350 Agricultural Water Use

The size of each territory indicates the annual water use, in cubic metres, for agricultural purposes, including irrigation and rearing livestock.

Between 1987 and 2003 an annual average of 2.4 trillion cubic metres of water were used around the world for agricultural purposes.

A large fraction of agricultural water use occurs in Asian territories such as India, Pakistan, Nepal, China, the Philippines, Indonesia, Japan and Viet Nam. There is also high per-person water use in the Middle Eastern and Central Asian territories of Kyrgyzstan, Tajikistan, Turkmenistan, Uzbekistan, Kazakhstan and Afghanistan.

Agricultural consumption of water resources is lower where rainfall is regular and temperatures are moderate or low, so that less irrigation is needed. It is also lower in places where fewer crops are grown or fewer animals reared.

AGRICULTURAL WATER USE
cubic metres per person per year, 1987–2003

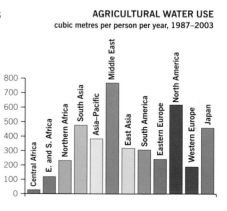

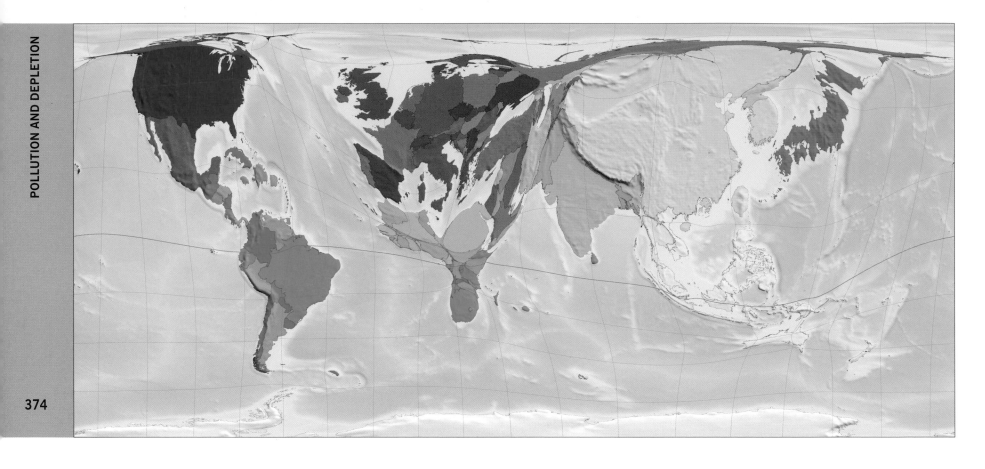

FEWEST WATER CONNECTIONS

Rank	Territory	%[a]
1	Somalia	1
2	Uganda	1
3	Afghanistan	4
4	Central African Republic	4
5	Chad	4
6	Togo	4
7	Ethiopia	5
8	Burundi	5
9	Guinea-Bissau	5
10	Burkina Faso	6
11	Angola	6
12	Bangladesh	6
12	Madagascar	6
14	Myanmar	6
15	Malawi	7
16	Niger	8
16	Mozambique	8
18	Rwanda	8
19	Equatorial Guinea	8
20	Cambodia	9

[a] % of households connected to water supplies, 2004

351 Piped Water

The size of each territory indicates the numbers of households with piped water supplies.

In Western Europe almost all households are supplied with piped water. In Central Africa only 10% are. Having water piped to your house saves time and energy that would otherwise have to be spent collecting water for drinking, cooking and washing. It also greatly facilitates the operation of sewerage systems.

'Clean water is essential for life and something most of us in the United Kingdom take for granted. But over a billion people in the world do not have access to it...' Barbara Frost, CEO of the charity WaterAid, 2007

HOUSEHOLDS WITH TAP WATER
% of households connected to tap water, 2004

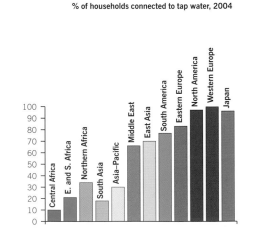

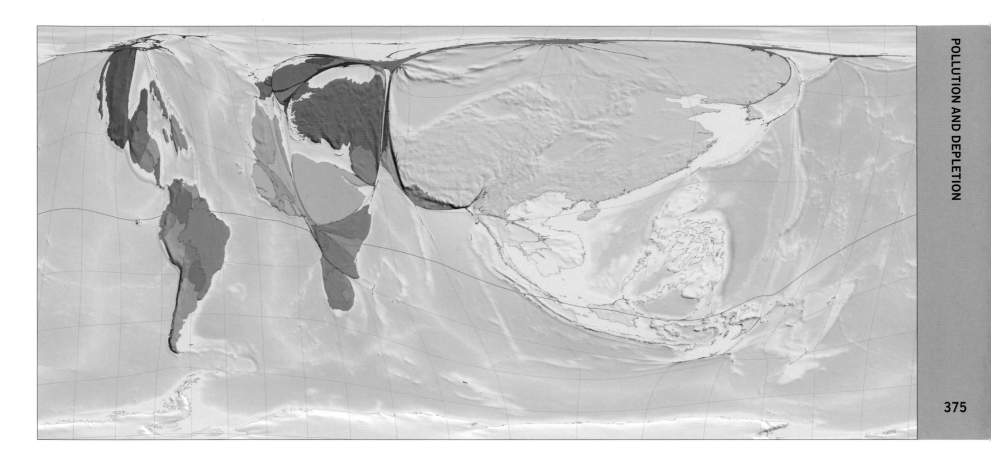

HIGHEST RATE OF NEW WATER CONNECTIONS

Rank	Territory	%ᵃ
1	Turkey	24
2	Paraguay	18
3	Guatemala	17
4	Senegal	16
5	Egypt	15
6	Syria	14
6	Philippines	14
8	Viet Nam	13
9	China	12
10	Lesotho	12

ᵃ **% of all households connecting to water supplies, 1995–2004**

TOTAL NEW WATER CONNECTIONS

Rank	Territory	People connectingᵃ
1	China	155
2	Turkey	17
3	Philippines	11
4	Egypt	11
5	Viet Nam	10
6	Indonesia	9
6	Mexico	8
8	Brazil	7
9	Russia	4
10	Thailand	4

ᵃ **million people in households connecting to water supplies, 1995–2004**

352 Increase in Piped Water

The size of each territory indicates the number of new households connected to piped water supplies between 1995 and 2004.

New water connections are one measure of improvement in basic living conditions. Only a few territories have a decreasing proportion of connected households. Territories in which the number of households with piped water fell during this interval have size zero on this map and so do not appear.

An estimated 289 million extra households were connected to piped water supplies between 1995 and 2004.

Most of these (54%) were in China. Other territories with high numbers of newly connected households were Turkey, the Philippines, Egypt and Viet Nam.

Of the territories with few newly connected households, some simply lack the resources to improve water distribution, while others had high rates of connection to begin with, so that there were few households left to be connected.

CHANGE IN WATER CONNECTION RATE
change in % of people in households connected, 1995–2004

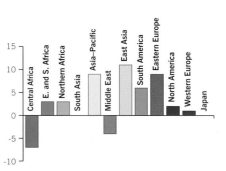

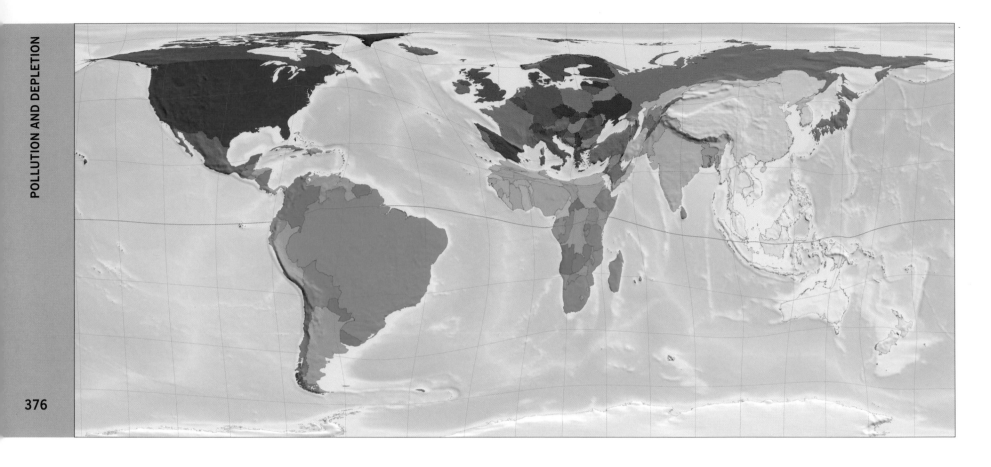

HIGHEST AND LOWEST BIOCAPACITY

Rank	Territory	Biocapacity[a]
1	Malta	699
2	Mauritius	687
3	Germany	429
4	Denmark	427
5	United Kingdom	403
6	Luxembourg	378
6	Belgium	378
8	Czech Republic	362
9	Netherlands	357
10	France	351
191	Botswana	14
192	Mali	13
193	Afghanistan	11
194	Niger	11
195	Somalia	11
196	Saudi Arabia	11
197	Namibia	11
198	Algeria	8
199	Libya	3
200	Iraq	2

[a] productivity of land measured as standardized
biologically productive hectares per sq km

353 Biocapacity

The size of each territory indicates how biologically productive it is. Biocapacity
may change as a result of both natural and human-induced effects.

Biologically productive land includes
cropland, pasture, forests and fisheries.
Biocapacity is measured in 'global
hectares', which is the productivity of the
world average hectare of land. Thus a piece
of land, of any size, with a biocapacity of 1
global hectare is as biologically productive
as the average single hectare worldwide.

The biocapacity of a territory is affected
both by physical conditions and by people's
actions. A pertinent example is Iraq, where

the Mesopotamian marshes were once part
of the Fertile Crescent, an area rich in
biodiversity and environmental conditions
conducive to the development of ancient
civilizations. Much of the marshland,
however, has now been drained and
become desert. Iraq's land is now
considered to be the least productive
in the world.

Today, 16% of the world's biocapacity
is in Brazil.

*'...land is not only the ground, it is not only a means of production
and it is not only the material reality that one knows...'*

Claudia Briones, professor of anthropology, University of Buenos Aires, 2006

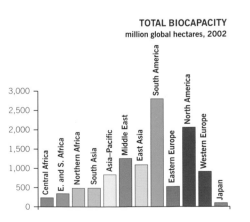

TOTAL BIOCAPACITY
million global hectares, 2002

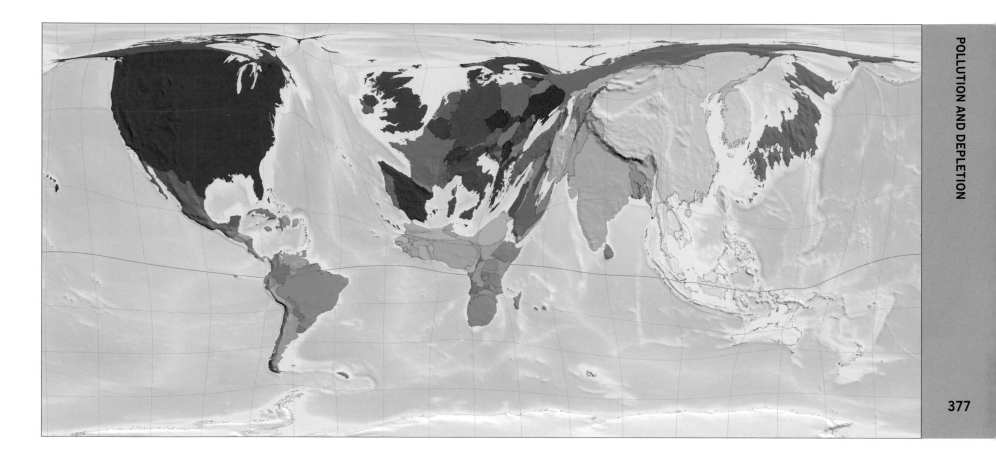

354 Ecological Footprint

The size of each territory indicates the land area needed to support its population in its accustomed mode of life.

The ecological footprint is a measure of the area needed, measured in 'global hectares' of averagely biologically productive land, to support a population's lifestyle, including their consumption of food, fuel, wood and fibres. The pollution this lifestyle generates, such as carbon dioxide emissions, is also counted as part of the footprint.

The United States, China and India have the largest ecological footprints. The footprints of China and India are large principally because these territories have large populations: the footprint per person in both territories is below average. The United States, on the other hand, has a much smaller population but a per-person footprint almost 5 times the world average.

'People consume resources and ecological services from all over the world, so their footprint is the sum of these areas, wherever they may be on the planet.' WWF, *The Living Planet Report*, 2006

AVERAGE ECOLOGICAL SHOE SIZE
person's ecological footprint in global hectares

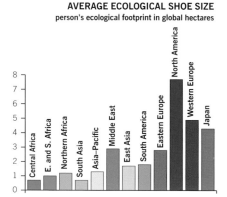

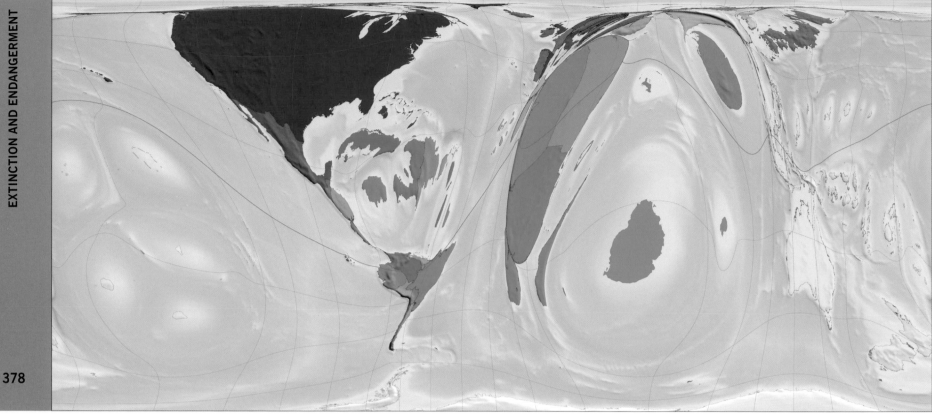

355 Extinct Species

The size of each territory indicates the numbers of species living there that became extinct between the years 1500 and 2004.

A species is considered extinct when no reasonable doubt remains that the last remaining representative of that species has died. This map shows the last known locations of the 783 now extinct species identified in the 'Red List' compiled by the International Union for Conservation of Nature and Natural Resources (IUCN). These species include mammals, birds, fish, reptiles, amphibians, invertebrates and plants.

Among other territories, the United States, Tanzania, Uganda and Mauritius all figure prominently on this map as the former homes of large numbers of now extinct species. Many island territories also appear large, because islands are often home to species that live nowhere else, putting them at greater risk of extinction than species that are more widely distributed. Island species are also particularly vulnerable to invasion by introduced predators.

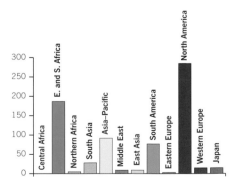

EXTINCTIONS
reported worldwide extinctions by 2004

'...current extinction is being precipitated by the widespread loss of habitats because of human activity...remaining habitats are small and fragmented, and their quality has been degraded...'

Anil Ananthaswamy, *New Scientist*, 2004

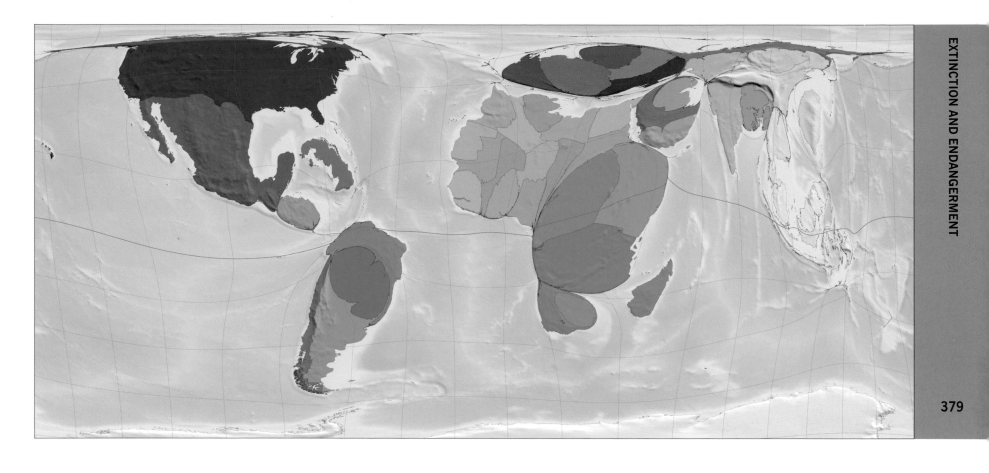

MOST AND FEWER SPECIES EXTINCT IN THE WILD BUT LIVING IN ZOOS

Rank	Territory	Species[a]
1	Western Sahara	4.4
2	United States	3.5
3	Poland	3.1
4	Paraguay	3.1
5	Mexico	3.0
6	Belarus	2.7
7	Lithuania	2.6
8	Kuwait	2.6
9	Uganda	2.6
10	Tanzania	2.4
38	Russia	0.83
39	Honduras	0.82
40	Sudan	0.76
41	China	0.71
42	Nigeria	0.66
43	Thailand	0.61
44	South Africa	0.59
45	Brazil	0.53
46	Malaysia	0.39
47	Indonesia	0.29

[a] extinctions in the wild of species surviving in zoos per 1,000 local species, 2004 (153 countries recorded no such extinctions)

356 Species Existing Only in Zoos

The size of each territory indicates the number of species that are extinct in their natural habitats and now exist only in captivity or under cultivation.

In 2004, 38 species of animal and 27 species of plant were extinct in the wild but survived in captivity or under cultivation. Most of the territories shown represent just one surviving species, but there are 12 in the United States and 8 in Mexico.

Of the 38 animal species on the map 13 are molluscs, 13 fish, 4 mammals, 4 birds, 1 crustacean, 2 amphibians and 1 insect. Of the 27 plant species, 21 were flowering plants; the other 6 were cycad, monocotyledon and algae. Continuing assessment of the risk of extinction may enable more species to be preserved in this way.

SURVIVING SPECIES EXTINCT IN THE WILD
species extinct except in captivity or cultivation by 2004

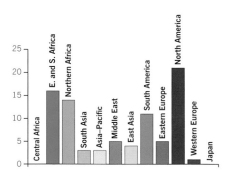

MOST AND FEWEST SPECIES AT RISK OF EXTINCTION

Rank	Territory	Species[a]
1	Mauritius	471
2	Madagascar	443
3	Ecuador	431
4	Sri Lanka	395
5	Jamaica	382
6	Cuba	373
7	Malaysia	349
8	Philippines	345
9	Fiji	341
10	United States	335
191	Estonia	30
192	Gambia	30
193	Central African Republic	30
194	Luxembourg	27
195	Liechtenstein	26
196	Niger	24
197	Botswana	19
198	Burkina Faso	18
199	San Marino	0
200	Vatican City	0

[a] species at risk from extinction per 1,000 local species, 2004

357 Species at Risk of Extinction

The size of each territory indicates the numbers of animal and plant species considered to be at risk of local extinction.

The IUCN classifies species according to their risk of dying out in a given local habitat. This map shows the distribution of the 24,495 species classified as 'critically endangered', 'endangered' or 'vulnerable' in at least one habitat in 2004. Note that this figure reflects the total number of possible local extinctions, so that a species at risk in two different territories is counted twice.

The largest numbers of species at risk are in Ecuador, the United States and Malaysia. In Ecuador 2,151 species, including many plants and 163 amphibians, are considered at risk.

Measured as a proportion of species assessed, the islands of Mauritius and Madagascar have the highest percentages of species at risk of extinction.

SPECIES LOCALLY AT RISK
species locally at risk of extinction, 2004

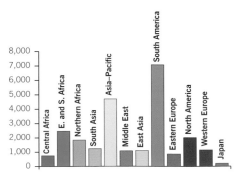

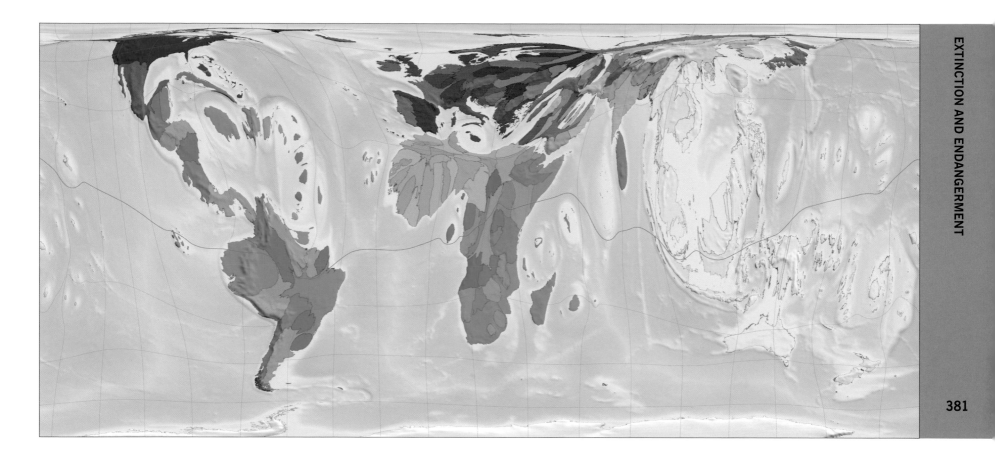

MOST AND FEWEST SPECIES AT LOWER RISK OF EXTINCTION

Rank	Territory	Species[a]
1	Tuvalu	356
2	Gaza and West Bank	281
3	Nauru	233
4	Samoa	220
5	Timor-Leste	203
6	Monaco	196
7	Kiribati	195
8	Brunei	178
9	Vanuatu	164
10	Marshall Islands	163
191	Iceland	40
192	Suriname	38
193	Bhutan	36
194	Niger	35
195	Lesotho	34
196	Burundi	33
197	Mali	33
198	Greenland	27
199	Andorra	19
200	Vatican City	0

[a] species at lower risk of extinction per 1,000 local species, 2004

358 Species at Lower Risk of Extinction

The size of each territory indicates the numbers of animal and plant species that are not under immediate threat of extinction but are considered to be at moderate risk in the future.

Worldwide, 12,639 animal and plant species are considered close to being 'threatened species' or are predicted to be at risk of extinction in the future in at least one habitat.

Indonesia is home to the largest number of species in this category: 471 in addition to the 833 seriously threatened species there that appear on Map 357. There were another 402 near-threatened species in Ecuador, in addition to the 2,151 threatened species there. Malaysia had 401 near-threatened and 892 threatened species.

Whereas Map 357, showing species at immediate risk of extinction, is dominated by just a few territories, most territories are visible on this map because the species in this lower-risk group are far more geographically widespread.

'Tylototriton shanjing is a crocodile newt...over-harvesting for use in traditional Chinese medicine is becoming a serious threat. It is also becoming popular in the international pet trade.' World Conservation Union, 2004

SPECIES LOCALLY AT LOWER RISK
species locally at a lower risk of extinction, 2004

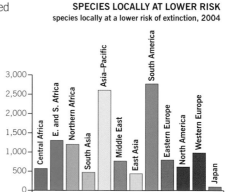

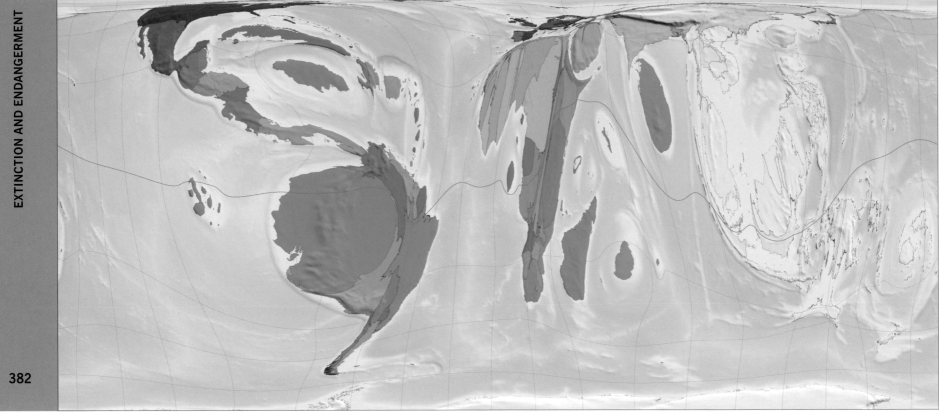

359 Plants at Risk

The size of each territory indicates the number of plant species at risk of local extinction.

In assessments carried out in 2004, 10,335 plant species were found to be at risk of becoming extinct in specific locales. This map shows where those species live. On average each territory is home to 53 locally threatened species. 8,321 species were also at risk of extinction globally.

The territory with the highest number of threatened species is Ecuador: of the 2,467 Ecuadorian species considered in the assessments, 74% were classified as being at risk. The second highest number of threatened species is in Malaysia, where 58% are considered at risk, followed by China, then Indonesia.

PLANT SPECIES AT RISK
plant species locally at risk of extinction, 2004

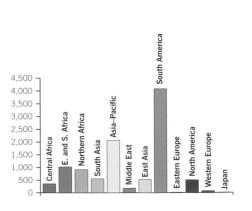

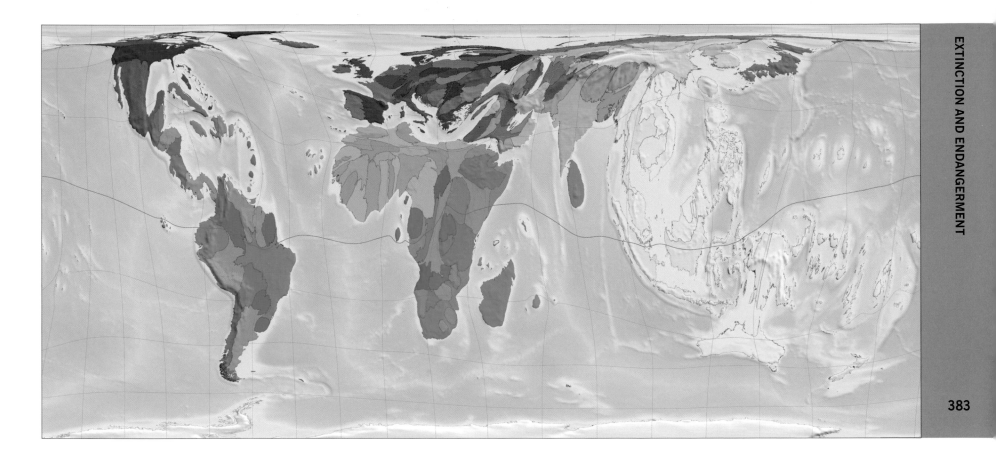

MOST MAMMAL SPECIES AT RISK OF EXTINCTION

Rank	Territory	Species[a]
1	Solomon Islands	48
2	Indonesia	42
3	Madagascar	41
4	Philippines	38
5	India	36
6	Papua New Guinea	36
7	Russia	36
8	Japan	34
9	Samoa	33
10	Laos	30
11	Micronesia	29
12	Turkmenistan	29
12	Cambodia	29
14	Ethiopia	28
15	Vanuatu	28
16	China	28
16	Iran	28
18	Australia	28
19	Bhutan	28
20	Mexico	27

[a] mammal species at risk of extinction per 1,000 local animal and plant species, 2004

360 Mammals at Risk

The size of each territory indicates the number of mammal species at risk of local extinction.

In 2004, 2,807 species of mammal were considered to be at risk of extinction in specific areas and 1,101 were also at risk of global extinction. On average, each territory was home to 14 locally threatened mammals.

Threatened mammal species include Verreaux's Sifaka lemur of Madagascar, whose deciduous forest habitat is being cleared for timber, firewood and charcoal production, and the channel islands fox of the California Channel Islands in the United States, which is hunted by golden eagles.

There are mammal species at risk in almost all territories; those with no threatened species are generally very small. The territory with the highest count of threatened species in 2004 was Indonesia, with 146. There were 85 species at risk in India and 80 in China.

MAMMALS AT RISK
mammal species locally at risk of extinction, 2004

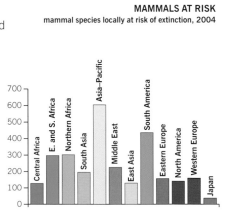

MOST BIRD SPECIES AT RISK OF EXTINCTION

Rank	Territory	Species[a]
1	Niue	182
2	Cook Islands	172
3	New Zealand	138
4	Gaza and West Bank	125
5	Timor-Leste	119
6	Samoa	77
7	Nauru	67
8	Kiribati	65
9	South Korea	60
10	Comoros	54
11	Philippines	53
12	Solomon Islands	51
13	Hong Kong (China)	49
14	Japan	49
15	Fiji	44
16	North Korea	43
17	Sao Tome and Principe	42
18	Mauritius	42
19	Mongolia	41
20	Qatar	40

[a] bird species at risk of extinction per 1,000 local animal and plant species, 2004

361 Birds at Risk

The size of each territory indicates the number of bird species at risk of local extinction.

In 2004, 3,371 species of birds were considered to be at risk of extinction in specific locales. 1,213 of these were also at risk of global extinction. In Indonesia, 121 bird species were at risk of local extinction; in Brazil, 120 species. On average there were 17 bird species at risk of local extinction in each territory.

Threatened bird species include the yellow-eared parrot, which is believed to survive only in the Central Andes of Colombia and is considered critically endangered. This species, threatened by the degradation and loss of its habitat, has been considered at risk since 1988.

'...species such as the house sparrow, snipe, starling, lapwing and corn bunting have been listed as birds of European concern, but these species have been declining in the United Kingdom's countryside for decades.' Mark Avery, conservation director of the RSPB, 2004

BIRD SPECIES AT RISK
bird species locally at risk of extinction, 2004

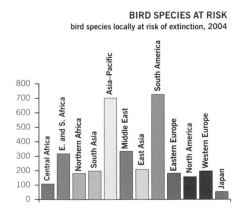

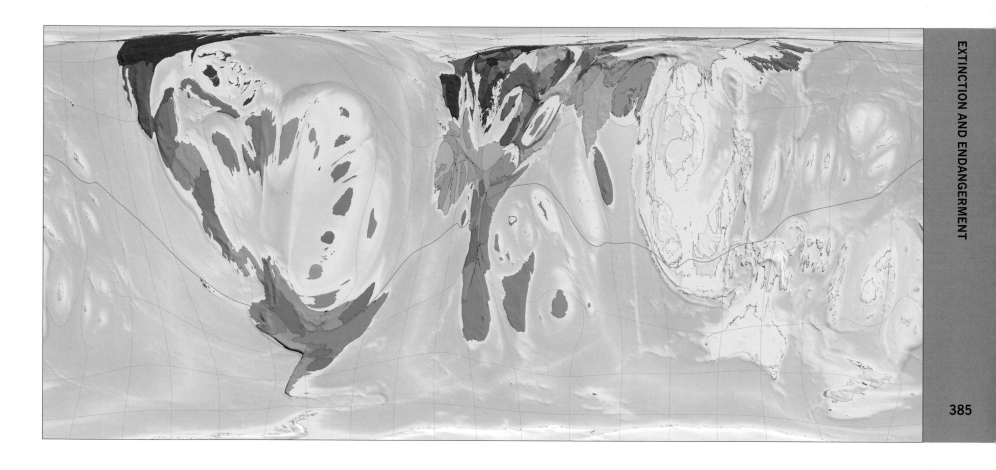

MOST REPTILE SPECIES AT RISK OF EXTINCTION

Rank	Territory	Species[a]
1	Tonga	26
2	Bangladesh	24
3	Dominican Republic	24
4	Cook Islands	23
5	Niue	23
6	New Zealand	22
7	Tuvalu	22
8	Marshall Islands	22
9	Fiji	20
10	Haiti	20
11	Grenada	20
12	St Lucia	19
13	Bahrain	18
14	Puerto Rico	18
15	Timor-Leste	17
16	Australia	17
17	Georgia	16
18	Mauritius	16
19	Antigua and Barbuda	16
20	Viet Nam	15

[a] reptile species at risk of extinction per 1,000 local animal and plant species, 2004

362 Reptiles at Risk

The size of each territory indicates the number of reptile species at risk of local extinction.

Reptiles are cold-blooded vertebrates and include tortoises, turtles, snakes, lizards, crocodiles and alligators. The word 'reptile' derives from the Latin *reptilis*, meaning 'creeping'.

In 2004, 888 reptile species were considered at risk of extinction in at least one local habitat and 304 were at risk of extinction globally. On average, each territory contained 5 reptile species in danger of local extinction, but some territories had considerably more than the average: there were 38 threatened reptile species in Australia, for example, 28 in Indonesia and 31 in China. Among the regions, South America and the Asia–Pacific and Australasia account for half of all locally threatened reptile species. Within South America most threatened species are in the northern part of the continent. Central American territories also have many threatened reptile species.

REPTILES AT RISK
reptile species locally at risk of extinction, 2004

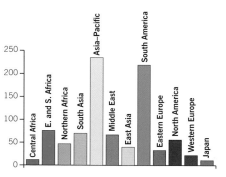

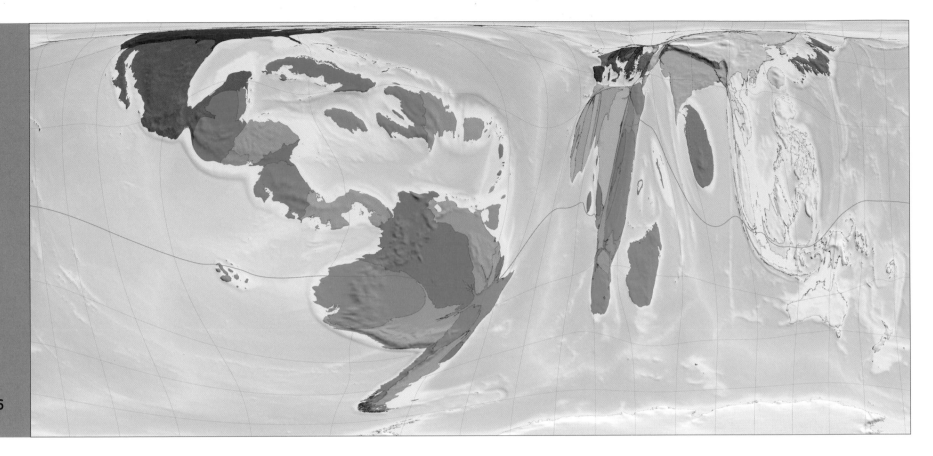

MOST AMPHIBIAN SPECIES AT RISK OF EXTINCTION

Rank	Territory	Species[a]
1	Haiti	103
2	Dominican Republic	73
3	Mexico	72
4	Cuba	64
5	Colombia	63
6	Guatemala	60
7	Madagascar	46
8	Sri Lanka	44
9	Honduras	43
10	Costa Rica	40
11	Philippines	36
12	Ecuador	33
13	Venezuela	30
14	China	30
15	Panama	30
16	Puerto Rico	29
17	India	28
18	Cameroon	26
19	Peru	26
20	Chile	25

[a] amphibian species at risk of extinction per 1,000 local animal and plant species, 2004

363 Amphibians at Risk

The size of each territory indicates the number of amphibian species at risk of local extinction.

Amphibians are vertebrates that live both in and out of water. The word in Greek, *amphibion*, means 'both lives', referring to this versatility. There are 3 orders of amphibians: frogs and toads, newts and salamanders, and the rarely seen caecilians.

In 2004 there were 2,083 species of amphibians considered at risk of extinction in at least one local habitat and 1,770 at risk of global extinction. Most endangered amphibians are in South America, where on average each territory had 33 species at risk of local extinction in 2004. The largest number, 208, were in Colombia; Ecuador came next, with 163. Six other South American territories had over 50 each. In Europe and in the Middle East and Central Asia no territory had more than 5 amphibian species at risk and worldwide there were 102 territories that recorded no amphibians at risk at all.

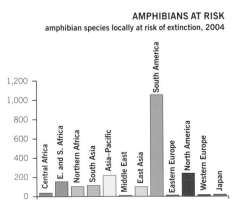

AMPHIBIANS AT RISK
amphibian species locally at risk of extinction, 2004

MOST FISH SPECIES AT RISK OF EXTINCTION

Rank	Territory	Species[a]
1	Monaco	161
2	Tuvalu	111
3	Nauru	100
4	Marshall Islands	76
5	Cape Verde	70
6	Niue	68
7	Grenada	60
8	Madagascar	55
9	Kiribati	52
10	Tonga	51
11	Timor-Leste	51
12	Croatia	51
13	Western Sahara	49
14	Cook Islands	46
15	United States	45
16	Samoa	44
17	Mexico	40
18	Greece	40
19	Maldives	39
20	Turkey	38

[a] fish species at risk of extinction per 1,000 local animal and plant species, 2004

364 Fish at Risk

The size of each territory indicates the number of fish species at risk of local extinction.

There are about 29,300 known species of fish. Of these, about 2,545 are considered at risk of local extinction and 800 at risk of global extinction. Both fish that live in fresh water and those that live in salt water are included in this map: salt-water species are assigned to the territories in whose ocean waters they swim.

There are endangered fish species in almost all territories. The largest numbers are in the United States, Mexico, Indonesia and Australia. There are a variety of threats facing fish species. The humphead wrasse, bowmouth guitarfish and largetooth sawfish, for example, are all threatened by fishing, getting caught inadvertently in nets intended to catch other species, while the damba mipentina is threatened by competition from invasive species and habitat loss, as well as by fishing.

'...cyanide and heavy metal-contaminated liquid spilled into the Lupus stream, reaching the Szamos, Tisza, and finally Danube rivers and killing hundreds of tonnes of fish...' Eva Kaszala, 'The Tisza River Spill', 2002

FISH SPECIES AT RISK
fish species locally at risk of extinction, 2004

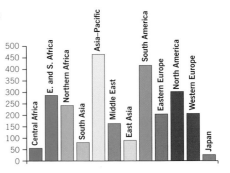

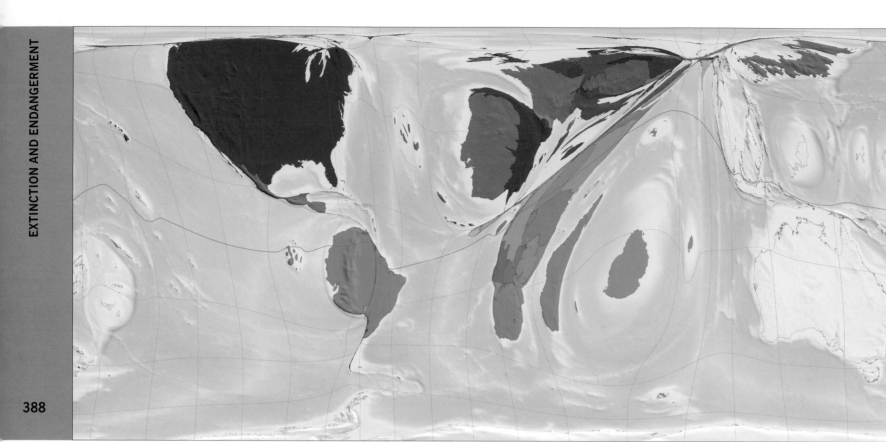

365 Molluscs at Risk

The size of each territory indicates the number of mollusc species at risk of local extinction.

Molluscs include slugs and snails, oysters, squid, octopus and cuttlefish. These creatures live mainly in the sea, although some live in fresh water or on land. There are roughly 70,000 recognized species of mollusc. 970 species are considered to be at risk of extinction in at least one local habitat. Over half of the endangered mollusc species are in the United States, Australia, Portugal and Ecuador. One critically endangered mollusc is the Dlinza Forest pinwheel snail, which lives only in the very small Dlinza Forest in South Africa; its livelihood is endangered by the effects of extreme weather and climate change.

MOLLUSC SPECIES AT RISK
mollusc species locally at risk of extinction, 2004

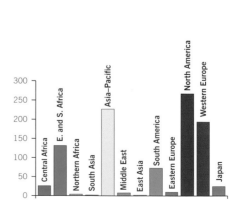

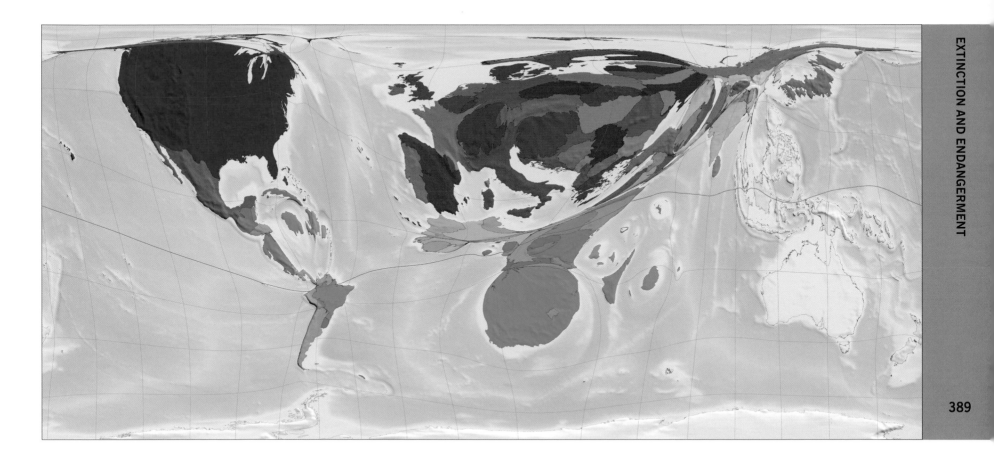

MOST OTHER INVERTEBRATE SPECIES AT RISK OF EXTINCTION

Rank	Territory	Species[a]
1	United States	88
2	Slovenia	79
3	South Africa	64
4	Italy	52
5	Switzerland	52
6	Australia	48
7	Hungary	42
8	Spain	37
9	Romania	36
10	France	34
11	Austria	34
12	Serbia and Montenegro	31
13	Czech Republic	29
14	Germany	29
15	Russia	24
16	Slovakia	24
17	Ukraine	24
18	Bosnia and Herzegovina	23
19	Georgia	23
20	Poland	22

[a] other invertebrate species at risk of extinction per 1,000 local animal and plant species, 2004

366 Invertebrates at Risk

The size of each territory indicates the number of invertebrate species (not including molluscs) at risk of local extinction.

Invertebrates are animal species with no backbone. (Some have an exoskeleton or shell.) Ants, spiders, dragonflies, crustaceans and stick insects are all examples of invertebrates. Molluscs are also invertebrates but are not included on this map, being shown separately on Map 365.

Worldwide, 1,496 species of invertebrates, excluding molluscs, are considered at risk of extinction in at least one locale, and 1,018 are at risk of extinction globally. The territory with the largest number of invertebrate species at risk is the United States, with 300 threatened species, followed by South Africa (109) and Australia (107).

'Invertebrates play important roles in nutrient cycling and in creating and maintaining biological diversity...audiences may overlook their significance because many invertebrate species are small or cryptic.'

R. T. Ryti, 'An overview of the role of invertebrates in risk assessment', 2000

INVERTEBRATES AT RISK
invertebrate species locally at risk of extinction per 1,000 animal and plant species, 2004

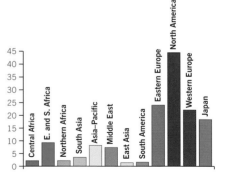

World Beliefs

Faiths and Beliefs

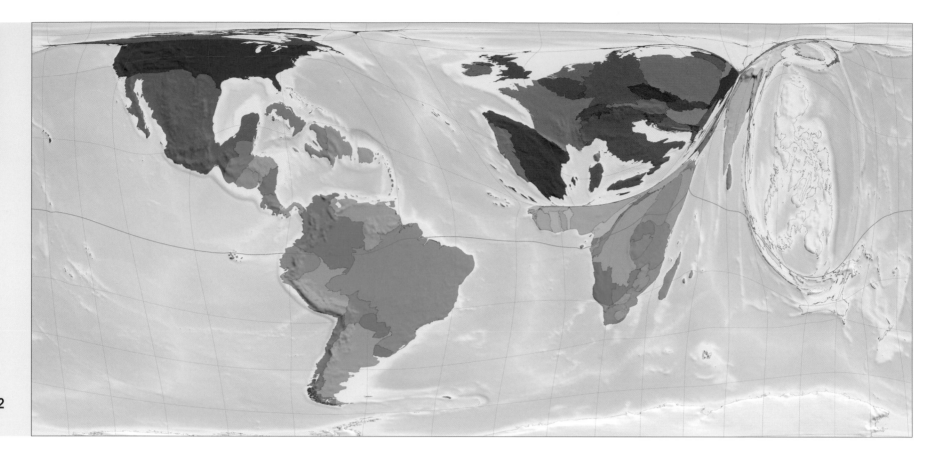

Rank	Territory	%ᵃ
1	Vatican City	100.0
2	Malta	96.9
3	Poland	91.8
4	Andorra	91.1
5	San Marino	90.6
6	Ireland	90.4
7	Spain	89.3
8	Colombia	89.2
9	Ecuador	88.5
10	Luxembourg	87.6
11	Mexico	86.2
12	Paraguay	85.7
13	Sao Tome and Principe	85.6
14	Portugal	85.5
15	Dominican Republic	85.1
15	Venezuela	85.1
17	Cape Verde	84.8
17	Slovenia	84.8
19	Seychelles	84.6
20	Monaco	83.7

ᵃ **% of population who are Catholic Christians**

367 Catholic Christians

The size of each territory indicates the number of adherents
to the Catholic denomination of Christianity.

Just under a billion people are believed to
be adherents to the Catholic denomination
of the religion of Christianity, half of them
living in North or South America.

An adherent to Christianity is known as
a Christian, meaning one who follows the
teachings of Jesus Christ. Christianity is the
largest religion in the world and its largest
denomination is the Catholic Church, also
called the Roman Catholic Church after the
location of its central authority. The word
'Catholic' means 'universal', denoting the

undivided Christian church; the term was
retained after the Reformation in Europe
in the 16th century to distinguish the
Roman church from the various Protestant
denominations that were formed at this time
and subsequently. The collected sacred
texts of Christianity are known as the Bible.

Catholic Christians account for some
70% of the population in South America,
50% in Central Africa, 40% in North
America, 50% in Western Europe and
25% in Eastern Europe.

CATHOLIC CHRISTIANS
% of population

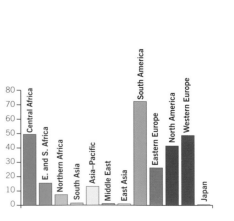

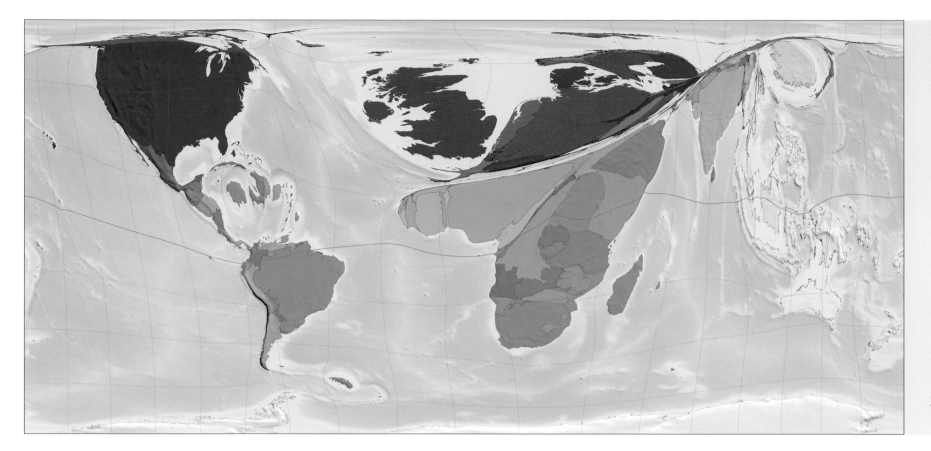

368 Protestant Christians

The size of each territory indicates the number of adherents to the Protestant denominations of Christianity.

Over 430 million people are believed to be adherents to the various Protestant denominations of Christianity. These denominations include Anglicans, Jehovah's Witnesses, Lutherans, Mormons and Quakers. Groups of Christians that became Protestant denominations first split from the Catholic Church in the 16th century, and new churches have continued to be created by further splits since then.

Almost a third of the populations of Central Africa and East and Southern Africa adhere to various Protestant Christian religions, as do a quarter of the populations of Western Europe and North America.

'No one after lighting a lamp puts it under the bushel basket, but on the lampstand, and it gives light to all in the house. In the same way, let your light shine before others, so that they may see your good works and give glory to your Father in heaven.'

The Bible, Matthew 5: 15–16

PROTESTANT CHRISTIANS
millions

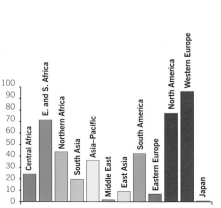

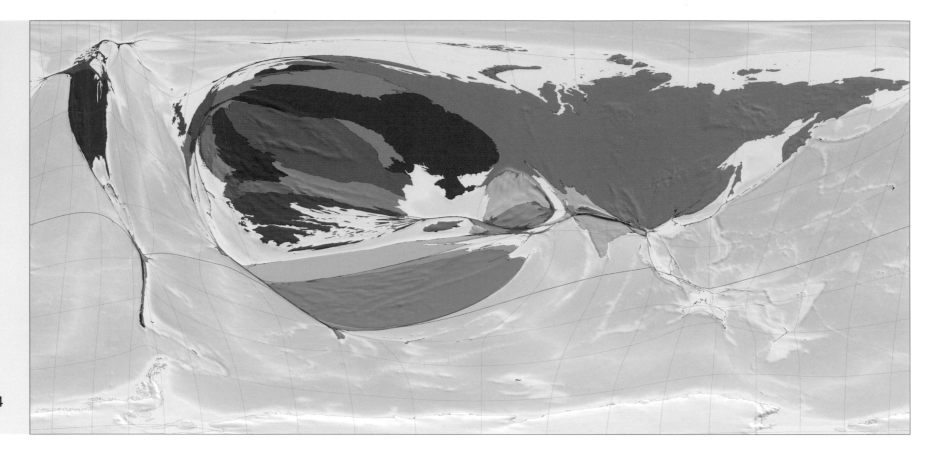

ᵃ % of the population who are Orthodox Christians

369 Orthodox Christians

The size of each territory indicates the number of adherents
to the Orthodox denominations of Christianity.

Around 230 million people are recorded
as being adherents to the various Orthodox
Christian denominations that by the 11th
century had established themselves
separately from the Western Catholic
Church. This map includes adherents of
both the main denominations of Orthodox
Christianity, the Eastern (Chalcedonian)
and Oriental, as well as smaller
denominations such as Nestorianism.

A quarter of all people in Eastern Europe
and the Middle East and Central Asia
(including Russia) adhere to Orthodox
Christian denominations, as do a tenth of
those in East and Southern Africa and 3%
of those in Western Europe. Some 78%
of Orthodox Christians in Western Europe
live in Greece, where 88% of the
population are adherents

*'God is the only eternal Being. Beyond time, space and all
limitations, He abides without a beginning and without an end.'*

The Ethiopian Orthodox Tewahedo Church Faith and Order, 2003

ORTHODOX CHRISTIANS
% of population

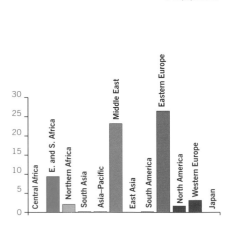

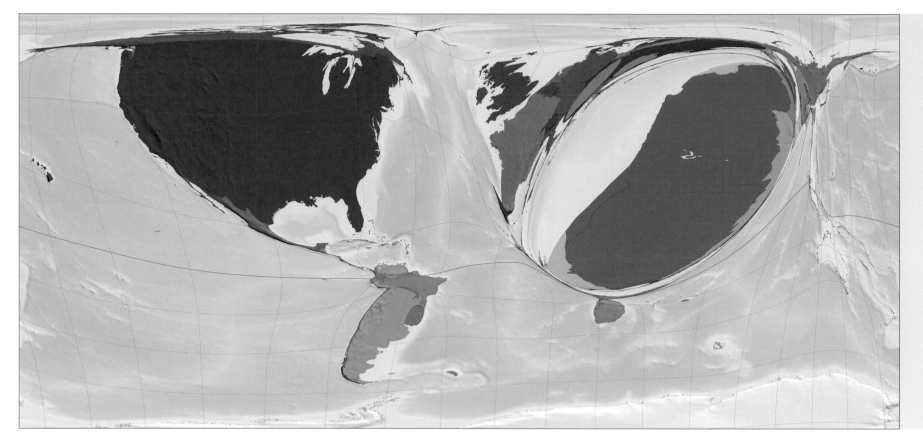

Rank	Territory	%ᵃ
1	Israel	71.39
2	Gaza and West Bank	11.82
3	United States	1.92
4	Monaco	1.70
5	Argentina	1.32
6	Canada	1.30
7	Uruguay	1.23
8	Moldova	1.11
9	Belize	1.08
10	Hungary	0.97
11	France	0.96
12	Australia	0.49
13	United Kingdom	0.47
14	Georgia	0.44
15	Latvia	0.41
16	Ukraine	0.39
17	Andorra	0.34
18	Azerbaijan	0.32
19	Belgium	0.27
19	Germany	0.27

ᵃ % of the population who are Adherents to
Judaism

370 Adherents to Judaism

The size of each territory indicates the number of adherents
to the Jewish faith, also known as Judaism.

There are just over 14 million adherents to
the Jewish faith, a religion originating with
the Israelites of the ancient Middle East.
Judaism is the oldest of the Abrahamic
religions, the other major ones being
Christianity and Islam. The most holy of the
writings of Judaism are known as the Torah.

The highest proportion of adherents to
Judaism per head of the population is
found in Israel (71%), followed by Gaza
and the West Bank (12%). Note that in
creating this cartogram Israel and Gaza
were rescaled as a single unit and so are
coloured the same, although the border
between them is shown. The West Bank,
on the other hand, is treated as a separate
territory.

Just 2% of the population of North
America and 1% of the population of the
Middle East adhere to Judaism.

*'When God began to create heaven and earth – the earth being
unformed and void, with darkness over the surface of the deep and
a wind from God sweeping over the water – God said, "Let there be
light"; and there was light.'* The Torah, Genesis 1. 1–3

ADHERENTS TO JUDAISM
millions

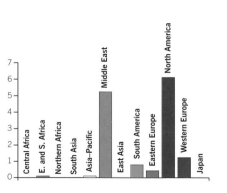

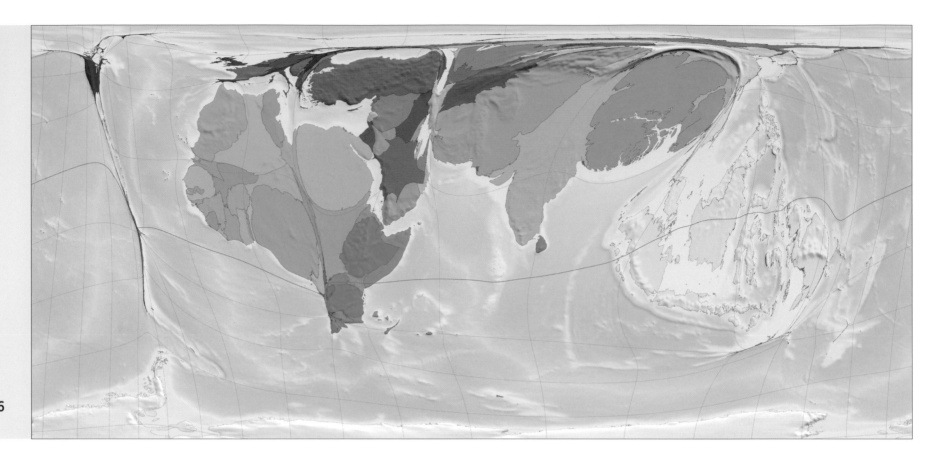

HIGHEST PROPORTIONS OF SUNNI MUSLIMS

Rank	Territory	%[a]
1	Mauritania	98.1
2	Somalia	98.0
3	Comoros	97.3
4	Tunisia	97.0
5	Morocco	96.6
6	Western Sahara	96.4
7	Djibouti	96.0
8	Libya	95.5
9	Maldives	93.5
10	Algeria	93.0
11	Jordan	92.0
12	Niger	87.7
13	Turkmenistan	84.7
14	Gambia	84.5
15	Egypt	84.1
16	Senegal	83.3
17	Saudi Arabia	83.0
18	Bangladesh	82.7
19	Afghanistan	81.6
20	Tajikistan	79.8

[a] % of the population who are Sunni Muslims

371 Sunni Muslims

The size of each territory indicates the number of adherents to the Sunni denomination of Islam.

Just over a billion people are believed to be adherents to (practising followers of) Sunni Islam. An adherent to Islam is known as a Muslim. Islam is the second largest religion in the world, after Christianity, and Sunni Muslims form the largest denomination. The word 'Islam' means 'submission', or the total surrender of oneself to God (Allah). The word 'Sunni' is derived from 'Sunnah', meaning 'words and actions', in this case those of Muhammad, the Prophet of Islam. The central religious text of Islam is the Qur'an.

Sunni Muslims account for two-thirds of the population in Northern Africa, 35% in the Middle East, 30% in the Asia–Pacific and Australasia region, 25% in both South Asia and Eastern Europe (including Turkey), and 22% in East and Southern Africa.

'And do not mix up the truth with the falsehood, nor hide the truth while you know [it]. And keep up prayer and pay the poor-rate and bow down with those who bow down.' The Holy Qur'an, 2. 42–3

SUNNI MUSLIMS
millions

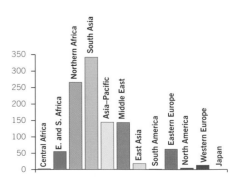

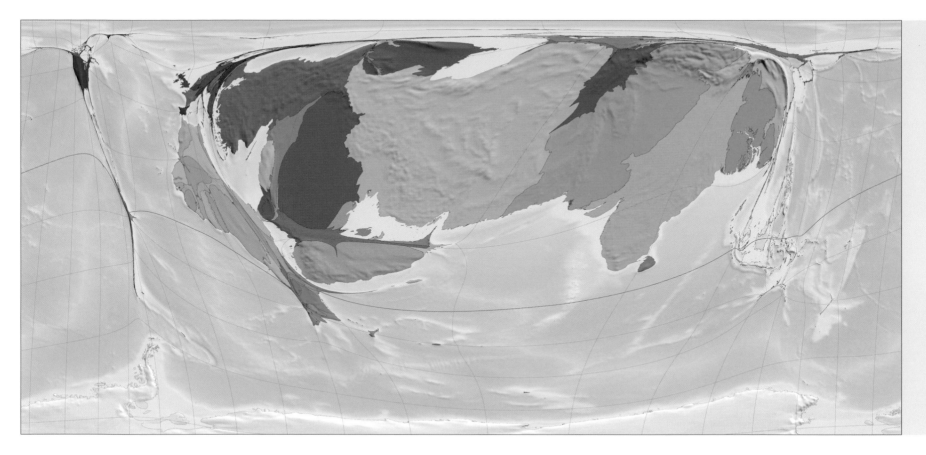

HIGHEST PROPORTIONS OF SHIA MUSLIMS

Rank	Territory	%ᵃ
1	Iran	87.5
2	Azerbaijan	69.6
3	Iraq	62.9
4	Bahrain	58.5
5	Yemen	41.6
6	Lebanon	35.6
7	Kuwait	30.0
8	Turkey	19.5
9	Pakistan	19.2
10	Afghanistan	17.9
11	Syria	13.9
12	Gaza and West Bank	12.8
13	United Arab Emirates	11.4
14	Albania	9.7
15	Chad	9.2
15	Saudi Arabia	9.2
17	Brunei	8.8
18	Togo	8.3
19	Qatar	8.2
20	Mali	8.1

ᵃ % of the population who are Shia Muslims

372 Shia Muslims

The size of each territory indicates the number of adherents to the Shia denomination of Islam.

Just over 200 million people are estimated to be adherents to the Shia denomination of Islam, half of them living in the Middle East. Shia Islam is the second largest denomination of the Islamic faith, after Sunni Islam, and comprises some 15% of Muslims worldwide. Shia Muslims adhere to the teachings of the Prophet Muhammad, but differ from Sunni Muslims in also following the religious guidance of his family, who are referred to as the Ahl al-Bayt, and certain of his descendants, known as Shia imams. Shia Muslims constitute a quarter of the population of the Middle East and Central Asia, but no more than 6% in any other region. Just over half of all Shia Muslims in the Middle East live in the Islamic Republic of Iran.

SHIA MUSLIMS
% of population

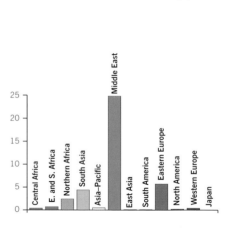

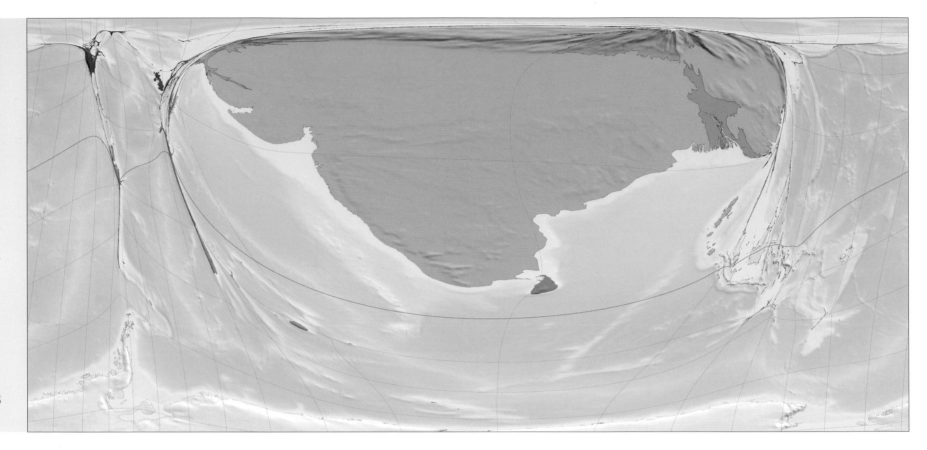

373 Hindus

The size of each territory indicates the number of adherents to Hinduism.

There are over 820 million Hindus, almost all of them in India or Nepal. Hinduism is one of the oldest major religions that is still practised. It is a conglomerate of diverse beliefs and traditions, and has no single founder. It is the world's third largest religion after Christianity and Islam, and the largest of the Dharmic religions, which originated in the Indian subcontinent and also include Jainism, Buddhism and Sikhism.

Hindus comprise almost 60% of the population of South Asia, but no more than 2% of the population of any other region. Outside India and Nepal, the largest numbers of Hindus live, in descending order, in Bangladesh, Indonesia, Sri Lanka, Pakistan, Malaysia, the United States and South Africa.

HINDUS
% of population

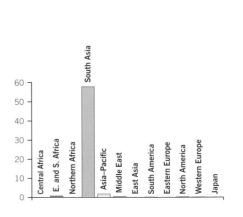

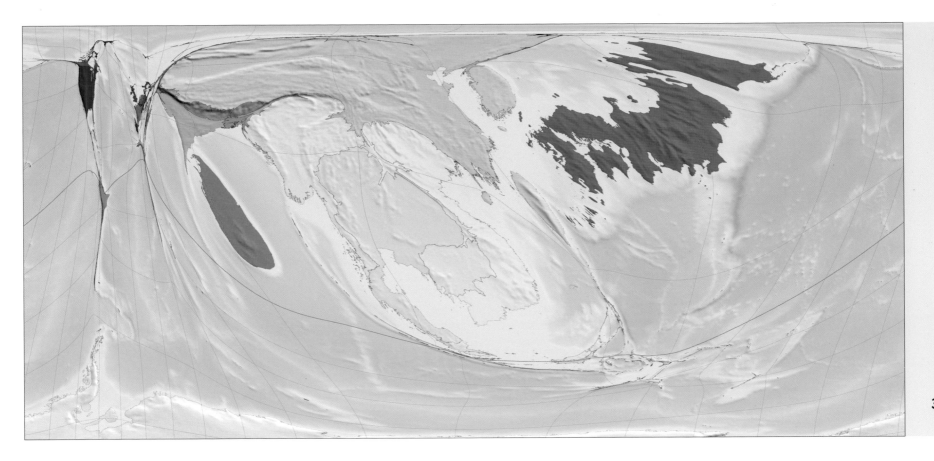

HIGHEST PROPORTIONS OF BUDDHISTS

Rank	Territory	%[a]
1	Cambodia	85.3
2	Thailand	83.0
3	Myanmar	73.7
4	Sri Lanka	68.4
5	Bhutan	65.9
6	Japan	55.0
7	Viet Nam	48.2
8	Laos	43.1
9	Mongolia	22.5
10	Taiwan	20.8
11	South Korea	15.1
12	Singapore	14.1
13	Hong Kong (China)	10.0
14	Brunei	9.7
15	Nepal	9.5
16	China	8.5
17	Malaysia	6.8
18	Australia	2.1
18	Lebanon	2.1
20	United Arab Emirates	2.0

[a] % of the population who are Buddhists

374 Buddhists

The size of each territory indicates the number of adherents to Buddhism.

Around 370 million people are adherents to various Buddhist traditions, making Buddhism collectively the fourth largest religion in the world. There are, however, many variations in teaching within Buddhism.

Buddhism began with the teachings of Siddhartha Gautama (the Buddha) around 2,500 years ago. Buddhist traditions focus on personal spiritual development.

Some 55% of people in Japan, 26% in the Asia–Pacific and Australasia region, and 8% in East Asia are adherents to Buddhist traditions. The largest number of Buddhists is found in China (107 million), followed by Japan (70 million), Thailand (52 million), Viet Nam (39 million), Myanmar (36 million) and Sri Lanka (13 million).

BUDDHISTS
% of population

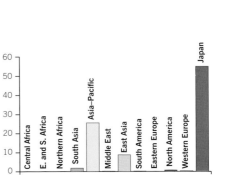

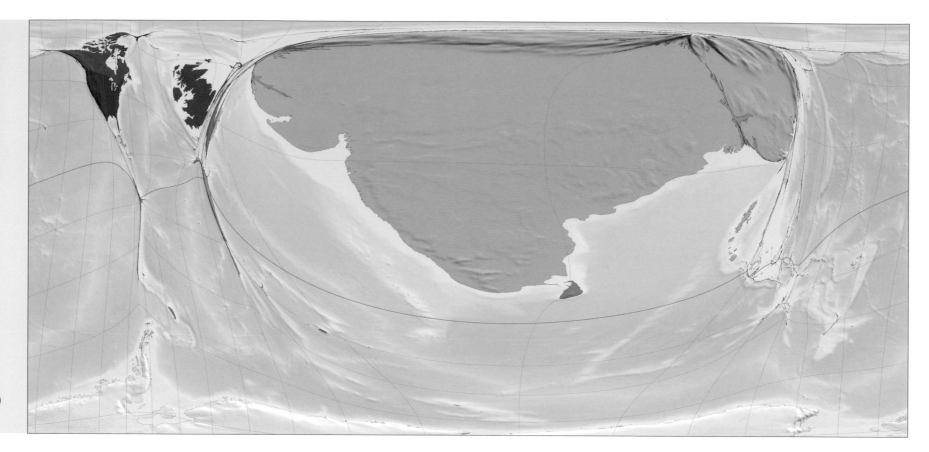

375 Sikhs

The size of each territory indicates the number of adherents to Sikhism.

There are estimated to be about 24 million Sikhs, 95% of whom live in India, although they constitute only about 2% of the Indian population. Sikhism is the fifth largest religion in the world after Christianity, Islam, Hinduism and Buddhism.

Sikhism was founded on the teachings of Guru Nanak Dev and succeeding gurus in northern India between the 15th and early 18th centuries. The word 'Sikh' comes from one of two roots in Sanskrit, either *śisya*, meaning 'disciple' or 'learner', or *śiksa*, meaning 'instruction'. Of the relatively small number of Sikhs not living in India, just over a quarter live in the United Kingdom, just under a quarter in Canada and about a fifth in the United States.

'Sikhism preaches a message of devotion and remembrance of God at all times, truthful living, equality of mankind and denounces superstitions and blind rituals.'

www.sikhs.org, 2003

SIKHS
millions

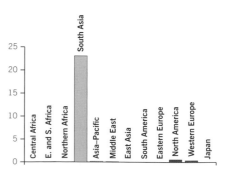

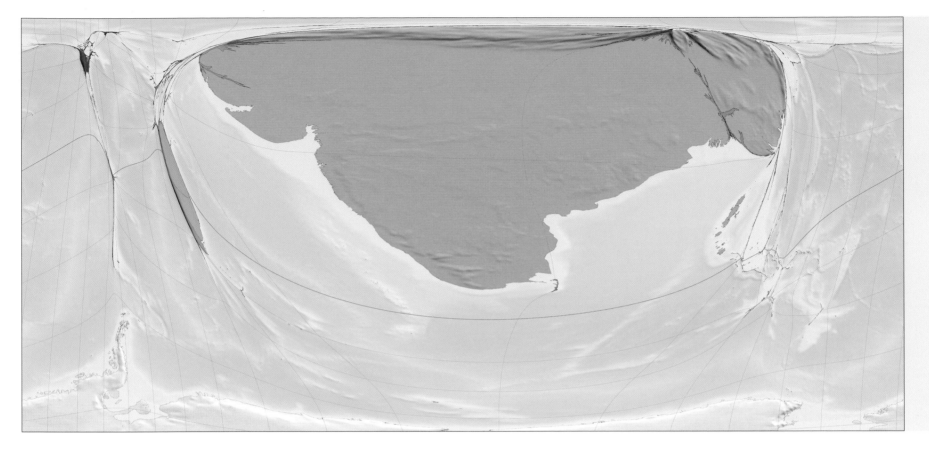

HIGHEST PROPORTIONS OF JAINS

Rank	Territory[a]	%[b]
1	India	0.517
2	Kenya	0.192
3	Seychelles	0.041
4	Nepal	0.025
5	Tanzania	0.023
6	Uganda	0.009
6	Myanmar	0.006
8	Australia	0.004
8	United States	0.003
10	France	0.002
11	Yemen	0.001

[a] In the sources to which we had access,
Jains were counted in only these 11 territories
[b] % of the population who are Jains

376 Jains

The size of each territory indicates the number of adherents to Jainism.

There are about 5.5 million Jains in the world, 98% of them in India. Jains follow one of the oldest of the world's religions, based on the teachings of the 24 Jinas (conquerors), also known as Tirthankaras. The 23rd Tirthankara is the earliest who can be dated and lived about 3,000 years ago.

Jainism advocates non-violence towards all living beings and self-effort directed at personal spiritual development. Jainism influenced the development of Buddhism and Hinduism, in turn influencing many other religions up to the youngest found today. The fundamental principles on which it is based are known as *tattva*.

Outside India, the largest groups of Jains are found in Kenya (61,000), Tanzania (8,000), the United States (7,000), Nepal (6,000), Myanmar (3,000) and Uganda (2,000).

'By undertaking these wholesome activities, we acquire punya or good karmas. ... When punya matures, it brings forth worldly comfort and happiness.'

Third Tattva, undated

JAINS
% of population

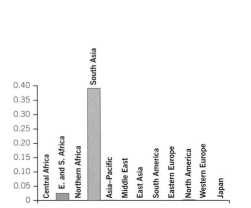

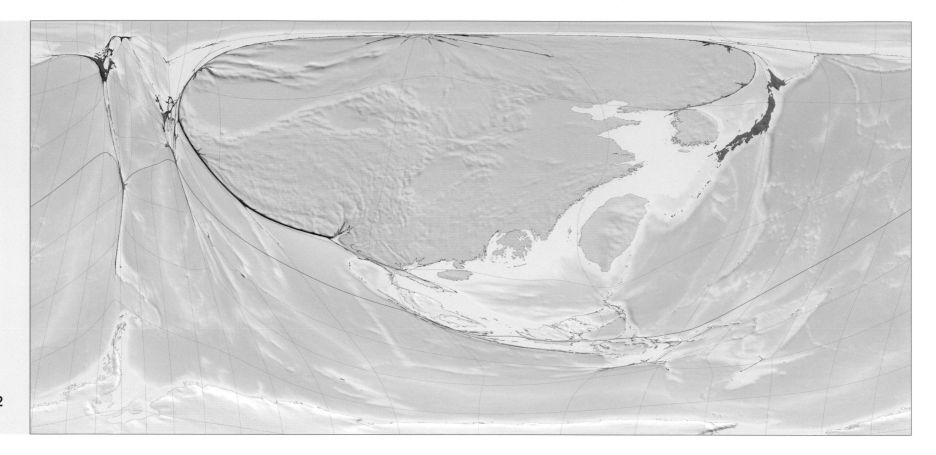

377 Taoic Religions

The size of each territory indicates the number of adherents
to various Taoic religions.

About 390 million people are adherents to
the various Taoic religions, which include
Confucianism, Shintoism, Taoism itself and
Chinese Universalism. Nine out of every
ten adherents to Taoic religions live in
China, though most of the 2.8 million
adherents to Shintoism live in Japan.

Taoic religions focus on the East Asian
concept of Tao, which can be translated
roughly as 'the path' or 'the way'. According
to the Reform Taoist Congregation, Tao

'refers to a non-sentient, impersonal
power that surrounds and flows through
all things, living and non-living'. Taoism
itself was founded by Lao-Tse, a
contemporary of Confucius (5th–6th
centuries BC), and is most common in
Taiwan. Confucianism is concentrated in
China, as is Chinese Universalism, which
has by far the most adherents among the
Taoic religions (380 million).

ADHERENTS TO TAOIC RELIGIONS
millions

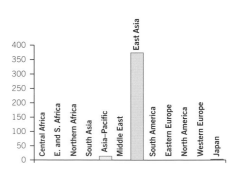

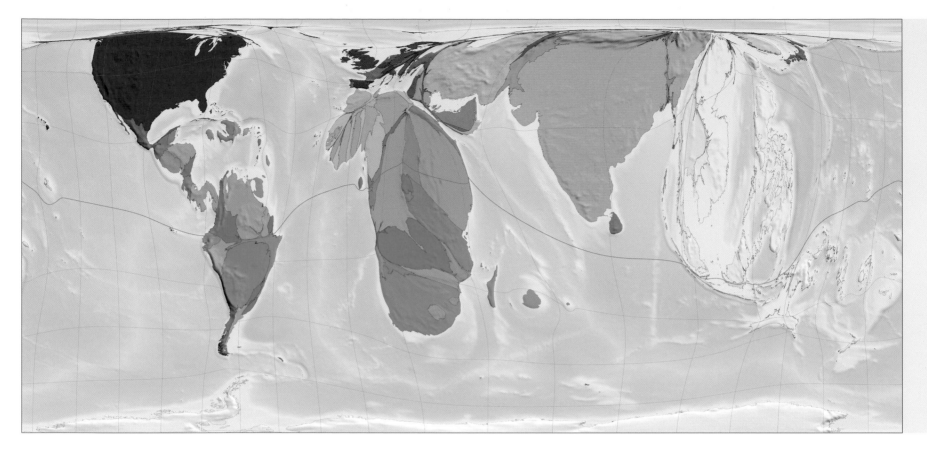

HIGHEST PROPORTIONS OF BAHA'I

Rank	Territory	%ᵃ
1	Nauru	9.6
2	Tonga	6.7
3	Kiribati	5.2
4	Tuvalu	5.0
5	Vanuatu	2.8
6	Belize	2.5
7	Sao Tome and Principe	2.4
8	Samoa	2.3
8	United Arab Emirates	2.3
10	Bolivia	2.2
11	Zambia	2.0
12	Mauritius	1.9
13	Dominica	1.7
13	Guyana	1.7
15	Marshall Islands	1.6
15	Micronesia	1.6
17	Niue	1.5
17	St Vincent and the Grenadines	1.5
17	Suriname	1.5
20	Panama	1.3

ᵃ % of the population who are Baha'i

378 Baha'i

The size of each territory indicates the number of adherents to the Baha'i religion.

There are just over 7 million adherents to the Baha'i faith worldwide. Baha'i is one of the youngest of the world's major religions, founded in 19th-century Persia (now Iran) by Bahá'u'lláh (1817–92). Baha'i teachings include abandoning all prejudice, eliminating extremes of poverty and wealth, and realizing universal education.

By region the Baha'i faith is most popular in Central Africa, where 0.5% of the population are adherents. By country, however, the largest numbers of followers are found in India (1.8 million), the United States (800,000), Iran (400,000), Viet Nam (350,000), Kenya (320,000), South Africa (300,000), the Philippines (230,000), the Democratic Republic of Congo (220,000), Zambia (210,000), Bolivia (190,000), Tanzania (170,000) and Venezuela (150,000). Overall the Baha'i religion is one of the most evenly distributed worldwide.

'The earth is but one country and mankind its citizens.'

Bahá'u'lláh, undated

BAHA'I
thousands

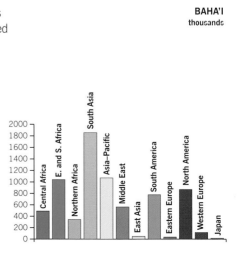

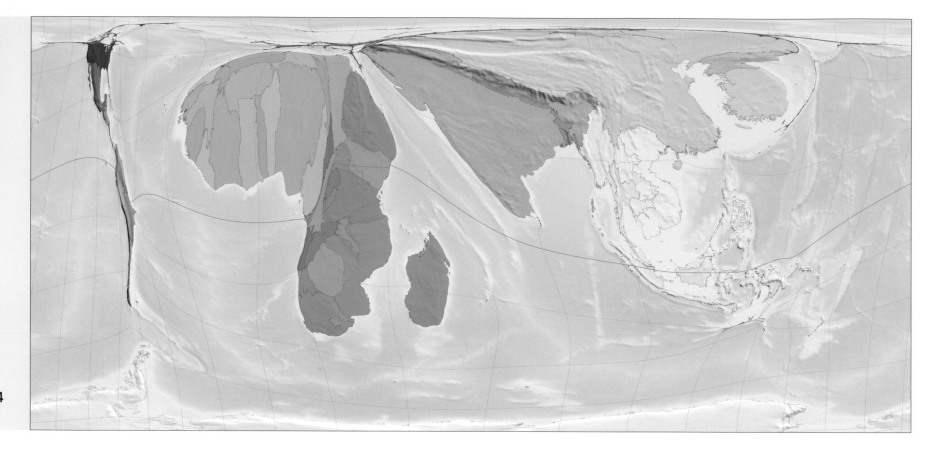

404

HIGHEST PROPORTIONS OF PAGANS

Rank	Territory	%ª
1	Benin	50
1	Mozambique	50
3	Laos	49
4	Madagascar	47
5	Guinea-Bissau	45
6	Liberia	42
7	Sierra Leone	39
8	Côte d'Ivoire	37
9	Botswana	35
9	Togo	35
11	Burkina Faso	32
11	Mongolia	32
13	Zimbabwe	29
14	Guinea	28
15	Cameroon	22
15	Ghana	22
17	Central African Republic	19
18	Mali	16
18	South Korea	16
20	Tanzania	15

ª % of the population who are pagans

379 Pagans

The size of each territory indicates the number of adherents to various pagan religions.

There are thought to be about 240 million people worldwide who adhere to pagan religions, including animists, spirit-worshippers, ancestor-worshippers, shamanists, polytheists, pantheists, and members of cargo cults and tribal messianic movements. Many of these religions are confined to a single tribe or people.

Virtually all pagans live in Africa and Asia. They constitute 1 in 7 people in East and Southern Africa, 1 in 9 in Northern Africa, 1 in 18 in Central Africa, 1 in 21 in the Asia–Pacific and Australasia region and in East Asia and 1 in 31 in South Asia. By contrast, they account for only 1 in 167 people in North America, 1 in 189 in South America, 1 in 2,500 in Western Europe, 1 in 12,000 in Eastern Europe and 1 in 13,000 in Japan.

'There is a respect for all of life and usually a desire to participate with rather than to dominate other beings.'
Prudence Jones, The Pagan Federation, 2000s

PAGANS % of all pagans

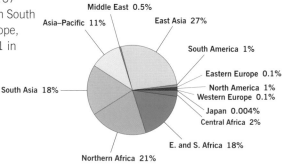

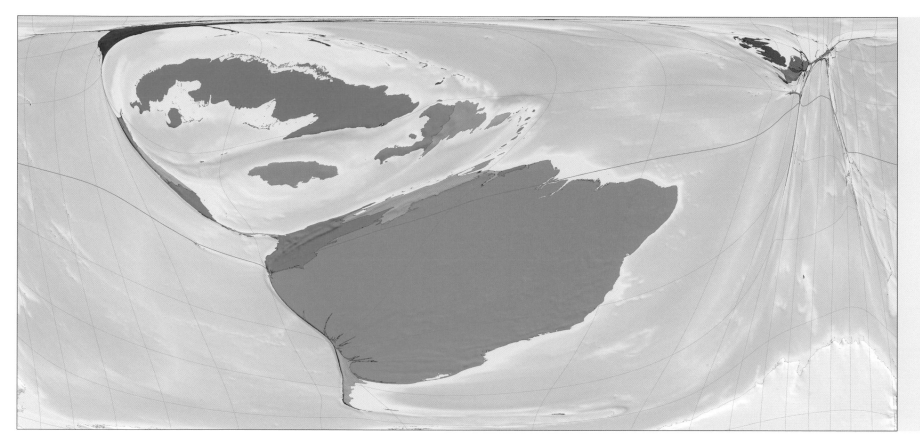

HIGHEST PROPORTIONS OF SPIRITUALISTS

Rank	Territory	%[a]
1	Cuba	17.2
2	Jamaica	10.1
3	Brazil	4.9
4	Antigua and Barbuda	3.6
4	Suriname	3.6
6	Haiti	2.7
7	Dominica	2.6
8	Dominican Republic	2.2
9	Bahamas	1.9
10	St Vincent and the Grenadines	1.8
11	St Lucia	1.7
12	Nicaragua	1.5
13	Trinidad and Tobago	1.4
14	Grenada	1.3
14	Guyana	1.3
14	St Kitts–Nevis	1.3
17	Venezuela	1.1
18	Belize	1.0
18	Colombia	1.0
20	Honduras	0.9

[a] % of the population who are spiritualists

380 Spiritualists

The size of each territory indicates the number of adherents to spiritualism.

Spiritualism has some 13 million adherents. Spiritualism is a philosophical doctrine, established in France in the mid-19th century on the basis of a series of books written by the French educator Hippolyte Léon Denizard Rivail (1804–69) under the pseudonym Allan Kardec. On the lid of Rivail's burial chamber in Paris is written: 'Naître, mourir, renaître encore et progresser sans cesse, telle est la loi' (translation below).

Over 97% of all spiritualists live in South America, the majority in Brazil (8 million), followed by Cuba (2 million). Outside South America, most live in the United States (140,000), the United Kingdom (82,000) and France (31,000).

'To be born, die, still to reappear and progress unceasingly, such is the law.'

Tombstone of Rivail / Kardec, 1869

SPIRITUALISTS
% of population

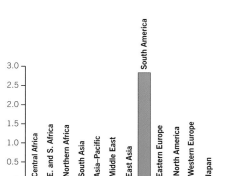

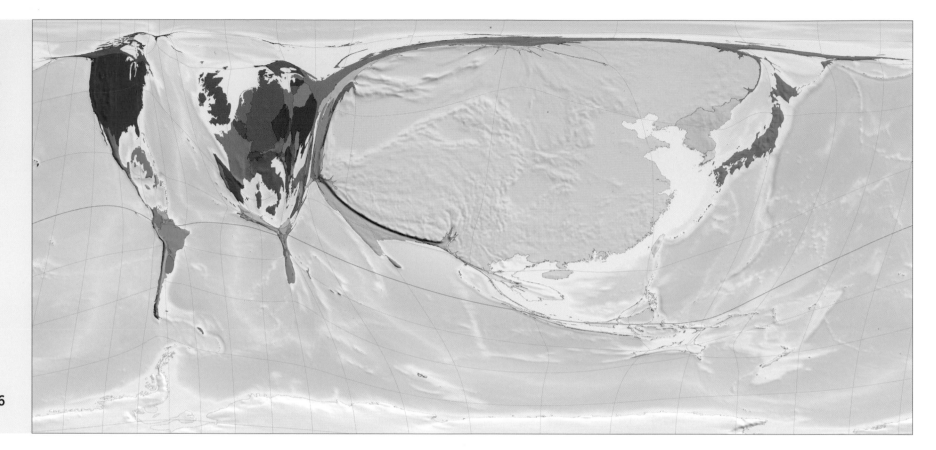

HIGHEST PROPORTIONS OF AGNOSTICS

Rank	Territory	%ᵃ
1	North Korea	55.7
2	China	41.5
3	Czech Republic	30.6
4	Mongolia	30.2
5	Uruguay	27.5
6	Kazakhstan	26.3
7	Latvia	25.4
8	Belarus	24.6
9	Estonia	24.4
10	Kyrgyzstan	21.5
11	New Zealand	20.0
12	Netherlands	19.0
13	Cuba	18.7
14	Moldova	18.6
15	Germany	18.4
16	Sweden	18.1
16	Uzbekistan	18.1
18	Russia	18.0
19	France	15.4
20	Australia	15.2

ᵃ % of the population who are agnostics

381 Agnostics

The size of each territory indicates the number of people living there who are believed to be agnostic as concerns religion.

Some 770 million people are thought to be agnostics, 70% of them living in China. Agnosticism is not a religion. Agnostics are people who believe either that it is not possible to have certain knowledge of the existence of a god or gods, or alternatively that while individual certainty may be possible, they personally have no such certainty. Agnostics do not deny the possibility of the existence of a god or gods but are not themselves adherents to any faith.

After China (which has 540 million agnostics), the territories with the largest numbers of agnostics are the United States (28 million), Russia (26 million), Germany (15 million), India (13 million) and Japan (13 million).

'...it is wrong for a man to say he is certain of the objective truth of a proposition unless he can provide evidence which logically justifies that certainty.'

Thomas H. Huxley, 1889

AGNOSTICS
% of all agnostics

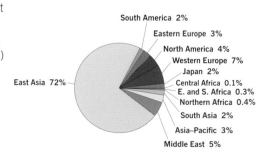

South America 2%
Eastern Europe 3%
North America 4%
Western Europe 7%
Japan 2%
Central Africa 0.1%
E. and S. Africa 0.3%
Northern Africa 0.4%
South Asia 2%
Asia–Pacific 3%
Middle East 5%
East Asia 72%

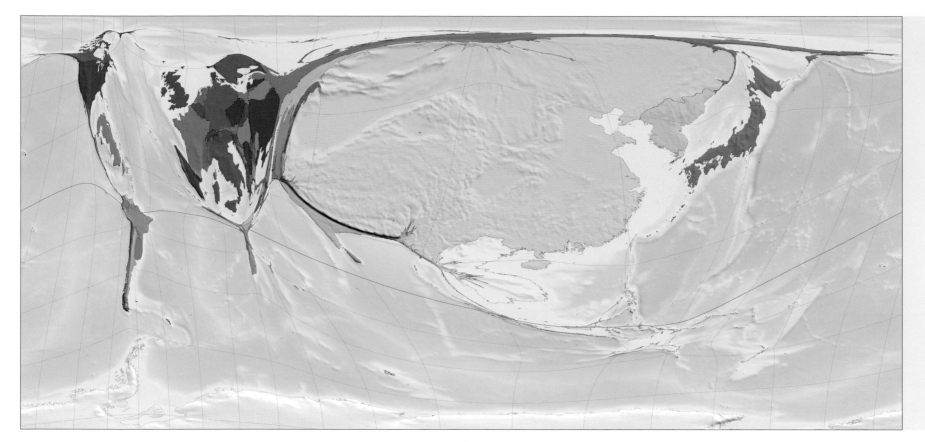

382 Atheists

The size of each territory indicates the number of people living there who are believed to be atheistic as concerns religion.

Some 150 million people worldwide are thought to be atheists, almost three-quarters of them living in East Asia. Atheism is not a religion but an absence of belief in the existence of a god or gods, or a dismissal of such beliefs as illogical. There are many different kinds of atheism. Indeed, some religions do not advocate belief in any god and their followers could therefore be considered atheists. Certain forms of Buddhism, for example, are sometimes classified as atheistic, although they are not so classified in this book.

Around 8% of people in East Asia, 3% in Japan, 2% in the Middle East and Central Asia, and 2% in Europe are thought to be atheists, though it should be noted that not all who define themselves as atheists do so with the same degree of assertiveness or commitment to atheist beliefs.

'If anything, an atheist has to be more morally responsible precisely because we don't blame a god for our own actions.'

Ivan Ratoyevsky, contemporary commentator

ATHEISTS
% of population

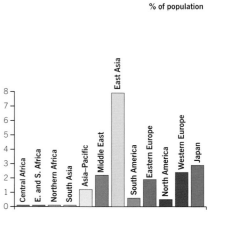

Glossary of Terms and Abbreviations

For additional abbreviations used in quotations, see p. 412.

AIDS acquired immune deficiency syndrome

billion one thousand million

CEAR Spanish Commission for Refugee Assistance

CFCs chlorofluorocarbons

ECE United Nations Economic Commission for Europe

EU European Union

FAO Food and Agriculture Organization of the United Nations

GDP gross domestic product: the total market value of all goods and services produced within a territory in a given year

GNI gross national income: total income received in return for goods and services produced in a territory, plus income from abroad such as that received by domestic companies operating elsewhere or from foreign investments

HIV human immunodeficiency virus

IDP internally displaced person

IMF International Monetary Fund

ITU International Telecommunication Union

IUCN International Union for Conservation of Nature and Natural Resources

local purchasing power see PPP

MIT Massachusetts Institute of Technolocy

MSF Médecins sans Frontières: medical charity

net exports Exports minus imports. On a 'net exports' map, territories that export less than they import have zero size and so do not appear.

net imports Imports minus exports. On a 'net imports' map, territories that import less than they export have zero size and so do not appear.

NGO non-governmental organization

OECD Organization for Economic Cooperation and Development

PPP Purchasing power parity: a way of adjusting amounts of money to show their local purchasing power, that is, what they would purchase in different countries. US$10 buys more in, say, Indonesia than it does in the United States; PPP US$10 means a sum that can buy the equivalent in Indonesia of what US$10 would buy in the United States.

person-hour the amount of work, travel or other activity performed by an average person in one hour

RSPB Royal Society for the Protection of Birds

SAARC South Asian Association for Regional Cooperation

STI sexually transmitted infection

TB tuberculosis

TEU 20-foot equivalent unit: a cargo container measure

tonne 1,000 kilograms; roughly the same weight as an imperial ton

tonne-kilometre 1 metric tonne travelling 1 kilometre

trillion one million million

UN United Nations

UN DHA United Nations Department of Humanitarian Affairs

UNDP United Nations Development Programme

UNEP United Nations Environment Programme

UNESCO United Nations Educational, Scientific and Cultural Organization

UNFPA United Nations Population Fund

UN-HABITAT United Nations Human Settlements Programme

UNICEF United Nations Fund for Children

WHO World Health Organization

WIPO World Intellectual Property Organization

WMO World Meteorological Organization

WSSD World Summit on Sustainable Development

WTO World Trade Organization

Sources of Data

Notes

Seven key data sources are used for many of the maps in this atlas. To save space, their names are reduced in the notes to the standard acronyms; the full names, abbreviated forms and the URLs for the websites of these organizations are listed at the head of the notes.

At least four different dates may be associated with each data source used for a map. (1) The date or dates of the events that gave rise to the data represented in the map; usually these events occurred a few years before the data were collected and compiled. (2) The date given in the title of the report used as the data source; often this corresponds with the collection of the data. (3) The date of first publication of the report, given in parentheses before the report title. (4) The date of last access to the URL; all URLs were checked in early 2008, so access dates are not given unless they are significant. The most useful date for an understanding of the data is that of the events represented in the map; this is therefore provided in the text for each map or in the notes to the table or graph. Dates given here thus refer only to the titles and publication of data sources and access to their websites.

To account for the exact provenance of the data shown in the maps would require a paper publication too large to be bound. The sources listed here often lack some data, so statistics have been estimated, using parts of the other large datasets referred to below where appropriate. For readers wishing to replicate our methods, we give a numeral in angle brackets (<1>) at the end of the title of each map; this is the refererence number for the map on the Worldmapper website (www.worldmapper.org), where a spreadsheet and technical notes on the data sources may be found. The website provides several hundred thousand additional details. The spreadsheets constitute a frozen record of the data at the time it was originally sourced – a necessary procedure because many of the data sources are online and are constantly being updated. Where the table numbers in an online source were altered between the time of initial access and the time when the map was prepared for this volume, the original and current numbers are cited below.

The rankings shown in the tables and discussed in the texts accompanying the maps include only the 200 territories for which statistics are measured in the data sources listed below, though 204 territories are shown in the maps themselves (see the Introduction for more details). Further, the rankings usually omit estimated values for missing data; the numbers shown thus accord most closely with the original source data. Rankings that include estimated values are shown on the Worldmapper website and may usefully be compared with those in the book.

In the tables of data, the figures shown are almost always rates rather than the absolute values depicted in the maps; the tables list the territories where the highest and lowest rates of the mapped data are to be found. The measure is usually a rate per capita using population data from the year 2002; see the sources of data for map 002 Total Population.

For the new maps, original data sources have been slightly altered to make data consistent with the Worldmapper territory populations for 2002.

Abbreviations and URLs

CIA Central Intelligence Agency (USA): https://www.cia.gov/library/publications/the-world-factbook/index.html

UNCTAD United Nations Conference on Trade and Development (various websites, individually cited)

UNDP United Nations Development Programme: http://hdr.undp.org/en/reports/global/hdr2004/

UNEP United Nations Environment Programme: http://geodata.grid.unep.ch/

UNICEF United Nations Children's Fund (various websites, individually cited)

WHO World Health Organization (various websites, individually cited)

World Bank World Bank: http://web.worldbank.org/data (time series are shown as a sequence of capital letters separated by full points: these are needed to access the original data)

001 Land Area <1> Economic and Social Research Institute (ESRI), *Digital Chart of the World* (1991); http://www.maproom.psu.edu/dcw/dcw_about.shtml. National boundaries are shown here on an equal-area projection. In the table and pie chart, land area excludes land covered by inland water and ice.

002 Total Population <2> UNDP (2004), *Human Development Report, 2004*, Tables 5, 18, 33; http://hdr.undp.org/en/reports/global/hdr2004/. • UNEP (2006) *Annual Report, 2005*; http://geodata.grid.unep.ch/. • CIA (2007) *The World Factbook, 2007*; https://www.cia.gov/library/publications/the-world-factbook/index.html.

003–006 Population in the Year 1 <7>, Population in the Year 1500 <8>, Population in the Year 1900 <9>, Population in the Year 1960 <10> A. Maddison (2007), *Historical Statistics for the World Economy: 1–2003 AD*; http://www.ggdc.net/maddison/.

408

007 Projected Population in the Year 2050 <11> United Nations Population Division (2006), *World Population Projections: The 2006 Revision*; http://www.un.org/esa/population/unpop.htm.

008 Projected Population in the Year 2300 <12> United Nations Population Division (2004), *World Population in 2300: Proceedings of the United Nations Expert Meeting on World Population in 2300 United Nations Headquarters New York*, Economic and Social Affairs Secretariat, Population Division, Working Paper 187, Revision 1 http://www.un.org/esa/population/publications/longrange2/2004worldpop2300reportfinalc.pdf; http://www.un.org/esa/population/publications/longrange2/Country_Tables.xls.

009–010 Births <3>, Attended Births <4> WHO (2005), *World Health Report, 2005*, Table 8; http://www.who.int/whr/2005/en/.

011–012 Number of Children <5>, Number of Elderly <6> UNDP (2004), *Human Development Report, 2004*, Table 5.

013–014 Right to Vote <359>, Voter Turnout <360> UNDP (2004), *Human Development Report, 2004*, Table 29; http://www.idea.int/vt/survey/voter_turnout_pop2-2.cfm. • International Institute for Democracy and Electoral Assistance (1998) and International Foundation for Election Systems (IFES, 2007), Election Guides; http://www.electionguide.org/. • Tianjian Shi (1999), 'Voting and nonvoting in China: Voting behavior in plebiscitary and limited-choice elections', *Journal of Politics*, 61, 1115–39.

015–018 International Emigrants <16>, International Immigrants <15>, Net Emigration <18>, Net Immigration <17> World Bank (2005), *World Development Indicators 2005*, Table 6.13; http://web.worldbank.org/data.

019–020 Refugee Origins <14>, Refugee Destinations <13> UNDP (2004), *Human Development Report, 2004*, Table 22.

021–028 Tourist Origins <20>, Tourist Destinations <19>, Net Outgoing Tourism <22>, Net Incoming Tourism <21>, Spending on Tourism <24>, Income from Tourism <23>, Net Income from Tourism <25>, Net Loss from Tourism <26> World Bank (2005), *World Development Indicators 2005*, Table 6.14.

029 Aircraft Departures <27> World Bank (2005), *World Development Indicators 2005*, time series on aircraft departures (IS.AIR.DPRT).

030–031 Aircraft Travel <28>, Air Passengers <29> UNEP (2006), *Annual Report, 2005*, time series on civil aviation traffic.

032–041 Rail Travel <30>, Cars <31>, Mopeds and Motorcycles <32>, Public Transport <142>, Commuting Time <141>, Roads <35>, Freight Vehicles <33>, Railway Lines <36>, Rail Freight <34>, Container Ports <38> World Bank (2005) *World Development Indicators 2005*, time series on railways, passengers carried (passenger-km) (IS.RRS.PASG.KM); passenger cars per 1,000 people (IS.VEH.PCAR.P3); two-wheelers (IS.VEH.2CYL.P3); Table 3.11 'Urban environment'; roads (total network km) (IS.ROD.TOTL.KM); freight vehicles (IS.VEH.NVEH.P3 less IS.VEH.PCAR.P3); vehicles (IS.VEH.NVEH.P3); rail lines (total route-km) (IS.RRS.TOTL.KM); railways, goods carried (tonne-km) (IS.RRS.GOOD.KM); container port traffic (twenty foot equivalent units, TEU) (IS.SHP.GOOD.TU).

042–043 Cargo Shipping <40>, Oil Tankers <39> UNCTAD (2005), *Handbook of Statistics On-line*, Table 8.6 (5.4 as at 18/01/2008) 'World merchant fleet by flag of registration and by type of ship'; http://www.unctad.org/Templates/Page.asp?intItemID=1890&lang=1; http://stats.unctad.org/Handbook/TableViewer/tableView.aspx?ReportId=1303.

044 Air Freight <37> UNEP (2006), *Annual Report, 2005*, time series on civil aviation traffic (total freight).

045–048 Rainfall <101>, Water Resources <102>, Groundwater Recharge <103>, Water Use <104> UNEP (2006) *Annual Report, 2005*, time series on rainfall by political units; dataset of internal renewable water resources (IRWR), recharge and use.

049–052 Forests 1990 <105>, Forests 2000 <106>, Forest Growth <107>, Forest Loss <108> World Bank (2005), *World Development Indicators 2005*, time series on forest area (AG.LND.FRST.ZS).

053–054 Fuel Use <119>, Fuel Increase <120> UNDP (2004), *Human Development Report, 2004*, Table 21.

055–056 Electricity Production <117>, Increase in Electricity Production <118> World Bank (2005), *World Development Indicators 2005*, time series on electricity production (EG.ELC.PROD.KH).

057 Traditional Fuel Consumption <109> UNDP (2004), *Human Development Report, 2004*, Table 21.

058–062 Hydroelectric Power <110>, Oil Power <111>, Gas Power <112>, Coal Power <113>, Nuclear Power <114> World Bank (2005), *World Development Indicators 2005*, time series on electricity production from hydroelectric sources (EG.ELC.HYRO.ZS); oil (EG.ELC.PETR.ZS); natural gas (EG.ELC.NGAS.ZS); coal (EG.ELC.COAL.ZS); nuclear sources (EG.ELC.NUCL.ZS).

063–072 Primary Exports in 1990 <347>, Primary Exports in 2002 <348>, Secondary Exports in 1990 <349>, Secondary Exports in 2002 <350>, High-Tech Exports in 1990 <351>, High-Tech Exports in 2002 <352>, Decline in Terms of Trade <353>, Improvement in Terms of Trade <354>, Debt Service in 1990 <355>, Debt Service in 2002 <356> UNDP (2004), *Human Development Report, 2004*, Table 15.

073 Demonstrations against the 2003 War in Iraq <361> A large number of sources were used to create this map. These were collected by John Pritchard and Lindsay Pritchard and are documented in full on the Worldmapper website. Three key sources are: D. Reynie (2004), *La Fracture occidentale: naissance d'une opinion européenne*, Paris, Éditions de La Table Ronde, pp. 161–204; D. Gordon (2003), 'The cost, consequences and morality of war in Iraq', *Radical Statistics*, 84; http://www.radstats.org.uk/no084/Gordon84.pdf; and Y. Kimmons and B. Williams (2003), 'How many Americans said no to war? A detailed study of 287 demonstrations for peace in the United States on February 15', *Liberal Oasis*; http://www.liberaloasis.co/peacereport.htm.

074 International Justice <362> International Committee of the Red Cross (2007), *États parties aux principaux traités de DIH*; http://www.icrc.org/Web/fre/sitefre0.nsf/iwpList103/ABFF592C8B3D842CC1256E350054B3D0.

075 International Food Aid <363> World Food Programme (WFP) (2005), *WFP Annual Report, 2005*, Annex 5; http://www.wfp.org/policies/Annual_Reports/index.asp?section=6&sub_section=3.

076 International Fast Food <364> McDonalds Corporation (2006), *Current Number of Restaurants per Country*; http://www.mcdonalds.com/corp/news/media.html.

077 Votes in the International Monetary Fund <365> International Monetary Fund (IMF) (2006), *Personal Correspondence with IMF Officials*; http://www.worldmapper.org/technotes.php?selected=365.

078 Who's Looking at Us? <366> The count of people looking at the Worldmapper website records the number of 'hits' not the number of visitors; during the period 1 January to 31 October 2006, represented in the map, there were 24.9 million hits; http://www.worldmapper.org/technotes.php?selected=366.

079–120 Fruit Exports to **Alcohol and Tobacco Imports <41>–<56> Ore Exports <61>, Ore Imports <62>, Metals Exports <71> Metals Imports <72> Steel and Iron Exports <63> Steel and Iron Imports <64> Fuel Exports <115>, Fuel Imports <116>, Crude Petroleum Exports <75> Crude Petroleum Imports <76> Refined Petroleum Exports <65> Refined Petroleum Imports <66> Chemicals Exports <85> Chemicals Imports <86> Gas and Coal Exports <59> Gas and Coal Imports <60> Wood and Paper Exports <73> Wood and Paper Imports <74> Natural Products Exports <81> Natural Products Imports <82> Exports of Machinery** to **Computer Imports <87>–<92>** UNCTAD (2005), *Handbook of Statistics On-line*, Table 4.2 (3.2 as at January 2008) 'International merchandise trade by product and by country'; http://www.unctad.org/Templates/Page.asp?intItemID=1890&lang=1; http://stats.unctad.org/TableViewer/tableView.aspx?ReportId=1298. World Bank (2005), *World Development Indicators 2005*, time series on fuel exports and imports as a percentage of merchandise exports (TX.VAL.FUEL.ZS.UN).

121–140 Transport and Travel Exports to **Royalty and Licence Fee Imports <57> <58> <69> <70> <77>–<80> <83> <84> <93>–<100> <167> <168>** UNCTAD (2005), *Handbook of Statistics On-line*, Table 5.2 'Trade in services by sector and country'; http://www.unctad.org/Templates/Page.asp?intItemID=1890&lang=1; http://stats.unctad.org/Handbook/TableViewer/tableView.aspx?ReportId=1304. UNDP (2004), *Human Development Report, 2004*, Table 12.

141–142 Research and Development Expenditure <165>, Research and Development Employees <166> UNDP (2004), *Human Development Report, 2004*, Table 12.

143–148 Wealth in the Year 1 <159>, Wealth in the Year 1500 <160>, Wealth in the Year 1900 <161>, Wealth in the Year 1960 <162>, Wealth in the Year 1990 <163>, Projected Wealth in the Year 2015 <164> A Maddison (2007), *Historical Statistics for the World Economy: 1–2003 AD*; http://www.ggdc.net /maddison/. Maddison does not include the data for the last map in this sequence; these were made by projecting the average annual rate of GDP growth for 1975–2002 forwards to cover the period to 2015.

149–160 Earnings of the Poorest Tenth of the Population to Number of People Living on more than US$200 a Day <149>–<158> <179> <180> UNDP (2004), *Human Development Report, 2004*, Tables 3, 4, 13, 14, and modelling assuming log-normal distributions.

161–164 Unadjusted Wealth <169>, Adjusted Wealth <170>, Growth in Wealth <171>, Decline in Wealth <172> UNDP (2004), *Human Development Report, 2004*, Table 13.

410

165–168 National Income <309>, National Savings <310>, Adjusted Savings Rate <317>, Negative Adjusted Savings <318> World Bank (2005), *World Development Indicators 2005*, Table 1.1 'World view: Size of the economy'.

169–172 Human Development <173>, Human Poverty <174>, Increase in Human Development <175>, Decrease in Human Development <176> UNDP (2004), *Human Development Report, 2004*, Table 2.

173–174 Undernourishment in 1990 <177>, Undernourishment in 2000 <178> UNDP (2004), *Human Development Report, 2004*, Table 7.

175 Underweight Children <182> UNDP (2004) *Human Development Report, 2004*, Tables 3 and 4.

176–178 Equality of the Sexes <181>, Males' Income <148>, Females' Income <147> UNDP (2004), *Human Development Report, 2004*, Tables 24 and 25.

179–182 Cereals Production <123>, Meat Production <125>, Vegetable and Vegetable Product Consumption <124>, Meat and Meat Product Consumption <126> UNEP (2006),

Annual Report, 2005, time series on cereal production (metric tonnes per year); meat production; average total calorie supply less supply from animal products (per capita per day); calorie supply from animal products.

183–184 Tractors Working <121>, Increase in Tractors <122> World Bank (2005), *World Development Indicators 2005*, time series on agricultural machinery, tractors (AG.AGR.TRAC.NO).

185–190 Male Agricultural Workers <128>, Female Agricultural Workers <127>, Males in Industry <130>, Females in Industry <129>, Service Sector Males <132>, Service Sector Females <131> UNDP (2004), *Human Development Report, 2004*, Table 27.

191–192 Female Domestic Labour <137>, Male Domestic Labour <138> UNDP (2004), *Human Development Report, 2004*, Table 28 (and 27).

193 Teenage Mothers <136> World Bank (2005), *World Development Indicators 2005*, Table 1.5.

194 Child Labour <135> World Bank (2005), *World Development Indicators 2005*, time series on labour force, children 10–14 (percentage of age group) (SL.TLF.CHLD.ZS).

195–196 Market Hours Worked by Males <140>, Market Hours Worked by Females <139> UNDP (2004), *Human Development Report, 2004*, Table 27 (and 28).

197–198 Male Managers <134>, Female Managers <133> UNDP (2004), *Human Development Report, 2004*, Tables 14 and 25.

199–202 Unemployed People <143>, Long-Term Unemployed <144>, Male Youth Unemployed <146>, Female Youth Unemployed <145> UNDP (2004), *Human Development Report, 2004*, Table 20.

203–204 Trade Unions <357>, Strikes and Lockouts <358> UNDP (2004), *Human Development Report, 2004*, Table 31. International Labour Organization (2002), *World Labour Report, 1997–1998*; http://www.ilo.org/public/english /dialogue/ifpdial/publ/wlr97/annex/tab11.htm.

205–206 Urban Population <189>, Projected Increase in Urban Population <190> UNDP (2004), *Human Development Report, 2004*, Table 5.

207–210 Number of Households <191>, Housing Prices <194>, People Living in Overcrowded Homes <192>, Durable Dwellings <193> World Bank (2005), *World Development Indicators 2005*, Tables 3a and 3.11.

211–212 Urban Slums <187>, Slum Growth <188> United Nations Human Settlements Programme (UN-HABITAT) (2006), *Slum Population in Urban Areas*, Millennium Development Goal; http://mdgs.un.org/unsd/mdg/Data.aspx.

213–216 Minimal Sanitation <183>, Basic Sanitation <184>, Sewerage Sanitation <185>, Poor Water <186> UNDP (2004), *Human Development Report, 2004*, Table 7. • UNEP (2006), *Annual Report, 2005*, time series on population connected to public waste-water treatment plants.

217–228 Children in Primary Education to Growth in Tertiary Education Spending <199>–<204> <207>–<212> UNDP (2004), *Human Development Report, 2004*, Tables 10 and 26. • UNDP (2005), *Human Development Report, 2005*, Table 27.

229–232 Youth Literacy <195>, Gender Balance of Illiterate Youth <197>, Adult Literacy <196>, Gender Balance of Illiterate Adults <198> UNDP (2004), *Human Development Report, 2004*, Tables 11 and 26.

233–234 Science Research <205>, Growth of Science Research <206> World Bank (2005), *World Development Indicators 2005*, time series on scientific and technical journal articles (IP.JRN.ARTC.SC).

235–236 Cost of Universal Telephone Service <330>, Telephone Faults <329> World Bank (2005), *World Development Indicators 2005*, time series on telephone revenue per mainline in current US$ (IT.MLT.REVN.CD); telephone faults per 100 mainlines (IT.MLT.FALT.M2).

237–242 Telephone Lines in 1990 <331>, Telephone Lines in 2002 <332>, Mobile Phones in 1990 <333>, Mobile Phones in 2002 <334>, Internet Users in 1990 <335>, Internet Users in 2002 <336> UNDP (2004), *Human Development Report, 2004*, Table 12.

243–247 Personal Computers <337>, Cable Television <338>, Televisions in Use <339>, Radios in Use <340>, Daily Newspaper Circulation 341> World Bank (2005), *World Development Indicators 2005*, time series on personal computers per 1,000 people (IT.CMP.PCMP.P3); cable television subscribers per 1,000 people (IT.TVS.CABL.P3);

television sets per 1,000 people (IT.TVS.SETS.P3); radios per 1,000 people (IT.RAD.SETS.P3); daily newspapers per 1,000 people (IT.PRT.NEWS.P3).

248–251 Weekly Newspaper Circulation <342>, Books Published <343>, Books Borrowed <344>, Films Watched <345> United Nations Educational, Scientific and Cultural Organization (UNESCO) online statistics, http://www.uis.unesco.org/ (accessed 10/10/2006).

252 Electricity Access <346> International Energy Agency (2002), *World Energy Outlook*, Annex 13.1; http://www.iea.org/textbase/nppdf/free/2000 /weo2002.pdf.

253–254 Public Health Spending <213>, Private Health Spending <214> UNDP (2004), *Human Development Report, 2004*, Table 6.

255–259 Working Midwives <215>, Working Nurses <216>, Working Pharmacists <217>, Working Dentists <218>, Working Physicians <219> WHO (2006), *Global Atlas of the Health Workforce* (as at January 2008 *Global Health Atlas*; http://www.who.int/globalatlas/).

260 Basic Healthcare Provision <220> WHO (2004), *World Health Report, 2004:Changing History*, Web Annex Table 9; http://www.who.int/whr/2004/en/

261 Hospital Beds< 221> World Bank (2005), *World Development Indicators 2005*, time series on hospital beds per 1,000 people (SH.MED.BEDS.ZS).

262–269 Affordable Drugs to HIV Prevalence <67> <68> <222>–<227> UNDP (2004), *Human Development Report, 2004*, Tables 6 and 8.

270–274 Tuberculosis Cases <228>, Malaria Cases <229>, Deaths from Malaria <230>, Cholera Cases <231>, Deaths from Cholera <232> WHO (2004), *Human Resources for Health: Basic Data* (as at 21/03/2005, Table 'Communicable Diseases'); for tuberculosis: http://www.who.int/tb/en/; for malaria and cholera: http://www.who.int/globalatlas/.

275 Childhood Diarrhoea <233> World Bank (2005), *World Development Indicators 2005*, time series on diarrhoea prevalence (percentage of children under 5) (SH.STA.DIRH.ZS).

276–280 Blinding Disease <234>, Polio Cases <235>, Yellow Fever <236>, Deaths from Rabies <237>, Outbreaks of Influenza <238> WHO (2004), *Human Resources for Health, Basic Data* (as at 21/03/2005, Table 'Communicable Diseases');

http://www.who.int/globalatlas/. For rabies and influenza follow the links to RabNet and FluNet from this site.

281 Prevalence of Diabetes <239> World Bank (2005), *World Development Indicators 2005*, Table 2.18 'Health: risk factors and future challenges'.

282 Alcohol Consumption <240> WHO (2006), *Behavioural and Risk Factor Indicators: Global Information System on Alcohol and Health (GISAH)*; http://www.who.int/research/en/.

283–285 Men Smoking <242>, Women Smoking <241>, Road Deaths <243> World Bank (2005), *World Development Indicators 2005*, Table 2.18 'Health: risk factors and future challenges'.

286 Years of Poor Health <244> UNEP (2006), *Annual Report, 2005*, time series on healthy life expectancy (HALE).

287–290 Life Expectancy <255>, Increase in Life Expectancy <256>, Women Outliving Men <257>, Maternal Mortality <258> UNDP (2004), *Human Development Report, 2004*, Tables 1 and 33.

291–292 Stillbirths <259>, Early Neonatal Mortality <260> WHO (2005), *World Health Report, 2005*, Table 8; http://www.who.int/whr/2005/en/.

293–296 Infant Mortality <261>, Decrease in Infant Mortality <262>, Child Mortality <263>, Decrease in Child Mortality <264> UNDP (2004), *Human Development Report, 2004*, Tables 1 and 33.

297–298 Mortality of Males Aged 15–60 Years <266>, Mortality of Females Aged 15–60 Years <265> WHO (2005), *World Health Report, 2005*, Table 8; http://www.who.int/whr/2005/en/.

299–308 People Affected by Disasters to **People Affected by Infestations <245>–<254>** UNEP (2006), *Annual Report, 2005*, time series on disasters; supplemented in the case of insect infestation with information from WHO Collaborating Centre for Research on the Epidemiology of Disasters (CRED), *Emergency Events Database* (EM-DAT); http://www.em-dat.net.

309–312 Military Spending 1990 <280>, Military Spending 2002 <279>, Arms Exports <281>, Arms Imports <282> UNDP (2004), *Human Development Report, 2004*, Tables 19 and 22.

313 Nuclear Weapons <289> Centre for Defence Information, Washington, DC, online statistics; http://www.cdi.org/issues/nukef&f/database/ nukearsenals.cfm (accessed October 2006).

314 Landmine Deaths and Injuries <290> Landmine Monitor (2006), *Landmine Monitor Report 2006: Toward a Mine-Free World*; http://www.icbl.org /lm/2005/intro/survivor.html#Heading4.

315–316 Armed Forces in 1985 <283>, Armed Forces in 2002 <284> UNDP (2004), *Human Development Report, 2004*, Table 19.

317–318 Armed Forces at War 1945–2004 <285>, Population at War 1945–2004 <286> Data for 1945–82: M. Kidron and D. Smith (1983), *The War Atlas: Armed Conflict, Armed Peace*, London: Pan Books. Data for the years 1983–89: Centre for the Study of Civil War at the International Peace Research Institute, Oslo (PRIO), and Uppsala Conflict Data Program (UCDP) at the Department of Peace and Conflict Research, Uppsala University (2006), *Armed Conflicts, 1946–2004*; http://www.prio.no/cwp/ armedconflict/. • Data for 1990–2004 (including data for wars that began earlier and continued into this period): 'Table of wars, 1990–2004', in D. Smith and A. Bræin, (2005), *The Atlas of War and Peace*, London: Earthscan, pp. 116–21.

319 War Deaths 1945–2000 <287> M. Leitenberg (2001), 'Deaths in wars and conflicts between 1945 and 2000', paper presented to the conference Data Collection in Armed Conflict, held in Uppsala, Sweden, 8–9 June 2001; http://www.einaudi.cornell.edu /peaceprogram/publications/occasional_papers/Death s-Wars-Conflicts3rd-ed.pdf (accessed October 2006).

320–322 War Deaths in 2002 <288>, Violent Deaths <291>, Suicides <292> WHO (2004), *World Health Report, 2004*, Web Annex Table 'Burden of disease'; http://www.who.int/evidence/bod/en/.

323–324 Prisoners <293>, Prisoners Awaiting Trial <294> International Centre for Prison Studies (ICPS) (2006), *World Prison Brief Online*; http://www.kcl.ac.uk/depsta/rel/icps/worldbrief/highest _to_lowest_rates.php (accessed 14.09.2006).

325–328 Carbon Dioxide Emissions in 1980 <296>, Carbon Dioxide Emissions in 2000 <295>, Increase in Emissions of Carbon Dioxide <297>, Decrease in Emissions of Carbon Dioxide <298> UNDP (2004), *Human Development Report, 2004*, Table 21. UNEP (2006), *Annual Report, 2005*; dataset: all_co2_mdg_total.

329 Projected Carbon Dioxide Emissions in 2015 <315> Energy Information Administration (EIA) (2006), *Annual Energy Outlook, 2006*, Table A10, 'World carbon dioxide emissions by region, reference case, 1990–2030'; http://www.eia.doe.gov/oiaf/ieo/pdf /ieoreftab_10.pdf . The source is a statistical agency of the U.S. Department of Energy; its regional projections have been used to rescale the World Bank projected totals for 2000 (see Worldmapper files for map 315) to 2015.

330 Particulate Damage <316> World Bank (2005) *World Development Indicators 2005*, Table 1.1 'World view: Size of the economy'.

331–334 Greenhouse Gases <299>, Agricultural Methane and Nitrous Oxide <300>, Sulphur Dioxide <301>, Nitrogen Oxides <302> United Nations Statistics Division (UNSD), Department of Economic and Social Affairs (2005), *Main Environmental Indicators*; http://unstats.un.org /ENVIRONMENT/q2004indicators.htm.

335–338 Chlorofluorocarbons <303>, Nuclear Waste <304>, Hazardous Waste <305>, Sewage Sludge <306> UNEP (2006), *Annual Report, 2005*; datasets: all_ozone_cfc used: Consumption of Ozone-Depleting Substances - Chlorofluorocarbons (CFCs); all-waste_nuclear: tonnes of heavy metal per year; all-waste_haz_ generated; and all-waste_sludge_prod.

339–340 Waste Generated <307>, Waste Recycled <308> United Nations Statistics Division (UNSD), Department of Economic and Social Affairs (2005), *Main Environmental Indicators*; http://unstats.un.org /unsd/ENVIRONMENT/q2004indicators.htm.

341–342 Maintenance Costs <311>, Fossil Fuel Depletion <312> World Bank (2005), *World Development Indicators 2005*, Table 1.1 'World view: Size of the economy'.

343–344 Gas Depletion<319>, Oil Depletion<320> British Petroleum (BP) (2005), *Statistical Review of the World*; http://www.bp.com/statisticalreview.

345–346 Mineral Depletion <313>, Forest Depletion <314> World Bank (2005), *World Development Indicators 2005*, Table 1.1 'World view: Size of the economy'.

347–350 Excess Water Use <323>, Domestic Water Use <324>, Industrial Water Use <325>, Agricultural Water Use <326> World Bank (2005), *World Development Indicators 2005*, Table 3.5 'World view: Freshwater'.

351–352 Piped Water <327>, Increase in Piped Water <328> WHO and UNICEF (2006), *Joint Monitoring Programme on Water and Sanitation Worldwide*; http://www.wssinfo.org/en/sanquery.html.

353–354 Biocapacity <321>, Ecological Footprint <322> World Wide Fund for Nature (WWF) (2006), *Living Planet Report*; http://www.footprintnetwork.org; http://www.panda.org/news_facts/publications/ living_planet_report/lp_2006/index.cfm.

355–366 Extinct Species <267>, Species Existing Only in Zoos <268>, Species at Risk of Extinction <269>, Species at Lower Risk of Extinction <270>, Plants at Risk <271>, Mammals at Risk <272>, Birds at Risk <273>, Reptiles at Risk <274>, Amphibians at Risk <275>, Fish at Risk <276>, Molluscs at Risk <277>, Invertebrates at Risk <278> International Union for Conservation of Nature and Natural Resources (IUCN) (2006), *Red List of Threatened Species 2004*. Tables 5, 6a, 6b; http://www.iucnredlist.org/.

367–382 Catholic Christians <555>, Protestant Christians <557>, Orthodox Christians <556>, Adherents to Judaism <563>, Sunni Muslims <566>, Shia Muslims <565>, Hindus <569>, Buddhists <568>, Sikhs <571>, Jains <570>, Taoic Religions <572>, Baha'i <553>, Pagans <577>, Spiritualists <578>, Agnostics <581>, Atheists <582> The main source of data for these maps is the World Christian Database; http://worldchristiandatabase. org/wcd/. Data on the Religious Society of Friends (Quakers) comes from the Friends themselves, courtesy of Thecla Geraghty. We have made occasional use of figures from ARDA (http://www.thearda.com/) and adherents.com. Original data sources have been slightly altered to allow for a small amount of double-counting in countries where people are counted as adherents to more than one religion. Totals usually include children to whom the beliefs of their parents are attributed.

Sources of Quotations

For reasons of brevity, clarity and consistency quotations have been minimally revised. The original readings are detailed below.

ACP	African, Caribbean and Pacific Group of States
CICETE	China International Center for Economic and Technical Exchange
DFID	Department for International Development (UK)
ECE	United Nations Economic Commission for Europe
IRIN	Integrated Regional Information Networks
NACLA	North American Congress on Latin America
NDRC	National Development and Reform Commission
NIWA	National Institute of Water and Atmospheric Research (New Zealand)
UNDHA	United Nations Department of Humanitarian Affairs
UNDP	United Nations Development Programme
UNEP	United Nations Environment Programme
UNESCO	United Nations Educational, Scientific and Cultural Organization
UNFPA	United Nations Population Fund
UNICEF	United Nations Children's Fund
UNIFEM	United Nations Development Fund for Women
VSO	Voluntary Service Overseas
WHO	World Health Organization
WWF	World Wide Fund for Nature

412

001 Land Area <1> http://www.oxfam.org.uk/what_we_do/index.htm (accessed 16/01/2006)

002 Total Population <2> Hania Zlotnik (2005), *Statement to the Thirty-Eighth Session of the United Nations Commission on Population and Development, 4 April 2005*; www.un.org/esa/population/cpd/Statement_HZ_open.pdf (accessed January 2006)

004 Population in the Year 1500 <8> http://www.bbc.co.uk/scotland/education/geog/population/profile.shtml?country=mexico (accessed 14/03/2006)

005 Population in the Year 1900 <9> Anthony J. Parel, ed. (1997), *Ghandi: 'Hind Swaraj' and Other Writings*, Cambridge: Cambridge University Press, p. 36

007 Projected Population in the Year 2050 <11> UNFPA, Country Support Team for the South Pacific (1999), 'Y6B: World population hits 6 billion!', *SouthPac News*, 7/1 (June); http://www.un.org/popin/regional/asiapac/fiji/news/99jun/y6b.htm (accessed 12/12/2005)

009 Births <3> Aung San Suu Kyi (1997), *Letters from Burma*, Harmondsworth: Penguin, p. 56

010 Attended Births <4> http://www.who.int/goodwill_ambassadors/liya_kebede/un/en/index.html (accessed 06/01/2006)

012 Number of Elderly <6> HelpAge International, *State of the World's Older People 2002*, p. 8; http://www.helpage.org/Resources/Policyreports#1118337662-0-11 (accessed 17/12/2005)

017 Net Emigration <18> Dominic Bailey (2005), 'EU outposts turn into fortresses', BBC News (29 September); http://news.bbc.co.uk/1/hi/world/europe/4294426.stm (accessed 17/12/2005)

018 Net Immigration <17> Angela Partington, ed. (1996), *The Oxford Dictionary of Quotations*, rev. 4th edn, Oxford: Oxford University Press, p. 184, no. 11

020 Refugee Destinations <13> Alexandra Fouche (2001), interview with Habib Souaïdia, in *The Road to Refuge*, BBC News special report; http://news.bbc.co.uk/hi/english/static/in_depth/world/2001/road_to_refuge/foreign_land/testimony.stm (accessed 18/12/2005)

022 Tourist Destinations <19> Howard Rosenzweig (2003), 'Copan update', *Honduras This Week*, online edition, no. 37 (22 September); http://www.marrder.com/htw/ (accessed 15/01/2006)

024 Net Incoming Tourism <21> Angela Partington, ed. (1996), *The Oxford Dictionary of Quotations*, rev. 4th edn, Oxford: Oxford University Press, p. 667, no. 20

031 Air Passengers <29> Peter Haggett (2001), *Geography: A Global Synthesis*, Harlow: Prentice Hall, p. 652

033 Cars <31> Paul Harris (2006), 'How the US fell out of love with its cars', *The Observer* (29 January); http://observer.guardian.co.uk/world/story/0,,1697517,00.html (accessed 14/02/2006)

035 Public Transport <142> Yingling Liu (2006), 'Bus Rapid Transit: A step toward fairness in China's urban transportation', *Worldwatch* (9 March); http://www.worldwatch.org/features/chinawatch/stories/20060309-1 (accessed 05/05/2006)

036 Commuting Time <141> 'Bangkok's industrial ring road', *Bridges* (March–April 2005) http://www.bridgebuildermagazine.com/ME2/Audiences/dirmod.asp?sid=&nm=&type=Publishing&mod=Publications%3A%3ABB+Articles&mid=6E82BAC0CEF34C7F912D607986507460&AudID=FA0B7EC5E4FA43DB8CE50898D8DCA098&tier=4&id=5F6DB02AC6E34DBD8E8A2223933ED250 (accessed 07/05/2006)

037 Roads <35> Ian G. Heggie (1994), *Management and Financing of Roads: An Agenda for Reform*, Africa Technical Department Series, Technical Paper no. 275; http://www.worldbank.org/afr/findings/english/find32.htm (accessed 17/01/2006)

039 Railway Lines <36> Yukinori Koyama (1997), 'Railway construction in Japan', *Japan Railway and Transport Review* (December), pp. 36–41, at p. 39; http://www.jrtr.net/technology/index_technology.html (accessed 17/01/2006)

040 Rail Freight <34> http://www.monologues.co.uk/Childrens_Favourites/Night_Mail.htm (accessed 17/01/2006)

041 Container Ports <38> Jung Chang (1991), *Wild Swans*, London: Flamingo, p. 294 [referring to China in 1957]

042 Cargo Shipping <40> Alison Gee (2002), 'Shipping firms call up pirate busters', BBC News (6 August); http://news.bbc.co.uk/1/low/business/2175819.stm (accessed 17/01/2006)

044 Air Freight <37> Kamau Ngotho (2005), 'Kenya's in foreign hands', *Sunday Standard* (17 April); http://www.eastandard.net/archives/sunday/print/news.php?articleid=18216 (accessed 14/02/2006)

046 Water Resources <102> http://www.mre.gov.br/cdbrasil/itamaraty/web/ingles/divpol/norte/am/bamazon/index.htm (accessed 04/04/2006)

047 Groundwater Recharge <103> http://www.wrl.unsw.edu.au/groundwater/hscresources/841002.html (accessed 04/04/2006) [for 'groundwater' orig. reads 'ground water']

048 Water Use <104> Céline Dubreuil (2006), *The Right to Water: From Concept to Implementation*, World Water Council; http://www.worldwatercouncil.org/index.php?id=32 (accessed 04/04/2006)

049 Forests 1990 <105> The Rainforest Foundation (2004), *New Threats to the Forests and Forest Peoples of the Democratic Republic of Congo*, briefing paper (February); www.fern.org/pubs/briefs/DRC_RF.pdf (accessed 04/04/2006) [for 'Democratic Republic of Congo' orig. reads 'DRC']

050 Forest 2000 <106> http://www.forest.ru/eng/news/news.html?cmd[36]=i-37-82faf499701b61b88bf9fc8f3d65ce60 (accessed 04/04/2006)

051 Forest Growth <107> http://www.wisdomquotes.com/cat_children.html (accessed 17/02/2006)

052 Forest Loss <108> E. G. Togu Manurung (2006), 'Continuing problems remain unresolved', *Jakarta Post* (10 May); http://www.thejakartapost.com/Outlook2006/eco09b.asp (accessed 15/05/2006)

053 Fuel Use <119> Terry Costlow (2003), 'Fuel cell research moving "at light speed"', *Today's Engineer* (April); http://www.todaysengineer.org/2003/Apr/fuel_cell.asp (accessed 04/04/2006)

054 Fuel Increase <120> 'Preparation prevents higher oil price pain', *China Daily* (28 March 2006); http://www.chinadaily.com.cn/chinagate/doc/2006-03/28/content_554050.htm (accessed 04/04/2006)

055 Electricity Production <117> http://www.doe.gov/energysources/index.htm (accessed 04/04/2006)

060 Gas Power <112> Bernard A. Gelb (2006), *Russian Oil and Gas Challenges*, Washington, DC: Library of Congress, Congressional Research Service; fpc.state.gov/documents/organization/58988.pdf (accessed 04/04/2006)

066 Secondary Exports in 2002 <350> 'China doubles auto exports in 2006', *People's Daily Online* (1 January 2007); http://english.peopledaily.com.cn/200701/01/eng20070101_337593.html (accessed 15/01/2007)

067 High-Tech Exports in 1990 <351> http://www.somo.nl/html/paginas/pdf/ICT_China_report_NL.pdf (accessed 06/02/2007) [for 'Pearl River Delta' orig. reads 'Pearl River Delta (PRD)']

069 Decline in Terms of Trade <353> Graham Young (1990), 'Fair trading', *New Internationalist*, no. 204 (February); http://live.newint.org/issue204/fair.htm (accessed 06/02/2007)

072 Debt Service in 2002 <356> Ernest Harsch (2004), 'Brazil repaying its "debt" to Africa: President's tour of Southern Africa strengthens south–south ties', *Africa Recovery*, 17/4 (January), p. 3; http://www.un.org/ecosocdev/geninfo/afrec/vol17no4/174brazil.htm (accessed 08/01/2007) [referring to Luiz Inácio Lula da Silva, president of Brazil]

074 International Justice <362> http://www.quotationsbook.com/quotes/9187/view (accessed 08/01/2007)

075 International Food Aid <363> Martin Wroe and Malcolm Doney (2004), *The Rough Guide to a Better World: How You Can Make a Difference*, London: Rough Guides, p. 52

077 International Monetary Fund <365> Barbara Gunnell (**2004**), 'A conspiracy of the rich', *New Statesman* (24 May); http://www.newstatesman.com/200405240018 (accessed 08/01/2007) [for 'International Monetary Fund' orig. reads 'IMF']

078 Who's Looking at Us? <366> Horst Frenz, ed. **(1969)**, *Nobel Lectures…: Literature, 1901–1967*, **Amsterdam: Elsevier**; http://nobelprize.org/nobel_prizes/literature/laureates/1938/buck-speech.html (accessed 08/01/2007)

085 Cereals Exports <47> http://www.visit-laos.com/food/index.htm (accessed 17/02/2006)

087 Meat Exports <49> http://www.mongabay.com/brazil.html (accessed 17/02/2006)

089 Fish Exports <51> Helga Josupeit (2003), *Considerations of the Cofi Sub-Committee on Fish Trade in Relation to International Trade and Food Security*, Report of the Expert Consultation on International Fish Trade and Food Security, Casablanca, Morocco, 27–30 January 2003; http://www.fao.org/documents/show_cdr.asp?url_file=/DOCREP/006/Y4961E/y4961e05.htm (accessed 17/02/2006)

091 Groceries Exports <53> http://www.icco.org/ (accessed 17/02/2006)

093 Alcohol and Tobacco Exports <55> http://www.nps.gov/colo/Jthanout/TobaccoHistory.html (accessed 17/02/2006)

095 Ore Exports <61> Kelly Hearn (2005), 'South America's mining wars heat up' *Global Policy Forum* (28 June); http://www.globalpolicy.org/socecon/trade/2005/0628samines.htm (accessed 02/03/2006)

098 Metals Imports <72> http://www.world-aluminium.org/history/ (accessed 02/03/2006)

100 Steel and Iron Imports <64> Anand S. T. Das (2006), 'North India: A potential steel exporter to Pak [sic]', *Chandigarh Newsline* (23 February); http://cities.expressindia.com/fullstory.php?newsid=171003 (accessed 02/03/2006)

101 Fuel Exports <115> Melissa Dell (2004), 'The devil's excrement: The negative effect of natural resources on development', *Harvard International Review*, 26/3 (fall); http://hir.Harvard.edu/articles/1261/ (accessed 14/04/2006)

102 Fuel Imports <116> Kunio Anzai (2004), *City Gas Industry's Agenda in Japan*; www.gas.or.jp/english/letter/images/04/pdf/Agenda1.pdf (accessed 14/04/2006)

103 Crude Petroleum Exports <75> Rupert Wingfield-Hayes (2006), 'Satisfying China's demand for energy', BBC News (16 February); http://news.bbc.co.uk/1/hi/world/asia-pacific/4716528.stm (accessed 15/03/2006)

104 Crude Petroleum Imports <76> BBC News (2006), 'Japan in surprise trade deficit' (23 February); http://news.bbc.co.uk/1/hi/business/4742582.stm

(accessed 15/03/2006)

105 Refined Petroleum Exports <65> Yoko Kitazawa (1990), 'The Japanese economy and South-East Asia: The examples of the Asahan Aluminium and Kawasaki Steel projects'; http://www.unu.edu/unupress/unupbooks/80a04e/80A04E07.htm (accessed 15/03/2006)

106 Refined Petroleum Imports <66> Rupert Wingfield-Hayes (2006), 'Satisfying China's demand for energy', BBC News (16 February); http://news.bbc.co.uk/1/hi/world/asia-pacific/4716528.stm (accessed 15/03/2006)

107 Chemicals Exports <85> Patricia L. Short (2005), 'Currency knocks chemical industry's exports, but companies still look for pickup in domestic demand', *Chemical and Engineering News*, 83/2 (10 January), pp. 23–25; http://pubs.acs.org/cen/coverstory/83/8302wcoeurope.html (accessed 15/03/2006) [quotation attributed to Rein Willems]

108 Chemicals Imports <86> Rachel Carson (1962), *Silent Spring*, Penguin Classics, Harmondsworth: Penguin, 2000, p. 24

110 Gas and Coal Imports <60> Steven Eke (2005), 'Russia threatens Ukraine gas cut', BBC News (13 December); http://news.bbc.co.uk/1/hi/business/4526138.stm (accessed 15/03/2006)

111 Wood and Paper Exports <73> Eduardo Tadem (1990), 'Conflict over land-based natural resources in the Asean countries'; http://www.unu.edu/unupress/unupbooks/80a04e/80A04E04.htm#2..3%20The%20plunder%20of%20forest%20resources (accessed 15/03/2006)

112 Wood and Paper Imports <74> http://www.fscus.org/ (accessed 15/03/2006)

113 Natural Products Exports <81> John H. Drabble (2004), 'Economic history of Malaysia'; http://www.eh.net/encyclopedia/article/drabble.malaysia (accessed 15/03/2006)

114 Natural Products Imports <82> M. Ghazanfar Ali Khan (2004), 'Restrictions on trading in wildlife animals tightened', *Arab News* (21 October); http://www.arabnews.com/?page=1&section=0&article=53221&d=21&m=10&y=2004 (accessed 15/03/2006)

115 Exports of Machinery <87> H. G. Wells (1898), *The Time Machine*, ch. 12; http://www.bartleby.com/1000/12.html#19 (accessed 16/03/2006)

117 Exports of Electronics <89> Michael J. Kelly (1997), 'Introduction', *World Technology Evaluation Centre Panel Report on Electronics Manufacturing in the*

Pacific Rim (May); http://www.wtec.org/loyola/em/01_03.htm (accessed 17/03/2006)

119 Computer Exports <91> Chun Wei Choo (1997), 'Singapore's vision of an intelligent island', *Intelligent Environments*, ed. Peter Droege, Amsterdam and New York, Elsevier; http://choo.fis.utoronto.ca/FIS/ResPub/IT2000.html (accessed 17/03/2006)

120 Computers Imports <92> Tim Hirsch (2004), 'Computers must be greener', BBC News (8 March); http://news.bbc.co.uk/1/hi/technology/3541623.stm (accessed 17/03/2006)

121 Transport and Travel Exports <93> World Trade Organization, Services Trade section http://www.wto.org/english/tratop_e/serv_e/serv_e.htm (accessed 21/04/2006)

123 Vehicle Exports <79> BBC News (2005), 'Airbus unveils "superjumbo" jet' (18 January); http://news.bbc.co.uk/1/hi/business/4183201.stm (accessed 17/03/2006)

125 Car Exports <77> http://www.volkswagen.co.uk/company/wolfsburg (accessed 17/03/2006)

127 Clothing Exports <83> Roland Buerk (2005), 'Bangladesh garments aim to compete', BBC News (6 January); http://news.bbc.co.uk/1/hi/world/south_asia/4118969.stm (accessed 17/03/2006)

128 Clothing Imports <84> BBC News (2005), 'EU and China reach textile deal' (5 September); http://news.bbc.co.uk/1/hi/business/4214490.stm (accessed 17/03/2006)

129 Exports of Toys <57> Agence France-Presse (2000), 'McDonald's employing child labour to produce toys' (17 August); http://www.hartford-hwp.com/archives/55/336.html (accessed 18/02/2006)

133 Mercantile and Business Exports <95> Brian O'Rorke (1996), 'Management consultants are pathfinders in Britain's export drive'; http://www.bcctc.ca/managemt.html (accessed 05/04/2006)

134 Mercantile and Business Imports <96> http://www.museu.gulbenkian.pt/coleccionador.asp?lang=en (accessed 05/04/2006)

135 Finance and Insurance Exports <97> Jack Schofield (2006), 'Forget garages, coffee shops are where businesses get started', *Guardian Unlimited* (23 February), 'Technology Blog'; http://blogs.guardian.co.uk/technology/archives/2006/02/23/forget_garages_coffee_shops_are_where_businesses_get_started.html (accessed 05/04/2006)

136 Finance and Insurance Imports <98> Sherine Abdel-Razek (2005), 'The hardest nut to crack:

The insurance sector is next on the privatisation bloc', *Al-Ahram*, no. 757 (August); http://weekly.ahram.org.eg/2005/757/ec2.htm (accessed 05/04/2006)

137 Patents Granted <167> 'The right to good ideas', *The Economist* (21 June 2001); http://www.jubileeresearch.org/finance/right_good_idea.htm (accessed 10/07/2006)

138 Royalty Fees <168> Bryan Chaffin (2005), 'Hollywood unions want piece of iTunes video pie', *Mac Observer* (14 October); http://www.macobserver.com/article/2005/10/14.14.shtml (accessed 10/07/2006)

139 Royalty and Licence Fee Exports <99> 'World Intellectual Property Day, 2006', *WIPO Magazine*, no. 2 (April 2006), p. 3. http://www.wipo.int/freepublications/en/list.jsp?organisation=WIPO&collection=pubdocs&sub_col=mag&year=2006 (accessed 28/04/2006)

140 Royalty and Licence Fee Imports <100> 'New international treaty: The Singapore treaty on the law of trademarks', *WIPO Magazine*, no. 2 (April 2006), p. 2; http://www.wipo.int/freepublications/en/list.jsp?organisation=WIPO&collection=pubdocs&sub_col=mag&year=2006 (accessed 28/04/2006) [quotation attributed to Burhan Gafoor]

141 Research and Development Expenditure <165> Kenneth Walker (2004), 'Scholarship emerges in Africa', *Carnegie Reporter*, 3/1 (autumn), p. 1; http://www.carnegie.org/reporter/09/scholarship/index.html (accessed 15/07/2006)

143 Wealth in the Year 1 <159> http://www.un.int/bangladesh/gen/intro.htm (accessed 15/07/2006)

144 Wealth in the Year 1500 <160> Richard Effland (2003), 'The rest of the story about Africa'; http://www.mc.maricopa.edu/dept/d10/asb/anthro2003/legacy/africa/africa.html (accessed 15/07/2006)

145 Wealth in the Year 1900 <161> 'Industrialization and its consequences, 1750–1914 CE', World History for Us All [project of San Diego State University with the National Center for History in the Schools, University of California, Los Angeles]; http://worldhistoryforusall.sdsu.edu/dev/eras/era7.htm (accessed 15/07/2006)

146 Wealth in the Year 1960 <162> Luis Alberto Moreno (2006), 'Beyond macro reforms: Sound macroeconomies are necessary, but not sufficient, foundations for Latin America's development', *Latin Business Chronicle* (11 May); http://yaleglobal.yale.edu/display.article?id=7388 (accessed 15/07/2006)

148 Projected Wealth in the Year 2015 <164> Martin Wolf (2003), 'Asia's awakening', *Financial Times* (21 September); http://www.geog.psu.edu/courses/geog100/Asia's%20awakening.doc (accessed 15/07/2006)

151 Earnings of the Poorest Fifth of the Population
<151> Anthony Jay, ed. (1997), *The Oxford Dictionary of Political Quotations*, Oxford: Oxford University Press, p. 270, no. 11

153 Absolute Poverty (People Living on US$1 a Day)
<179> Frantz Fanon (1961), *Damnés de la terre*; Eng. trans. Constance Farrington as *The Wretched of the Earth*, London: MacGibbon & Kee, 1963 http://www.cooper.edu/humanities/classes/coreclasses/hss3/f_fanon.html (accessed 15/07/2006)

154 Abject Poverty (People Living on US$2 a Day)
<180> Anthony Jay, ed. (1997), *The Oxford Dictionary of Political Quotations*, Oxford: Oxford University Press, p. 147, no. 1

158 Number of People Living on US$50–100 a Day
<156> Arundhati Roy (2002), 'Shall we leave it to the experts? Enron's power project in India demonstrates who benefits from globalization', *The Nation* (18 February);http://www.ratical.org/co-globalize /AR021802.html (accessed 18/06/2006)

159 Number of People Living on US$100–200 a Day
<157> Adam Smith (1776), *An Inquiry into the Nature and Causes of the Wealth of Nations*; http://www.mdx.ac.uk/www/study/xsmith.htm (accessed 18/06/2006)

161 Unadjusted Wealth <169> Deanne Julius (2005), 'US economic power: waxing or waning?', *Energy*, 26/4 (winter); http://hir.harvard.edu /articles/1287/ (accessed 21/07/2006)

165 National Income <309> Robert Francis Kennedy, address given at the University of Kansas, Lawrence, Kansas, 18 March 1968; http://projectories.net/ (accessed 15/01/2007)

166 National Savings <310> Adam Smith (2006), interview with Muhammad Yunus, '…poverty in the world is an artificial creation' (13 October); http://nobelprize.org/nobel_prizes/peace/laureates/2006/ yunus-telephone.html (accessed 15/01/2007)

168 Negative Adjusted Savings <318> Lezak Shallat (2006), 'Give them credit', *New Internationalist*, no. 392 (August); http://www.newint.org/features/2006/08/01 /credit-card-system/ (accessed 15/01/2007)

170 Human Poverty <174> Angus Maddison (2002), 'Research objectives and results, 1952–2002'; http://www.ggdc.net/Maddison/content.shtml (accessed 03/06/2006)

172 Decrease in Human Development <176> Khaled Hosseini (2004), *The Kite Runner*, London: Bloomsbury Publishing, p. 38 [orig. pubd 2003]

173 Undernourishment in 1990 <177> Anthony Jay, ed. (1997), *The Oxford Dictionary of Political Quotations*,

Oxford: Oxford University Press, p. 276, no. 1

174 Undernourishment in 2000 <178> Lynn Brown (2004), 'No drop in world hunger deaths', BBC News (8 December); http://news.bbc.co.uk/2/hi/africa /4078003.stm (accessed 03/07/2006)

175 Underweight Children <182> BBC News (2006), 'Malnutrition hits economic growth' (2 March); http://news.bbc.co.uk/2/hi/business/4767864.stm (accessed 03/07/2006) [quotation attributed to Jean-Louis Sarbib]

176 Gender Empowerment <181> http://www.who.int/director-general/speeches /2000/english/20000609_newyork.html (accessed 03/07/2006)

178 Females' Income <147> Kimberly Goad (2006), 'Big-earning wives (and the men who love them)', *Redbook Magazine*; http://redbook.ivillage.com/sex /0,,98b05ztg,00.html (accessed 03/07/2006)

179 Cereals Production <123> John Travis (2002), 'Hot cereal: Rice reveals bumper crop of genes', *Science News* (6 April); http://www.sciencenews.org/articles /20020406/fob1.asp (accessed 16/05/2006)

180 Meat Production <125> George Monbiot (2005), 'The price of cheap beef: Disease, deforestation, slavery and murder', *The Guardian* (18 October); http://www.guardian.co.uk/comment/story/0,,1594653, 00.html (accessed 16/05/2006)

183 Tractors Working <121> Bharat Dogra (2000), 'India: "Green revolution" – bad news for poor labourers', *Corp Watch* (22 November); http://www.corpwatch.org /article.php?id=309 (accessed 16/05/2006)

186 Female Agricultural Workers <127> Jonathan Swift (1754), *Gulliver's Travels*; http://www.nri.org/ILUS/homepage.htm (accessed 16/05/2006)

188 Females in Industry <129> UNDP, CICETE, NDRC and UNIFEM (2003), *China's Accession to WTO: Challenges for Women in the Agricultural and Industrial Sectors* (July), p. 61

189 Service Sector Males <132> Stephan Khan (2003), 'Bombay calling…', *The Guardian* (7 December); http://www.guardian.co.uk/money/2003 /dec/07/internetphonesbroadband.phones (accessed 16/05/2006)

190 Service Sector Females <131> Simon Montlake (2006), 'China's factories hit an unlikely shortage: labor', *Christian Science Monitor* (1 May); http://www.csmonitor.com/2006/0501/p01s03-woap.html (accessed 16/05/2006)

191 Female Domestic Labour <137> Franz Kafka (1925), *Der Prozess*; Eng. trans. Willa Muir and Edwin Muir as *The Trial*, Penguin Modern Classics, Harmondsworth: Penguin, 1953

193 Teenage Mothers <136> Debbie A. Lawlor and Mary Shaw (2002), 'Too much too young? Teenage pregnancy is not a public health problem', *International Journal of Epidemiology*, 31, pp. 552–54, at p. 553

194 Child Labour <135> Ndabaninga Sithole (1959), *African Nationalism*, London: Oxford University Press, p. 7

201 Male Youth Unemployed <146> Matthew Schofield (2006), 'Youth unemployment in Europe is especially confounding in Greece', *The State* (South Carolina, 28 April) http://www.thestate.com/mld/thestate /news/world/14454313.htm (accessed 16/05/2006)

202 Female Youth Unemployed <145> ECE (2003), *Trends in Europe and North America: The Statistical Yearbook of the Economic Commission for Europe, 2003*, ch. 4; http://www.unece.org/stats/trend/ch4.htm (accessed 16/05/2006)

203 Trade Unions <357> Irwin Abrams, ed. (1997), *Nobel Lectures…: Peace, 1981–1990*, Singapore: World Scientific Publishing; http://nobelprize.org /nobel_prizes/peace/laureates/1983/walesa-acceptance.html (accessed 12/11/2006)

204 Strikes and Lockouts <358> Martin Wroe and Malcolm Doney (2004), *The Rough Guide to a Better World: How You Can Make a Difference*, London: Rough Guides, p. 9

206 City Growth <190> John Vidal (2004), 'Beyond the city limits', *The Guardian* (9 September); http://www.guardian.co.uk/life/feature/story/0,,1299798 ,00.html (accessed 22/07/2006)

207 Number of Households <191> Cat Lazaroff (2003), 'Smaller households lead to vanishing biodiversity', *Environment News Service* (13 January); http://forests.org/articles/reader.asp?linkid=19308 (accessed 22/07/2006)

211 Urban Slums <187> David Adam (2006), 'Urban population to overtake country dwellers for first time', *The Guardian* (16 June); http://www.guardian.co.uk /international/story/0,,1798774,00.html (accessed 22/07/2006)

213 Minimal Sanitation <183> Kevin Watkins (2006), 'We cannot tolerate children dying for a glass of water', *The Guardian* (8 March); http://www.guardian.co.uk/comment/story/0,,1725919, 00.html (accessed 22/07/2006)

214 Basic Sanitation <184> Bruce Gordon, Richard Mackay and Eva Rehfuess (2004), *Inheriting the World: The Atlas of Children's Health and the*

Environment, Brighton: Myriad Editions, p. 16

215 Sewerage Sanitation <185> Peter Bane (2006), 'Design for a solar moldering toilet'; http://www.sunnyjohn.com/toiletpapers2.htm (accessed 22/07/2006)

216 Poor Water <186> 'Ethiopia: Sanitation facilities severely lacking – UNICEF', *IRIN News*, UN Office for the Coordination of Humanitarian Affairs (7 September 2004); http://www.irinnews.org/report.asp? ReportID=43065 (accessed 22/07/2006)

217 Children in Primary Education <199> http://chinesefood.about.com/library/blquotationsch.htm (accessed 09/08/2006)

224 Growth in Primary Education Spending <208> Adriana Puiggros (1996), 'World Bank education policy: Market liberalism meets ideological conservatism', NACLA, *Report on the Americas*, reposted on the Cambridge Indymedia website; http://www.indymedia.org.uk/en/regions/cambridge/200 4/02/286118.html (accessed 04/08/2006)

226 Growth in Secondary Education Spending Growth <210> Sutapa Choudhury (2005), *Teacher Professionalism in Punjab: Raising Teachers' Voices*, Islamabad: VSO Pakistan and DFID; http://www.vso.org.uk/news/events/talking_teachers.asp (accessed 09/08/2006)

227 Tertiary Education Spending <211> Micahel C. Gonzales (1999), 'The World Bank and African primary education: Policies, practices, and recommendations', *Africa Today*, 46/1, 119–34; http://muse.jhu.edu/journals/africa_today/v046/46.1gon zales.html (accessed 09/08/2006)

228 Growth in Tertiary Education Spending <212> Andreas Schleicher (2006), 'Europe's university challenge', *OECD Observer* (May), repr. from *The Economics of Knowledge: Why Education is Key for Europe's Success*, Lisbon Council Policy Brief, Brussels; http://www.oecdobserver.org/news/fullstory.php/aid/188 9/ (accessed 09/08/2006)

229 Youth Literacy <195> Koïchiro Matsuura (2001), 'Message from the Director-General [of UNESCO] on…International Literacy Day, 8 September 2001'; http://www.unesco.org/education/literacy_2001/en_dg.s html (accessed 09/08/2006)

231 Adult Literacy <196> Stephen Jay Gould (1980), *The Panda's Thumb: More Reflections in Natural History*, New York: Norton; http://www.wisdomquotes.com/cat_knowledge.html (accessed 09/08/2006)

232 Gender Balance of Illiterate Adults <198> Bharati Silawal-Giri (2003), 'Literacy, education and women's empowerment', in *Literacy as Freedom*, Paris:

UNESCO; http://unesdoc.unesco.org/images/0013/001318/131823e.pdf (accessed 09/08/2006)

234 Growth of Science Research <206> Tharman Shanmugaratnam (2003), 'Speech…at the opening of the International Conference on Materials for Advanced Technologies…at Suntec Singapore International Convention and Exhibition Centre' (8 December); http://www.moe.gov.sg/speeches/2003/sp20031208.htm (accessed 09/08/2006)

237 Telephone Lines 1990 <331> Pekka Tarjanne (1997), keynote address, The 21st Century Global Information Society, Jakarta, 12 June; http://www.itu.int/itudoc/osg/ptspeech/chron/1997/jakarta.txt (accessed 16/02/2007)

239 Mobile Phones in 1990 <333> Anthony Zwane (2004), 'Good to talk', *New Internationalist*, no. 365 (March); http://live.newint.org/columns/currents/2004/03/01/talk/index.php (accessed 17/01/2007)

241 Internet Users in 1990 <335> Gabriela Tôrres Barbosa (2006), 'At the top of the hill', *New Internationalist*, no. 386 (January); http://www.newint.org/features/2006/01/01/brazil/ (accessed 17/01/2007)

242 Internet Users in 2002 <336> http://www.itu.int/partners/quotes.html (accessed 17/01/2007)

244 Cable Television <338> http://www.starhub.com/portal/site/CableTV/menuitem.227318645a7b5eb3f75eb1109b1000a0/?vgnextoid=2a572ec5ce53c010VgnVCM100000474114acRCRD#paytv (accessed 08/02/2007)

247 Daily Newspaper Circulation <341> BBC News (2006), 'Chechen war reporter found dead' (7 October); http://news.bbc.co.uk/1/hi/world/europe/5416218.stm (accessed 14/01/2007)

248 Weekly Newspaper Circulation <342> http://www.geocities.com/anita_job/quotes.html (accessed 17/01/2007)

251 Films Watched <345> Phurba Gyalzen (2005), 'Bollywood', *Green Cine*; http://www.greencine.com/static/primers/bollywood.jsp (accessed 17/01/2007)

254 Private Health Spending <214> Phineas Baxandall (2001), 'How U.S. health care stacks up internationally', *Dollars and Sense* (May–June); http://www.thirdworldtraveler.com/Health/How_USHealthCare_StacksUp.html (accessed 08/08/2006)

255 Working Midwives <215> http://www.unfpa.org/news/coverage/2006/may6-12-2006.htm (accessed 29/09/2006)

256 Working Nurses <216> Glenys Kinnock (2005), speech to the ACP–EU Joint Parliamentary Assembly, Edinburgh, 21 November; http://www.acpsec.org/en/jpa/edinburgh/kinnock_opening_spch_e.htm (accessed 29/09/2006)

257 Working Pharmacists <217> 'Togo: New malaria drugs beyond the reach of many', *IRIN News*, UN Office for the Coordination of Humanitarian Affairs (n.d. [2006]); http://www.irinnews.org/report.asp?ReportID=42112&SelectRegion=West_Africa&SelectCountry=TOGO (accessed 29/09/2006)

262 Affordable Drugs <222> Salil Tripathi (2004), 'Prevention is better than cure', *Apocalypse: The Truth about Aids*, pp. 108–19 [*Index on Censorship*, special issue]

263 Medical Exports <67> John Sulston (2001), 'Preface', *Patent Injustice: How World Trade Rules Threaten the Health of Poor People*, Oxfam briefing paper; http://www.oxfam.org.uk/what_we_do/issues/health/patent_injustice.htm (accessed 27/03/2006)

264 Medical Imports <68> Phil Bloomer (2005), 'Millions in developing countries threatened by bird flu pandemic', Oxfam press release (20 October); http://www.oxfam.org.uk/press/releases/birdflu201005.htm (accessed 27/03/2006)

266 Infants not at Risk from TB <224> Desmond Tutu (2006), *Advocacy, Communications and Social Mobilization to Fight TB: A 10-Year Framework for Action*, Geneva: WHO; http://www.who.int/tb/publications/2006/en/index.html (accessed 16/09/2006) [for 'TB' orig. reads 'tuberculosis']

267 Condom Use by Men <226> Dan Smith with Ane Bræin (2003), *The State of the World Atlas*, 7th edn, London: Earthscan, p. 62

268 Condom Use by Women <225> UNESCO (2005), *Teachers For All: What Governments and Donors Should Do*, Global Campaign for Education, policy briefing; http://www.vso.org.uk/Images/GCE_Teachers_For_All_tcm8-6873.pdf (accessed 16/09/2006) [quotation attributed to Clive Wing]

271 Malaria Cases <229> Jong-Wook Lee (2006), 'Report by the Director-General to the Fifty-ninth World Health Assembly, Geneva, 22 May'; http://www.who.int/mediacentre/events/2006/wha59/dr_lee_report/en/index.html (accessed 16/09/2006)

273 Cholera Cases <231> Pascale Harter (2005), 'Mauritania's deadly daily poverty', BBC News (8 September); http://news.bbc.co.uk/1/hi/world/africa/4219104.stm (accessed 16/09/2006)

274 Deaths from Cholera <232> 'Mozambique: Cholera death toll rises', *IRIN News*, UN Office for the Coordination of Humanitarian Affairs (24 March 2004); http://www.irinnews.org/report.asp?ReportID=40231&SelectRegion=Southern_Africa (accessed 16/09/2006)

277 Polio Cases <235> Matthieu Kamwa (2002), 'Polio eradication in a war situation: The experience of the Democratic Republic of Congo', WHO, *African Health Monitor*, 3/1 (July–December 2002) [for 'Democratic Republic of Congo' orig. reads 'DRC']

279 Deaths from Rabies <237> K. Sandeep (2002), 'Man's worst foe', *The Hindu* (6th May); http://www.hinduonnet.com/thehindu/mp/2002/05/06/stories/2002050601220200.htm (accessed 16/09/2006)

281 Prevalence of Diabetes <239> Unite for Diabetes (n.d. [2006]), 'The case for a UN resolution'; http://www.unitefordiabetes.org/campaign/resolution/ (accessed 16/09/2006)

283 Men Smoking <242> http://www.cancercouncil.com.au/editorial.asp?pageid=370 (accessed 01/10/2006)

284 Women Smoking <241> Simon Chapman (2001), '"Don't smoke" said the cigarette packet', *News in Science* (9 March); http://www.abc.net.au/science/news/stories/s256934.htm (accessed 16/09/2006)

286 Years of Poor Health <244> Jeffrey D. Sachs (2000), speech to WHO conference, Massive Effort Advocacy Forum, Winterthur, Switzerland, 5 October; http://www.winterthurhealthforum.ch/HtmlDoc/SpeechSachs031000.html (accessed 16/09/2006)

287 Life Expectancy <255> 'AIDS cuts life expectancy in Zambia to 33 years', *The Body* (6 May 2003); http://www.thebody.com/cdc/news_updates_archive/2003/may6_03/zambia_aids.html (accessed 12/11/2006)

289 Women Outliving Men <257> Miranda Hitti (2006), 'Why do women live longer than men? Researchers examine role of risky behavior in life expectancy', *WebMD Medical News* (11 May); http://www.webmd.com/content/article/122/114535.htm (accessed 12/11/2006)

292 Early Neonatal Mortality <260> 'Africa: 1.16 million newborns in sub-Saharan Africa die each year: UN report', United Nations News Service in New York (22 November 2006); http://allafrica.com/stories/200611220148.html (accessed 30/11/2006)

294 Decrease in Infant Mortality <262> Martin Wroe and Malcolm Doney (2004), *The Rough Guide to a Better World: How You Can Make a Difference*, London: Rough Guides, p. 32

296 Decrease in Child Mortality <264> Carol Bellamy (2004), Introduction, *Progress for Children: A Child Survival Report Card*, vol. 1, UNICEF, p. 3; http://www.unicef.org/publications/index_23557.html (accessed 12/11/2006)

297 Mortality of Males Aged 15–60 Years <266> 'Southern Africa, special report: New thinking needed on "AIDS orphans"', *IRIN News* (30 October 2003); http://www.irinnews.org/S_report.asp?ReportID=37548 (accessed 12/11/2006)

298 Mortality of Females Aged 15–60 Years <265> Susan Ratcliffe, ed. (2000), *Oxford Dictionary of Thematic Quotations*, Oxford: Oxford University Press, p. 102

299 People Affected by Disasters <245> Ian Timberlake (2006), 'Red Cross says preparation can mitigate the toll of disasters', *Terra Daily* (20 November); http://www.terradaily.com/reports/Red_Cross_Says_Preparation_Can_Mitigate_The_Toll_Of_Disasters_999.html (accessed 12/11/2006)

300 People Killed by Disasters <246> BBC News (2006), 'Hundreds dead in Ethiopian floods' (16 August); http://news.bbc.co.uk/1/hi/world/africa/4797415.stm (accessed 12/11/2006)

305 People Killed by Storms <251> Sazedur Rahman (2001), 'Tropical cyclone in Bangladesh', *Nirapad Newsletter*, no. 4 (September), p. 3; http://nirapad.org/care_nirapad/Home/Magazine/chronology/currentissue/html/news1.html#tc (accessed 17/11/2006) [orig. site unavailable; now http://nirapad.org/Magazine/news3.html]

306 People Killed by Avalanches and Landslides <252> UNDHA (1993), 'Ecuador: torrential rains, tremor and landslides in Nambija area', DHA Geneva Information Reports, nos. 1–2 (11 May); http://wwwnotes.reliefweb.int/w/rwb.nsf/6686f45896f15dbc852567ae00530132/18a30abc59801977c1256575004f0e70?OpenDocument (accessed 12/11/2006)

308 People Affected by Infestations <254> Exodus, 10: 14–15 (American Standard Version); http://www.biblegateway.com/passage/?book_id=2&chapter=10&version=8 (accessed 01/12/2006)

310 Military Spending 2002 <279> Anthony Jay, ed. (1997), *The Oxford Dictionary of Political Quotations*, Oxford: Oxford University Press, p.397, no. 1

313 Nuclear Weapons <289> Takashi Hiraoka (1995), Hiroshima Peace Declaration (6 August); http://www.cdi.org/nuclear/nukequo.html (accessed 05/12/2006)

316 Armed Forces in 2002 <284> Anthony Jay, ed. (1997), *The Oxford Dictionary of Political Quotations*, Oxford: Oxford University Press, p. 183, no. 1

320 War Deaths in 2002 <288> Sture Allén, ed. (1993), *Nobel Lectures…: Literature, 1981–1990*, Singapore: World Scientific Publishing; http://nobelprize.org/nobel_prizes/literature/laureates/1982/marquez-lecture-e.html (accessed 14/12/2006)

323 Prisoners <293> Susan Ratcliffe, ed. (2000), *The Oxford Dictionary of Thematic Quotations*, Oxford: Oxford University Press, p. 309

325 Carbon Dioxide Emissions in 1980 <296> Sunita Narain (2002), 'All said and done', *Equity Watch*, special edn, no. 3 (28 October); http://www.cseindia.org/html/cmp/climate/ew/ew_oct28/government.htm (accessed 19/01/2007)

327 Increase in Emissions of Carbon Dioxide <297> UNEP (2005), 'CO2 storage may minimize climate change', press release (26 September); http://www.unep.org/Documents.Multilingual/Default.asp?DocumentID=452&ArticleID=4960&l=en (accessed 19/01/2007)

332 Agricultural Methane and Nitrous Oxide <366> NIWA (2005), 'Polar ice reveals pre-industrial pollution', press release; http://news.mongabay.com/2005/0909-niwa_csiro.html (accessed: 18/01/2007)

333 Sulphur Dioxide <301> Sanjoy Hazarika (1994), 'Traffic makes New Delhi air most polluted in south Asia', *New York Times* (11 June); http://select.nytimes.com/gst/abstract.html?res=F70712FC3B5E0C728DDDAF0894DC494D81&n=Top%2fReference%2fTimes%20Topics%2fOrganizations%2fW%2fWorld%20Health%20Organization (accessed 19/01/2007)

336 Nuclear Waste <304> Richard Black (2006), 'Time for action on nuclear waste', BBC News (27 April); http://news.bbc.co.uk/2/hi/science/nature/4951522.stm (accessed 19/01/2007)

338 Sewage Sludge <306> Olena Tkachenko (2005), 'Dead water: Every year 1.9 billion m³ of crude sewage are dumped into the Sea of Azov', *Day, Weekly Digest* (31 May); http://www.day.kiev.ua/137967/ (accessed 19/01/2007)

340 Waste Recycled <308> Hilary Osborne (2006), 'Rich nations accused of dumping e-waste on Africa', *Guardian Unlimited* (27 November); http://environment.guardian.co.uk/waste/story/0,,1958353,00.html (accessed 19/01/2007)

342 Fossil Fuel Depletion <312> http://www.worldcoal.org/pages/content/index.asp?PageID=22 (accessed 19/01/2007)

344 Oil Depletion <320> http://www.karavans.com/peakoil.html (accessed 19/01/2007)

346 Forest Depletion <314> Chris Richards (2006), interview with Annie Kajir, *New Internationalist*, no. 391 (July); http://newint.org/columns/makingwaves/2006/07/01/annie-kajir/ (accessed 19/01/2007)

348 Domestic Water Use <324> Alex Kirby (2003), 'Why world's taps are running dry', BBC News (20 June); http://news.bbc.co.uk/1/hi/sci/tech/2943946.stm (accessed 19/01/2007)

349 Industrial Water Use <325> Ismail Serageldin (1999), speaking at the Water Forum, Netherlands, 30 November; http://ct.water.usgs.gov/EDUCATION/morewater.htm (accessed 21/01/2007)

351 Piped Water <327> WaterAid (n.d. [2007]); http://www.wateraid.org.uk/uk/get_involved/drink_more_water/3114.asp (accessed 21/01/2007)

353 Biocapacity <321> Tomás Bril Mascarenhas (2006), 'The privatization of Patagonia', *New Internationalist*, no. 392 (August); http://www.newint.org/columns/essays/2006/08/01/patagonia/ (accessed 23/01/2007)

354 Ecological Footprint <322> WWF (2006), *The Living Planet Report, 2006*, Gland, Switzerland: WWF, p. 14.

355 Extinct Species <267> Anil Ananthaswamy (2004), 'Earth faces sixth mass extinction', *New Scientist* (18 March); http://www.newscientist.com/article.ns?id=dn4797 (accessed 21/12/2006)

358 Species at Lower Risk of Extinction <270> http://www.iucnredlist.org/info/gallery2004 (accessed 21/12/2006)

361 Birds at Risk <273> Tim Radford (2004), 'Almost half of all Europe's bird species at risk', *The Guardian* (8 November); http://www.guardian.co.uk/uk_news/story/0,3604,1345844,00.html (accessed 21/12/2006)

364 Fish at Risk <276> Eva Kaszala (2002), 'The Tisza River spill, the Romanian gold mine, and international environmental implications', *Trade Environment Database*, case study no. 653 (January); http://www.american.edu/TED/tisza-spill.htm (accessed 21/12/2006)

366 Invertebrates at Risk <278> R. T. Ryti (2000), 'An overview of the role of invertebrates in risk assessment', paper presented to the annual meeting of the Society for Risk; http://www.riskworld.com/Abstract/2000/SRAam00/ab0ac313.htm (accessed 21/12/2006)

368 Protestant Christians <557> http://bible.oremus.org/?passage=Matthew+5 (accessed 12/06/2009)

369 Orthodox Christians <556> http://www.ethiopianorthodox.org/english/dogma/faith.html#salvation (accessed 12/06/2009)

370 Adherents to Judaism <563> http://www.literarytranslation.com/workshops/thebible/texts/ (accessed 12/06/2009)

371 Sunni Muslims <566> http://etext.lib.virginia.edu/koran.html (accessed 12/06/2009)

375 Sikhs <571> Sandeep Singh Brar, www.sikhs.org (accessed 22/05/2009)

376 Jains <382> Part of the Third Tattva or fundamentals of Jainism, at http://www.jainworld.com/philosophy/fundamentals.asp (accessed 22/5/2009)

377 Taoic Religions<572> The quotation 'refers to a non-sentient, impersonal power…living and non-living' is taken from http://www.reformtaoism.org/beliefs_m-z.php#tao

378 Baha'i <553> Quote from Bahá'u'lláh, believed to be the latest of God's divine messengers, at http://www.bahai.org/ (accessed 22/5/2009)

379 Pagans <577> http://www.paganfed.org/paganism.php (accessed 22/5/2009)

380 Spiritualists <578> Inscription on the tombstone of Allan Kardec (Hippolyte Léon Denizard Rivail), and his wife, Amélie Gabrielle Boudet; translation at http://www.absoluteastronomy.com/topics/Allan_Kardec (accessed 12/06/2009)

381 Agnostics <581> Thomas H. Huxley, 'Agnosticism' (1889), http://www.religioustolerance.org/agnostic.htm (accessed 22/05/2009)

382 Atheists <582> Ivan Ratoyevsky, at http://www.atheistbus.org.uk/atheism/ (accessed 22/05/2009)

416

Acknowledgments

This book is based on the authors' Worldmapper website and would not have been possible without the contributions of many colleagues. We especially thank John Pritchard for his work in creating and maintaining the website, for helping to source, check and correct the data we initially used, and for tirelessly responding to requests for help with this atlas. His assistance in collating and verifying the data on religion for this revised and expanded edition is deserving of renewed recognition. Our other collaborators on the Worldmapper project, Graham Allsopp and Ben Wheeler, gave advice generously. We also thank Michael Gastner, who was involved with the original development of the algorithm. Megan Barford helped early on in the project, and Lyndsay Pritchard later, entering and checking a huge amount of data. Bethan Thomas and David Dorling have each read through and helped to correct at least one set of draft texts for all the maps, and we are extremely grateful to them. The care and attention of staff at Thames & Hudson is greatly appreciated, and we thank Jamie Camplin, Andrew Sanigar and Drazen Tomic, as well as the copy-editor, Gillian Somerscales.

We are also very grateful to the following individuals and groups for their permission to reproduce cartograms in the Introduction: Waldo Tobler for a computer cartogram he generated 35 years ago; Candida Lacey of Myriad Editions for an image from *The State of the World Atlas* by Dan Smith and Ane Bræin; and the American Geographical Society and the National Council for Geographic Education (of the United States) for examples of the work of Erwin Raisz.

Finally, for making many suggestions as to what we could map, how we could map differently, spotting initial mistakes in our numbers or text on the web, and much more besides, we want to thank the thousands of Worldmapper viewers who have emailed us over the past few years.